Bionanodesign
Following Nature's Touch

RSC Nanoscience and Nanotechnology

Series Editors

Professor Paul O'Brien, *University of Manchester, UK*
Professor Sir Harry Kroto FRS, *University of Sussex, UK*
Professor Harold Craighead, *Cornell University, USA*

Titles in the Series:
1: Nanotubes and Nanowires
2: Fullerenes
3: Nanocharacterisation
4: Atom Resolved Surface Reactions: Nanocatalysis
5: Biomimetic Nanoceramics in Clinical Use: From Materials to Applications
6: Nanofluidics: Nanoscience and Nanotechnolgy
7: Bionanodesign: Following Nature's Touch

How to obtain future titles on publications:
A standing order plan is available for this series. A standing order will bring
delivery of each new volume immediately on publication.

For further information please contact:
Sales and Customer Care, Royal Society of Chemistry, Thomas Graham House,
Science Park, Milton Road, Cambridge, CB4 0WF, UK
Telephone: +44 (0)1223 432360, Fax: +44 (0)1223 420247, Email: sales@rsc.org
Visit our website at http://www.rsc.org/Shop/Books/

Bionanodesign
Following Nature's Touch

Maxim Ryadnov
Department of Chemistry, University of Leicester, Leicester, UK

RSCPublishing

RSC Nanoscience and Technology No. 7

ISBN: 978-0-85404-162-6
ISSN: 1757-7136

A catalogue record for this book is available from the British Library

Published by The Royal Society of Chemistry,
Thomas Graham House, Science Park, Milton Road,
Cambridge CB4 0WF, UK

Registered Charity Number 207890

For further information see our website at www.rsc.org

Preface

The progress of today's science and technology encounters an increasing demand for finer and more efficiently performing materials with properties superior to those of current and hence ageing devices. Whether this concerns electronics or drug delivery, cancer diagnostics or alternative energy sources the search for means of miniaturising the existing materials or devising fundamentally new components with higher capacities appears to be relentless.

A saving solution to this is widely proposed as the design and fabrication of nanostructures, molecular architectures with dimensions featured below 100 nm.

By convention, and as originally formulated by Richard Feynman, the challenge of constructing macroscopic structures through the manipulation of individual molecules or even atoms prompted the emergence of a rapidly evolving field – nanotechnology. By definition, nanotechnology mirrors complex organisation at the nanoscale and is underpinned by a variety of related physical events that are combined into one universal process – molecular self-assembly.

The phenomenon of self-assembling molecules is attractive from both academic and application perspectives. However, preferential attention is being given to approaches whereby nanostructured materials or their components can be produced, moreover, produced at whim; that is, designed.

The pursuit for routes that can lead to rational or at least predictable design strategies invoked the main objective of this publication – to bring together contemporary approaches for designing nanostructures that employ naturally derived self-assembling motifs as synthetic platforms.

Entitled *bioinspired nanoscale design* or *bionanodesign* the book is written in the shape of a review, referenced as fully as permissible within the context of biomolecular recognition and self-assembly, which forms a general trend throughout.

RSC Nanoscience and Nanotechnology No. 7
Bionanodesign
By M Ryadnov
© Maxim Ryadnov 2009
Published by the Royal Society of Chemistry, www.rsc.org

The volume is composed of three core chapters focusing on three prominent topics of applied nanotechnology where the role of nanodesign is predominant. Specific applications that arise from designed nanoscale assemblies as well as fabrication and characterisation techniques are of a much lesser focus and whenever they appear serve as progress and innovation highlights.

In this sense, the book takes a nonstandard approach in delivering the material of this kind. It does not lead straight to applications or methods as most nanotechnology titles tend to do, but instead it admits the initial and primary stress on "nano" rather than on "technology". The task is significantly eased by the cohort of brilliant bioinspired designs reported to date and complicated by the volume they create almost on a weekly basis. For this reason, the author apologises for the inevitable, but not necessarily deliberate, omission of examples, many of which may prove to be equally if not more influential in bionanodesign.

Maxim Ryadnov
November 2008

Contents

RSC Nanoscience and Nanotechnology No. 7
Bionanodesign
By M Ryadnov
© Maxim Ryadnov 2009
Published by the Royal Society of Chemistry, www.rsc.org

A designer knows he has achieved perfection not when there is nothing left to add, but when there is nothing left to take away.
Antoine de Saint-Exupery

There's plenty of room at the bottom
Richard Phillips Feynman

CHAPTER 1

Introductory Notes

1.1 Inspiring Hierarchical

It is becoming widely accepted that the decisive role in building nanostructures belongs to the hierarchical nature of molecular self-assembly, which renders the process a "bottom-up" strategy in accessing architectures of various complexities. The approach is thus reverse to the notion of miniaturising materials,[1] which assumes a top-down direction. Indeed, historically "top-down" methods such as photolithography were the first to be introduced into the practice of nanofabrication and processing. Yet, otherwise fairly efficient in nanoscale patterning and shaping on solid surfaces, the methods soon proved to be limited by the very basis of the technology – the use of devices that are considerably larger than the target materials. In this respect, hierarchical self-assembly, which allows for the spontaneous building of a target composite from the bottom up, *i.e.* from individual molecules up to microscopic functionally specialised shapes and morphologies, offers a promising alternative with practically unlimited capacities.

In principle, this is what reserves the potential to define and manipulate the properties of desired structures and materials at the nanoscale.[2]

Notably, a strong dependence on this is exhibited by biopolymers whose precise functional expressions necessarily determine the morphological diversity of biological structures. Conversely however, additional constraints are required to provide the accurate reproducibility of a given assembly by a certain biopolymer type, to which a gratifying provision is made by another intrinsic property of self-assembly characteristic of biological systems.

This is autonomous control over supramolecular propagations of individual molecules. The main mechanism here involves molecularly encoded folding, which enables correlation of each level of architectural hierarchy with the structural assignment of specialised self-assembly patterns.[3] Thus, assembling

RSC Nanoscience and Nanotechnology No. 7
Bionanodesign
By M Ryadnov
© Maxim Ryadnov 2009
Published by the Royal Society of Chemistry, www.rsc.org

biopolymer blocks such as proteins and nucleic acids at the subcellular level, often with a precision of a single nanometre,[3] becomes possible. However, one's ability to reproduce such a state of control and prediction remains to be demonstrated. Admittedly, this is due to incomplete understanding of molecular self-assembly *per se*, whilst gaining more insight into biomolecular hierarchies can lead to qualitatively new models and protocols in designing materials with otherwise unknown or unachievable properties.[4] Therefore, an explicit guidance to the fabrication of functional or specialist nanostructures is of paramount importance.

1.2   Encoding Instructive

Replicating Nature's designs faithfully reproduced over millions of years presents perhaps the most straightforward route to success. Nature shares examples of nanodefined self-assemblies in virtually all levels of biological organisation. These may include, but are not limited to, the repertoire of topologically infinite DNA structures, the wealth of viral forms, the functional elegance of enzyme machineries and protein cages, the architectural unification of extracellular matrices and biological membranes. Taken together these are soliciting for a robust design rationale that claims to be innate within the broadest possible spectrum of nanostructures.

But what are the ways of extracting or adapting this for engineering artificial systems?

Intriguingly, of different types as well as within every single type, natural designs are individually unique and especially in functions they carry or are assigned to. On the one hand, this creates precedents of conserved templates readily adaptable for synthetic designs. On the other, biopolymers universally obey the same assembly principle; they adopt three-dimensional secondary structures to build functional quaternary systems – natural nanoscale objects.

Synthetic designs reported to date take both routes. Protein or DNA structures based on preassembled native folds as well as systems designed from scratch, but unambiguously through the emulation of natural assembly elements, are peers. Therefore, a general approach to tackle the problem may focus on the assimilation of Nature's ways in creating macromolecular assemblies and specifically by employing and extending the structure–assembly relationship of existing examples. Eventually, this may constitute the sought essence of a structure-based strategy that specifically exploits biomolecular recognition for the generation of nanoscale composites. Steady progression in this direction revealed in the past decade states that systems shown as more advanced tend to result from better understood assembly elements. For instance, designs derived from DNA manifest precision and control to match, whereas unparalleled is also the representation of self-assembly elements in different biomolecular classes, with proteins and peptides giving the richest repertoire of self-assembling motifs.

1.3 Starting Lowest

Yet, irrespective of the chemical archetype or class of assembly, the synthesis of a discrete system that would span nano- to microscale dimensions is never a trivial task.[3,4] Monodispersity, an ability to maintain the internal order and morphology of resulting assemblies, reproducibility of prescribed assembly modes are amongst major hurdles to overcome towards functional nanostructures.

Naturally occurring systems are free of such obstacles. This is partly because there are no limitations in size and shape in choosing assembling components where complexity is not an issue and any is affordable, and partly because natural nanostructures are highly conserved sequential couplings of exquisitely fitted subunits that use spatially self-maintained molecular arrangements.

In principle, employing design assumptions offered by natural self-assembling motifs should be beneficial for engineering artificial systems or mimetics, which in this notation can be viewed as bioinspired. Logically, nanoscale objects generated in this way can lead to materials with predictable and tuneable properties that are frequently referred to as "smart" materials. However, this hardly proves to be the case and in particular for *de novo* nanoscale designs that, despite their impressive numbers, remain short of original examples.

Indeed, where the total number of particular designs may well have approached hundreds, rationally designed nanoscale morphologies are confined to a very few. Naturally, the latter is determined by applications, but possibly to a larger extent by the synthetic inaccessibility of large biomolecular subunits of natural assemblies.

As an inevitable consequence, the success of artificial designs is hampered by the need of finding efficient ways that would allow for control over assembly of smaller, simpler, albeit more entropy-dependent, self-assembling motifs. Therefore, very often identifying a suitable molecular candidate with high reproducibility and predictability in assembly, even with the admittance of more sophisticated chemistries, is critical.

1.4 Picturing Biological

Given Nature's preference for biopolymer precursors in constructing nanostructures a set of requirements can be identified for a potential self-assembling candidate as follows.

First, it must be synthetically accessible in a monodisperse form. This requirement is limiting and hence indispensable for any type of intended nanostructures. This also directly relates to the autonomous control of the nanoscale assembly.

Second, it has to adopt a recognition pattern ensuring minimised impact of entropy factors (*e.g.* inter- and intramolecular dynamics) on the assembly. This ensures the hierarchical order of the assembly and consequently presents a major morphology-specifying parameter.

Third, its assembly should obey the chosen mode of hierarchical ordering encoded and hence predetermined in primary sequences. This requirement is intrinsic for all biopolymers but can be waived for certain molecular mimetics that preferentially lean on bulk forces supporting self-assembly, *e.g.* the hydrophobic effect.

There are several biomolecular motifs that can meet such design criteria. With their encoding traits established empirically, all attest strong correlations between the chemistry and assembly. However, of notable advantage are those represented by two main classes. These are nucleic acids and proteins or rather their shortened versions, oligonucleotides and peptides, respectively. Other motifs developed and used over the course of the last several years can be seen as their derivatives or supplements.

Exemplified by just these two, the main factors underlying the functions of native nanostructures including monodispersity, consensus folding and environmental responsiveness provide inspirational impacts on artificial designs. The influence of such examples on scientific thought is immense and in conjunction with the growing body of synthetic develops and constantly improving analytical techniques is stimulative towards more systematic studies for elucidating main compatibility marks between structural principles behind native nanoscale designs and synthetic nanostructures.

All in all, this urges putting mainstream trends in nanofabrication, existing and probable, under the strong emphasis of design aspects. An attempt to address this or at least to touch some of the most design-responsive points in the prescriptive self-assembly is made in this volume.

References

1. G. E. Moore, Cramming more components onto integrated circuits, *Electronics Magazine*, 1965, **38**.
2. R. P. Feynman, There's plenty of room at the bottom, *Engineering and Science*, 1960, **23**, 22–36.
3. G. M. Whitesides, J. P. Mathias and C. T. Seto, Molecular self-assembly and nanochemistry: a chemical strategy for the synthesis of nanostructures, *Science*, 1991, **254**, 1312–1319.
4. J.-M. Lehn, Supramolecular Chemistry – Scope and Perspectives Molecules, Supermolecules, and Molecular Devices (Nobel Lecture), *Angew. Chem. Int. Ed.*, 1988, **27**, 89–112.

CHAPTER 2

Recycling Hereditary

It has been more than half a century since the year that defined the way biology is taught today. The big five – five research papers published in *Nature* within a span of three months in 1953[1–5] – hit the longstanding milestone in biology: the deciphering of the architectural code of DNA. The importance of the discovery has been stressed and recapped in numerous reviews and books that collectively put the matter into a dimension of the all-time scientific heritage of undisputable proof. Although it is difficult to identify a biologically relevant discipline that does not benefit from the knowledge of the DNA structure, none is likely more dependent in its essence on the accuracy with which the geometry and spatial organisation of DNA is predicted and described than biological nanotechnology. With the core characteristic of nanotechnology being the creation of diverse structures with nanoscale precision on the one hand and the refined specificity of binding interactions offered by Watson–Crick base pairing on the other, DNA has emerged as a leading instrument in nanodesign.[6,7]

In fact, according to various estimations, the human genome contains from thirty to hundred thousand genes, with all being based on the same molecular module, DNA. This clear statement for DNA as the central molecule of life has confirmed its central status in nanotechnology likewise within just a decade.[6–9]

Forged by Seeman[8] the notion of DNA nanotechnology – the term now widely accepted[10,11] – will be expanded in this chapter starting from the concepts of topological DNA variations pioneered by Seeman[8] to algorithmic DNA self-assembly conceived by Winfree[12] and applied to origami layouts of artificial DNA scaffolds developed by Rothemund.[13]

2.1 Coding Dual

DNA is termed by many as the language of genes, or, to put it another way, as a repository of the information genes carry and require passing to/over successive

RSC Nanoscience and Nanotechnology No. 7
Bionanodesign
By M Ryadnov
© Maxim Ryadnov 2009
Published by the Royal Society of Chemistry, www.rsc.org

generations. Invariably, such a function, which rationalises the very notion of DNA, dictates the structural parameters and folding paths of the molecule. These, apart from having to be conserved and independently exquisite (by default), need to be able to accommodate a simple and faithfully reproducible mode of self-replication that can be translated into the material of life – proteins. A set of rules that ensures this happening over and over again is termed the genetic code, with its specialisation established as the assignment of a codon, a triplet of nucleotides, to one of twenty proteinogenic amino acids.[14] Strictly speaking, there is more than one genetic code[15,16] as well as more than one mode of DNA base paring.[17] However, those are particular cases and can be ignored within the context of DNA structural reproducibility.

More important in this regard is the fact that (1) only a part of genetic information is encoded by the code, and (2) each cell type (except stem cells of course) specialises in expressing only one set of genes despite having the full copy of the genome. Furthermore, the genome is believed to contain the so-called "pseudogenes",[18] inactive and nonexpressible parts of the genome that are often thought of as an evolutionary artefact or "junk" with no functional purpose.[19] The term is admittedly provisional and debatable as "noncoding" DNA accounts for about 90% of the human genome and, for one instance, can be a stored material with an unidentified function.[20,21] This may prove to be very important from the standpoint of nanodesign as the functional uncertainty of junk DNA as opposed to translated DNA can relate to structural alleviations observed for noncoding DNA structures; that is, the requirement for protein-coding DNA, which is read from one end to another, to be a linear molecule can be waived for noncoding DNA.

In turn, this implies that DNA architecture is intrinsically amenable to different topologies and shapes, the repertoire of which, as can be judged by the recent progress in DNA-based designs, seems to be inexhaustible.[22] Whether the latter is predisposed by Nature or is imaginatively artificial, designing novel DNA structures comes down, if not to the detailed understanding of DNA chemistry then to at least the visionary acceptance of its postulated architectural and hierarchical expressions. This is the departing point in any DNA nanodesign that once taken may be and is often overlooked in subsequent complexed and advanced examples.

2.1.1 Deoxyribonucleic

2.1.1.1 *Building up in Two*

DNA or deoxyribonucleic acid is a monodisperse polymer composed of three types of repeating units – carbohydrate (deoxyribose, pentose monosaccharide); heterocyclic base that can be one of four: adenine (A), cytosine (C), guanine (G) or thymine (T); and phosphate – that together make up one DNA monomer, nucleotide, Figure 2.1. Therefore, an alternative name for DNA commonly used as its chemical rather than biofunctional definition is polynucleotide. The sequence of phosphates and carbohydrates (sugars) coupled

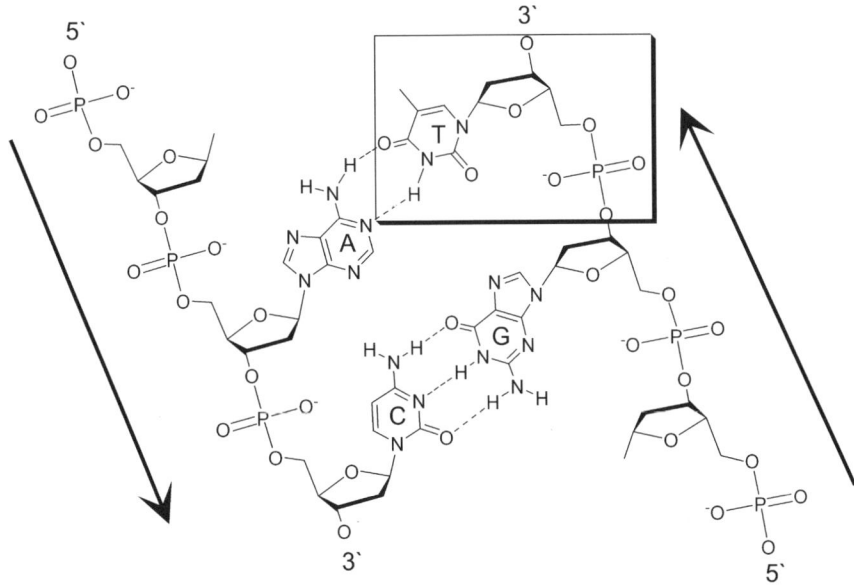

Figure 2.1 Chemical structure of DNA; two polynucleotide chains run antiparallel to form one double-stranded DNA via Watson–Crick base pairing, A–T and C–G. For simplicity only two pairs are shown. Dotted lines indicate hydrogen bonds. The square highlights a DNA monomer – nucleotide monophosphate consisting of a base (T), a five-carbon pentose ring, deoxyribose, and a phosphate group linking two pentoses. Arrows point in the $5' \rightarrow 3'$ direction of the asymmetric ends.

alternately constitutes a polynucleotide backbone that is decorated with bases linked to the first carbon atoms of the five-membered pentose rings, Figure 2.1.

Importantly, a phosphodiester bond furnished between two carbohydrates is asymmetrical as the bond links different (third and fifth) carbon atoms of the two. This renders a polynucleotide directional and underlies the signature of DNA – the memory of chain direction. For example, an individual poly-nucleotide chain is a single-stranded DNA (ssDNA) that can form relatively flexible structures. However, these are thermodynamically unstable and an ssDNA tends to intertwine with another ssDNA or with itself. Either way, this classical double helix DNA (dsDNA) arrangement ensues as an antiparallel assembly such that one strand is oriented oppositely to the other. The shape, stability and the very occurrence of the structure yet depends on the pattern and extension of the interstrand interactions predominantly provided by hydrogen bonding between the bases of the opposite strands – base pairing. The com-plementary base pairing as postulated by Watson and Crick[3] involves highly specific A–T and G–C interactions and has proved to be sufficient for pro-gramming a large and diverse set of nucleotide sequences confirming the robustness of this type of binding. The donor–acceptor patterns of hydrogen bonds are not identical for the base pairs and differ by geometry and the number of bonds per pair. The A–T pair is formed by two bonds, whereas the more stable G–C relies on three bonds. G and A belonging to a heterocyclic family of purines are larger molecules than C and T that are pyrimidine derivatives. Due to the size and the assumed geometry of binding, purines can only marry pyrimidines and *vice versa*. These two parameters sum up a simple mechanism that regulates appropriate pairings along polynucleotide sequences as selected against their relative stabilities and geometry, Figure 2.1.

2.1.1.2   Keeping in Shape

The behaviour of the backbone of dsDNA can be described as that of a polymer with conformational freedom considerably restricted by the regular stacking of sugar moieties. Taken together this underpins the conformational preference of dsDNA that folds as a right-handed helix with two distinctive grooves, the major and the minor, ~ 2.2 nm and 1.2 nm wide, respectively, Figure 2.2. This is the so-called B conformation and is the most common of three also including A,[5] a more compact or dehydrated form of B, and Z,[23,24] a transient left-handed zigzag structure. Combined the criteria favour an exclusive dsDNA conformation, the antiparallel B-form, Figures 2.1 and 2.2. An ingenious consequence of this design is the remarkable stability and reconstruction properties of dsDNA that allow it to survive and function under cellular conditions.

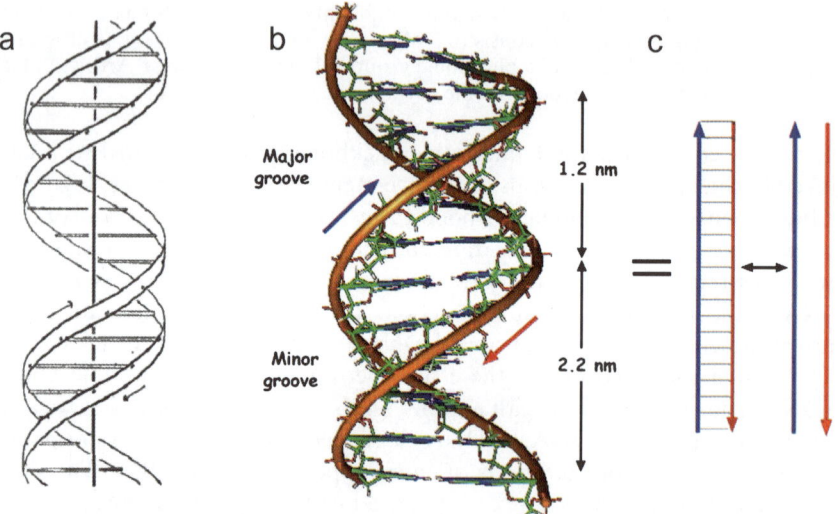

Figure 2.2   Diagrams of the double helix; (a) As originally proposed by Watson and Crick (reprinted by permission from Macmillan Publishers Ltd: Watson, J. D. and Crick, F. H. C. Molecular Structure of Nucleic Acids: A structure for Deoxyribose Nucleic Acid. *Nature*, **171**, 737–738. Copyright (1953)), (b) B-form of DNA as idealised from PDB, (c) 2D ladder-like structures with arrows denoting $5' \rightarrow 3'$ strands, with base pairs depicted as ladder steps (left) and, as equally used, omitted (right).

dsDNA is a very stable molecule (considerably more in physicochemical terms than other associated biopolymers, *e.g.* proteins), but the stability is at the expense of maintaining the complementarity constraint, which, if adequate, ensures the reconstruction of dsDNA from one of its strands. Rendering DNA materials self-replicating and self-repairable, which significantly increases the probability of error-free designs, this is consecutively supported by another strikingly simple mechanism. The binding energy of two strands generated by the described constraint can be approximated to k_BT per base pair, where k_B is Boltzmann's constant and T is a variable, *i.e.* temperature.[25–27] Given that under physiological conditions thermal changes exceeding $30k_BT$ are not common double helices with more than 30 base pairs are resistant to temperature-induced separations.[25,26] Instead, DNA is targeted by specific enzymes known as restriction endonucleases or restrictases. Each restrictase has a specific DNA cleavage site and therefore by cleaving DNA the enzymes reproducibly generate fragments of known sizes and sequences irrespective of whether DNA is from natural sources or of a synthetic origin. This property – unprecedented amongst other molecular types – offers tremendous potential for programming single-molecule constructions from nano- to microscales. Clearly, this is one of the implicit reasons behind the popularity and success of DNA in nanodesign. The property is structurally inherent to DNA and remains true for any topology the double helix can be shaped to. DNA is fairly adaptable to virtually any supramolecular algorithms, provided these are applicable to lengths exceeding the persistence length of the helix estimated as 150 base pairs or 50 nm.[28,29] Below this size DNA is considered as rod-rigid and inflexible, *i.e.* straight. However, owing to complementarity and considerable flexibility of ssDNA a common practice is to assign a given assembly mode to one strand that by being complemented by the other strand converts the mode into the desired morphology. This can vary depending on whether the desired assembly is a spatially fixed scaffold or a conformationally transient device, which is subject to the choice of a particular DNA motif, again be it natural or synthetic.

Base sequences also have their part in contributing to the flexibility of DNA.[30] For example, regions rich in more stable G–C pairs give stiffer helices than those made of A–T blocks. It can be presumed that the affinity of complementary sequences can be increased at the expense of higher contents of G–C pairs that would make the chosen assembly more specific and stable. This is as yet not as straightforward as it may seem and depends on the correlation of the hybridisation of desired and unwanted conformations. Hybridisation is a thermodynamic process of pairing two strands into a double helix. If the strands are complementary their hybridisation is said to be energetically favourable, if not, it is unfavourable. The event of folding two strands into a stable duplex is normally referred to as annealing to reflect its thermodynamic nature. Perfectly complementary strands anneal cooperatively at benign conditions. Their disassembly can be induced by thermal separation, but this would require transition temperatures, *i.e.* melting points, substantially higher than those needed to reverse annealing of strands with lower complementarity.

Thus, in designing DNA constructions the stabilities of intended and to-be-avoided hybridisations should be correlated and made maximised and minimised, respectively. This can be done by restricting similarities between complement sequences and within individual sequences containing regions that are not meant to anneal.[31] The latter is particularly important in designing topologically driven DNA constructions that are reliant on nonlinear motifs with higher requirements for complementarity. In this regard, thermal bending is another often quoted phenomenon contributive to different DNA shapes.[32,33]

Motion caused by temperature is the force behind the subtle deformations of the helix, subtle but constantly excited and sufficient to expose restrictaze sites.[32] This serves the biological purpose. However, structurally thermal bending can be accompanied by the generation of permanent bends.[33,34] These still ensue without untwisting or distorting the helix that can be attributive to sugar–phosphate backbones that by behaving as fixed strings retain their spiral shapes. The strings can be loosened only when fully stretched under an external force. Under normal conditions, thermal bending tends to fold helices that are substantially longer than the persistence length into random coil conformations. In this case the contribution of unusual shapes to the helix rigidity is minimised, which can constitute a mechanism facilitating in the meeting of distant sequences of the same molecule potentially useful in constructing extended closed systems.[35]

In one way or another, designing DNA-based devices and scaffolds feeds on DNA shapes available through adaptation of natural motifs or structure-defining phenomena such as thermal bending. The coiling of two complementary strands into a rope-like double helix is the native geometry of DNA, but not the only one.

Inspirations for others come in plenty from a repertoire of DNA assembly motifs found in both functional and in particular "junk" DNA whose lack of biological function can make it a principal source of geometrical and topological templates. Nonetheless, even restricted to motifs of functional DNA the number of design tips and hints is more than guiding. With the development of high-resolution microscopy techniques such as atomic force, transmission and cryo-electron microscopies it did not take long for artificial designs to fully appreciate this. Intermediate structural elements such as supercoils and knots, splits and junctions, crossovers and catenanes implicated in different DNA processes give the basis of the concept of a topology-driven DNA assembly[8–9,31,36–39] – a divergent point of the following section.

2.1.2 Priming Topological

With the identification of DNA assemblies supporting various biological functions alongside the emergence of original artificial designs, the representation of DNA as a double helix proves to have become a simplification.[6,7] This is true partly when nonlinear motifs are considered, partly when their relatively advanced complexity is taken into account.[39,40] Yet, all of the currently known, and probably most of the future, elements can be loosely viewed as variations of the double helix that retains its primary function as the main building block in the DNA architecture. Double-helical regions form the bulk of most DNA structures that essentially differ only in the number and type of structural elements, the ones that primarily direct assembly modes and define the final shapes of intended constructions. The careful choice of base sequences makes it possible to rationally design such elements or replicate their natural predecessors. For example, sticky ends – single-stranded binding overhangs of a duplex – which represent a common instrument in genetic processes such as transcription and recombination has become one of the most recognisable features of DNA assembly with the development of DNA recombinant technology.

Likewise, designing sticky ends or at least adapting their cohesion principles has been established as a prerequisite in engineered nanoarchitectures.[6] Other motifs involved in normal DNA processing being also common models in nanodesign include a variety of junctions, loops and hairpins.[8,9] The essence of building nanostructures based on these "primary" motifs is the reciprocal exchange of strands multiplied into crisscrossed or flexibly branched domains that is eventually realised through the duplex formation.[37] These can be used to generate individually discrete systems including catenated polyhedra,[41] single- and double-stranded knots,[38] and Borromean rings.[42,43] Subsequent assembly forms or "secondary" motifs such as crossover tiles,[44-46] edge sharing[47] and paranemic cohesion[48] target the next hierarchical level focused on the generation of nanoarchitectures with traceable and periodic morphological features that otherwise may not be acquired.

2.1.2.1 Resequencing Basic

2.1.2.1.1 Choosing the Fittest. Irrespective of the complexity of designed DNA nanostructures the initiating and hence defining step in their construction is the choice of base sequences. Apart from directional assignments of specific assemblies that eventually determine intended morphologies one point remains prioritised in designing discrete nanoscale structures. The realisation of prescribed hybridisation as precise as practical is primarily responsible for the high yields and monodispersity of assembled objects. Correlation between melting points of target and undesired constructions presents only a part of the solution that is also influenced by assembly kinetics. A key complication is addressed through the folding tendencies of ssDNA that for longer sequences become more autonomous and less predictable, which prompts a systematic and laborious conformational analysis or qualitatively more radical approaches. The latter appears to be more favourable in sequence selection, with several prediction programs being developed. These either operate on maximum-dissimilarity algorithms[31,49] or deduction approaches based on thermodynamic stabilities of designed folds.[50-52]

For example, DNA word-design strategies[53,54] make use of existing thermodynamic models reflecting the stabilities of formed duplexes and probabilities of secondary structures. In these algorithms each DNA "word" – typically one DNA molecule of 12 to 16 nucleotides – is a tandem unit concatenated with other "words" into longer sequences.[55] A key objective here is the generation of large libraries of individual DNA constructs by incorporating multiple "word" variants at each tandem position. The format is thus combinatorial and allows the assembly of individual noninteracting strands assembled from individual "words". Specific strands of the libraries can then be selected based on their predicted and experimentally verified hybridisation behaviours.

In a similar manner, SEQUIN – a sequence-symmetry minimisation algorithm proposed by Seeman[31] – selects sequences as constructed from subsequences of given lengths. The trick is to have each individual subsequence made available only once in all possible sequences with no occurrence of the complementary subsequence. To enable this, the subsequences are prescribed as overlapping and very short, typically 3–5-mers. The procedure is stepwise and implies the assembly of a target sequence from the collection of subsequences remaining in each step. Strictly speaking, the program presents a text-line interface, with sequence optimisation being only semiautomated and requiring the involvement of the user throughout the process.

Other popular and interrelated examples may include DNADesign and Spuriousdesign developed by Winfree and colleagues[56,57] that are of elementary interfaces to the Vienna RNA package,[58] the core of which is a set of routines for identifying the most optimised RNA secondary structures. The routines are accessed through C code libraries linked to independent programs (*e.g.*, RNAfold, RNAdistance).[58] In DNADesign MATLAB-based algorithms are employed for the analysis of complementary subsequences optimised with user-defined scoring functions. Spuriousdesign, or rather its compiled C program SpuriousC, offers a faster interface and typically uses negative design principles that are primarily focused on minimising "spurious" or unwanted interactions of undesirably complementary subsequences. The codes prove to be fairly efficient for designing relatively simple DNA folds and can be used for designing secondary blocks for more complex and extended assemblies. However, as for all procedures of the type each particular case normally requires the development of specific routines whose direct implementation on the interface may not be straightforward.

2.1.2.1.2 Evolving Diverse. Reif and coworkers sought to improve this by applying the codes into a graphical user interface that ideally should lead to a fully automated sequence optimisation. The proposed package, TileSoft,[59] features an evolutionary algorithm that chooses "the fittest DNA sequences" from randomly generated populations. These individual sequences are "mutated" to give new sequences that are reinserted into the same population. The process is repeated; that is, it evolves, until a termination point is reached. Algorithmically, TileSoft may be viewed as optimised SEQUIN and DNADesign – the same sequence-minimisation procedures that are exponentially weighted within a library through random to accepted (fittest) modifications – but its prime advantage is the integration of the user interface with its optimisation module, which allows direct visualisation of optimised sequences. In principle, the algorithm is amenable to designing more complex structures that can also be done in parallel and in multiplets.

Regarding the latter, a multimerisation method recently developed by Podtelezhnikov *et al.*[60,61] offers somewhat higher potential for designing nonlinear DNA motifs. The method derives from an assumption that short DNA fragments can be multimerised into different topologies of various fixed lengths. The approach was initiated by the task of finding an equilibrium in distributions of topological states of a nonlinear motif, namely circular DNA.[60] Circular DNA, which is described later, is a common natural topology and can be readily reproduced synthetically by ligation of linear fragments. Cyclisation primarily depends on the length and bending of DNA.[62,63] The former is to be set empirically, whereas the latter can be intrinsic or induced chemically or through binding interactions with protein ligands or other relevant substrata.[63] Therefore, apart from deriving estimates for standard structural parameters such as the persistence length, helicity and complementarity, an additional parameter, fragment bending or the bend angle, was introduced to establish the distribution of multimerisation products.[35,62] The bending of ligated fragments in this case is evaluated as caused by specific irregular structures or elements. This allows for a quantitative analysis of resultant structures or to put it more precisely for that of their structural features as obtained by the ligation of fragments into linear and nonlinear (circular) multimers. Assessing the efficiency of cyclisation can thus provide quantitative information on conformational preferences of the structural elements within the generated multimers. To realise this, the cyclisation efficiency was specified by the ring-closure probability, which is also referred to as a *j*-factor – an experimental value that can be directly measured for each individual multimer.[64,65]

The *j* factor can be defined as the ratio of two equilibrium constants,[66] one for cyclisation, K_1, and one for biomolecular association via two complementary sticky ends, K_2:

$$j = K_1/K_2$$

In this notation, a *j* factor equals the effective concentration of one end of a linear chain in the vicinity of the other end in the set angular and torsional orientations.[66] The end-joining reactions are typically faster than covalent closure of the sticky ends, and hence the value of a given *j* factor can be computed as directed by the conformational parameters of an individual DNA fragment. If known, which is possible because such parameters can be experimentally obtained; *j* factors can relate measured properties of DNA fragments with their conformational parameters. Consistent with this, the efficiency of the method was demonstrated not only for well-characterised linear duplexes[61] but also for more complex nonlinear topologies (*e.g.* double crossovers).[67]

Admittedly, the number of algorithms similar to those described is expected to be increasing particularly in relation to improved automation. In-depth analyses of specific concepts proposed to facilitate this are given in several excellent reviews that can also offer the reader a broader view of the problem.[68,69]

However, one issue shared by all algorithmic procedures developed to date is likely to stay. The problem is that although most of the programs are being shown efficient enough to convert the design process into interactively automatic procedures, virtually all of them are ultimately restricted to the sequence level. The reason behind this is the origin of the algorithms that essentially take start from the same point, which can be exemplified by the DNA computing principle.[70]

In his seminal work aimed at solving the directed Hamiltonian path problem with the help of the standard molecular biology protocols, Adleman postulated that a set of relatively simple rules can program DNA hybridisation reactions into self-assembling systems.[70] In the described model DNA was to encode a directed graph (digraph) in which different DNA constructs would give a set of all possible paths. The idea was to connect (assemble) the paths (DNAs) in a single combinatorial step (hybridisation) in such a way that different constructs constituting different vertices and edges of the digraph would give a traceable sequence of vertices. Such a directed propagation ensues spontaneously (self-assembly) as a result of hybridisation reactions with each giving a DNA duplex representing a valid path in the digraph. Thus, when connected DNA is expected to give mathematically predictable patterns. However, the method assumes canonical DNA polymerisation and therefore is a generalisation of the linear DNA assembly, which invariably limits the model to finite-state automation.[69] As a result, the outlined programs are not deemed universally reliable in supporting further or more sophisticated levels as any additional and more complex parameters such as spatial orientation and topological directionality are still to be prescribed or correlated intuitively and empirically; that is, by hand.

In this light, the importance of understanding of DNA structural motifs imposing additional requirements becomes even more apparent. For example, DNA can randomly form hairpins with high occurrence probability. These structures, particularly the shortest tri- and tetranucleotide loops, can be extremely stable to significantly inhibit hybridisation. The effect common in DNA assembly has been largely neglected in most software packages. Analogously, equally frequently occurring internal and bulge loops are not accounted for by most algorithms.[69] This is not surprising particularly given that the formation of internal loops can be accompanied by the stepwise stacking of noncanonical base pairs.[71] Furthermore, stacking interactions can be symmetry driven that is facilitated by helical twist angles[71] and impacts on the stability of loops of different symmetries but same lengths. Structure-destabilising DNA bulges – junction loops that can be as short as one nucleotide – have been studied more systematically;[72,73] however, their structural parameters can only be approximated using a series of assumptions.[74]

These are just a few examples that though constitute only a small fraction of known DNA motifs are yet to be considered in prediction algorithms. Meanwhile, different topological forms of DNA are by no means contingent and are truly essential components of biological systems.[6,22] DNA motifs are nonlinear elements and by definition are constrained molecules. Linear duplexes are seemingly less constrained, but again by nature are adjustable for topology control at the secondary structure level that eventually determines the whole variety of unusual DNA shapes. Consequently, constructing larger objects on the nanoscale compiled from such geometrically predefined units implies the use of coherent means to be self-assembly (noncovalent) and chemical conjugation (covalent). Although no strict boundary line exists between these two in the context of the fabrication of DNA nanoarchitectures the main accent here is made on DNA motifs as expressed through DNA self-assembling properties as identified by Seeman.[8,9]

As touched above, DNA motifs can be separated into two categories. For the purpose of this chapter, these are given as 2D orthogonal variations of linear DNA, "primary" motifs, and their posterior assembly forms or "secondary" motifs that are directly used as building blocks in the construction of discrete nanosystems.[6]

2.1.2.2 *Primary Motifs*

2.1.2.2.1 Gluing Universal. In describing primary motifs it is important to reinstate the role of sticky ends in DNA assembly, Figure 2.3(a).

Sticky ends that otherwise can be considered as the simplest primary element have a universally more important function as the major cohesion force. First of all, sticky ends support and transfer perhaps the most reliable and predictable set of intermolecular interactions available in self-associating systems. It is also very straightforward to implement their general complementarity and affinity principles into specific sequences. They can be as diverse as practical, and any of them will maintain the local structure of the antiparallel B-DNA,

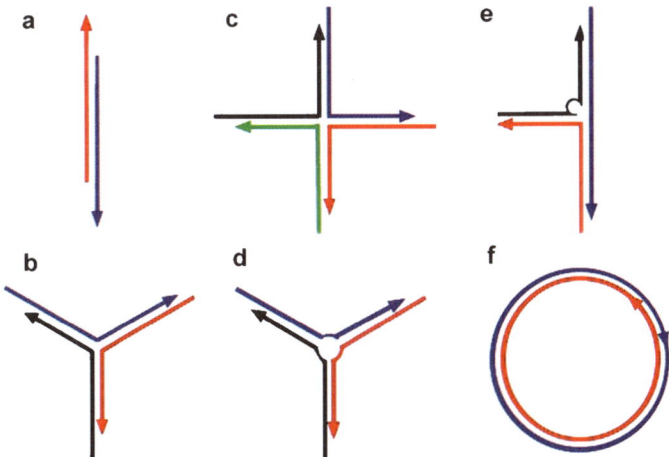

Figure 2.3 Schematic representation of "primary" DNA motifs; (a) Sticky ends, (b)–(e) Junctions with three (b) and four arms (c), internal loop (d), bulge loop (e), (f) circular DNA.

which always gives a building block with fixed parameters that are known with and featured to exquisite precision down to the subnanometre scale.

The length of sticky ends can be designed to significantly vary from a few nucleotides to tens or more than a hundred residues without affecting the affinity and binding properties of DNA. The range of routine synthetic and enzymatic techniques is available to construct target overhangs with achievable selectivity of a single residue that is not matched by any other molecular class or system. And, finally, sticky ends are compatible with other elements of prac-tically any geometry, symmetry, structural regularity and size. Altogether this makes the motif preceding other primary elements, each of which is held up by the sticky-end assembly.

2.1.2.2.2 Alienating Axial. The next simplest and remarkably influential primary motifs in line are axial DNA topologies – junctions and circular DNA, Figure 2.3.

As previously discussed, the double helix is a linear molecule; that is, axially straight. Nonetheless, as synthetically as well as naturally DNA junctions are readily generated. One of the most common examples are three- and four-arm junctions whose architectures rest on three and four complementary ssDNA respectively and can be modelled immobilely.[8] The arms of these structures, ephemeral in DNA replication and recombination,[75–77] are double-stranded DNA, each being made of halves of two different ssDNA. One ssDNA thus participates in two different arms, Figures 2.3(b) and (c). This allows for the connection of the helices through sharing one branching point. Higher gene-ration junctions with a greater number of arms can also be made, but these turn out to be significantly less stable and structurally unpredictable.[78,79] Extended assembly of immobile junctions that lack homologous sequence symmetry can be used in the fabrication of more complex networks and lattices.[8,9] This is done by terminating the arms in sticky ends to promote their self-assembly and following ligation, Figure 2.3. The procedure is thus identical to that used in standard linear DNA. The main feature of such motifs is their "connected-ness", defined as the number of edges (double-stranded arms) that merge at a vertex (branching point).[8,78,79] Hence, the number of arms of a junction is what limits the maximum connectedness of the vertex.

Another important feature is that the mobility of a given vertex can be set and controlled by the incorporation of smaller motifs that can be prescribed by sequence design. This can also determine mechanical properties of a junction. For example, the inclusion of unpaired bases into the strands at vertices would yield internal loops resulting in the junction becoming flexurally less stiff, Figure 2.3(d).

More impressively, the generation of a bulge loop in just one strand can fix the angle between two arms to 180° induced by edge-stacking interactions,[80,81] Figure 2.3(e). Branched junctions of the described types have been used as building blocks for a number of geometric constructs such as a quadrilateral,[82] a cube[83] and a trefoil knot.[84] These belong to the class of catenated and knotted motifs and are described below.

Noteworthy is the relative success in constructing these objects that inspired the search for other classes of multistranded and double-helical DNA in which each strand contributes to two different helices.[85] This led to identifying the so-called anti- and mesojunctions[86,87] – other generalisation models of multistranded DNA complexes such as the Holliday recombination intermediate.[77,86,88] Unlike in a conventional junction, only some and none of the axes in a mesojunction and an antijunction, respectively, point towards the centre of the junction. One consequence of this innate feature of the complexes is the different numbers of geometric populations for different values of maximum connectedness that grow with the increase in the number of arms. This makes designs based on the types difficult to predict and assign to specific nanoarchitectures, which in construction terms makes them somehow similar to crossings in knot-based designs.[38,89] At the same time, conventional junctions, however compelling building blocks they may appear, proved to be disadvantageous in the construction of regular grid-like nanoarchitectures such as lattices, matrices or dendrimers.[90,91] Structural irregularities typically observed in such cases are ascribable to the conformational instability of branching points that ultimately impede the physicomechanical properties of junctions.[90,91]

A seemingly simple primary element that deserves thorough attention is the circular DNA, Figure 2.3(f). This element is the one of choice for bacterial DNA molecules[92] that are normally referred to as plasmids and are primarily used in genetic engineering as gene vectors. It is also the form of mitochondrial DNA of eukaryotes and is a normal constituent of macro or "giant" DNA, protein-fused multiloop structures.[93] Biological functions of circular DNA are diverse, but structurally two forms of circular DNA are acknowledged.[94] These are designated as forms I and II. The former is a closed form; that is, each of the strands is closed on itself. It is the major type contributable to the very term of circular DNA and the one that is being considered here. Form II is a less compact derivative of form I occurring as a result of a single-stranded break.[95]

The formation of the closed form having both strands intact necessitates a specific topological state, and the cyclisation of a linear chain entails a conformational adjustment as set by the introduced constraint.

In polymer physics any polymer chain when cyclised can be described as a knot of one type or another.[96] Applicably to the conformationally fixed double helix, DNA knots are seen as formed by the helix axis. For example, with respect to its axis the closed double helix depicted in Figure 2.3(f) is unknotted and therefore is considered as a trivial knot. In this case, the topology of the axis is taken as the major determinant of conformational properties of circular DNA. The strands of unstressed or "relaxed" DNA are interwoven with a standard twist around the axis that can be altered by a strain.[97] Cyclisation itself provides sufficient stress for the double helix to twist out of the circular form that is achieved through supercoiling.[98]

Supercoiling is the process of global deformations of circular DNA that help DNA relax.[35] This ensues at the expense of an accompanying effect – an increase in the compactness of the closed cycle.[95] The deformations result in the helix being wound or rotated around its axis the number of times equivalent to the number of accommodated helical twists.[63] By convention, such rotations are called writhes.[99] One writhe is the number of the helix intersections or coils, whereas one twist is the number of helical turns. In this notation, supercoiling can be simplified as a sum of mutually convertible writhes and twists, which in topological terms allows the description of the cyclised double helix as a torus – a circle rotated around its axis.[99,100] The number of rotations in this case can then serve as the basis for classifications of the resultant knots.[101,102]

Furthermore, because in the closed form of the circular DNA both strands are linked, the link between them can give a quantitative parameter to model supercoiling or describe it mathematically.[33] This is normally done with the use of a "linking number", Lk,[99] which is the algebraic number of intersections between one strand and the edge of an imaginary surface created by the other strand.[100] Quantitatively, Lk is close to the sum of helical turns and writhes and depends only on the strand topology. Consequently, it remains unchanged through all conformational changes of a closed cycle and its value is always an integer.

The principles that underpin the rationale behind the knot theory[102] applied to DNA knots was proposed by the group of researchers led by Cozarrelli and Vinogradskii.[97–98,103] The approach is based on a further assumption suggesting that an individual knot has to be deformed to the standard form of its projection on a plane, deforming a knot does not change it.[98] The standard form is an image created by a minimum number of crossings with the total absence of self-intersections.[102] Using this description, any knot can be classified merely according to the lowest number of intersections of any of its possible diagrams, its crossing number,[102] Figure 2.4. The simplest knot, aside from the unknot that has no crossings, would have three intersections in its standard form and there would be only four types of knots with less than six crossings,[102] Figure 2.4.

However, the number of knot types increases dramatically with the increasing number of intersections. For example, there are 49 types of knots with 9 crossings, 165 with 10, 552 with 11, which in total mounts to 1 701 936 knots with less than 17 crossings.[102,104] Knots are catalogued by their crossing numbers with a subscript stating a particular knot shown out of many with the same number of intersections, Figure 2.4. These are simple or prime knots. The theory is also used to tabulate composite knots, which are created by cutting and then reconnecting arcs on two prime knots without losing their orientations.[102] However, these are of lesser relevance to predictable DNA nanodesign.

The interpretation of circular and supercoiled DNA as knots is not of purely mathematical interest and linking the experimentally proven occurrence of DNA knots[105] with topological methods may lead to qualitatively novel design rules. Adapting the knot theory for *de novo* DNA construction may follow a similar instrumental path as the revealing of enzymatic supercoiling of DNA[106,107] or the understanding of the mechanisms of DNA replication and recombination.[101] Furthermore, topological approaches prove to be directly contributive to establishing subtle differences in the behaviour of enzymes belonging to closely related classes.[108]

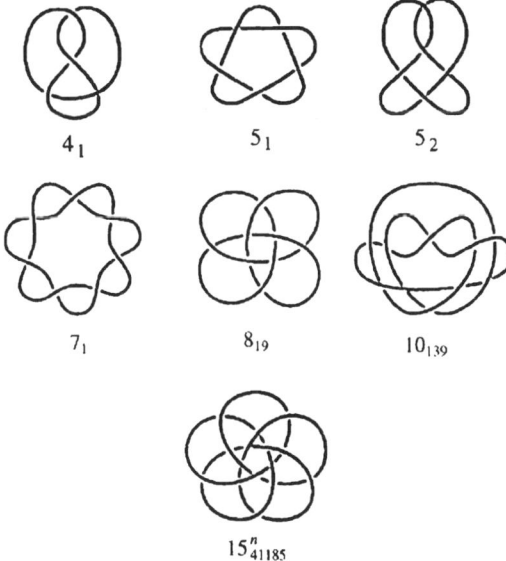

4_1 5_1 5_2

7_1 8_{19} 10_{139}

15^n_{41185}

Figure 2.4 Knotting DNA; standard forms of knots presented with different number of crossings in plane projection, (reprinted with permission from Podtelezhnikov, A. A. *et al.* Equilibrium distributions of topological states in circular DNA: Interplay of supercoiling and knotting. *Proc. Natl. Acad. Sci.*, **96**, 12974–12979. Copyright (1999) National Academy of Sciences, USA)

For example, in their classical work on topoisomerases – enzymes that regulate DNA supercoiling and hence act on the topology of DNA – Brown and Cozzarelli discovered that the unknotting of DNA by topoisomerase II (bacterial gyrase) is accompanied by a stepwise decrease in writhes.[107,108] The decrease was shown to consist of two steps of transient breaks in DNA reflected by the enzyme first cutting both strands on one side of a crossing point and then religating them on the reverse side.[107]

Interestingly, for type I topoisomerases that untangle DNA by cutting a single strand to rotate the generated free ends, and only then reannealing them, the decrease in crossing number consists of a single step.[109] Thus, both topoisomerase types change the linking number of DNA, but do it differently; type I changes by one, whereas type II by two.[110]

The potential of circular DNA in nanodesign is enormous, but may not have been exploited to the deserved extent. Ironically, one part of the problem is the topological properties of the element that are difficult to reproduce in experimental designs at will. As a result, the task of designing knotted DNA motifs is extremely challenging and demands ingenious solutions and solid understanding of both mathematics and chemistry of DNA assembly. In this regard, it perhaps is not surprising that the overwhelming majority of DNA circular motifs, as many others, come from designs by Seeman[10,38] – the initiator of DNA nanotechnology.

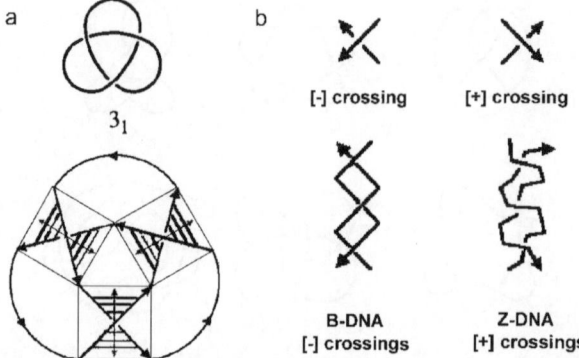

Figure 2.5   Design of single-stranded DNA knots; (a) The relationship between the crossings of a trefoil knot and B-DNA. Thick arrows indicate the path of the DNA. Each square shown in dotted lines as defined by a crossing between two arrows consists of four domains, two between parallel and two between antiparallel arrows. Antiparallel arrows within domains are linked by six thin lines denoting six base pairs corresponding to a half-turn of antiparallel B-DNA. Dotted double-headed arrows indicate helix axes, (b) Negative and positive crossings corresponding to right-handed B-DNA shown as one and a half turn, and its "mirror reflection" – left-handed zigzag Z-DNA, respectively, (reproduced with permission from Seeman, N. C. DNA Nanotechnology: Novel DNA Constructions. *Annu. Rev. Biophys. Biomol. Struct.*, **27**, 225–248 (1998)).

In the continuing string of newly introduced unusual species Seeman and colleagues[6,89,90] strongly advocate that nanostructures are best and predominately characterised by their topology (*e.g.* branching and linking) rather than their geometry. In particular, this serves the cause of applying DNA nanotechnology to the construction of motifs defined by their linking – knots.[38,85] The key requisite for this postulated as an ability to create a synthetic version of a crossing or a node in such motifs is readily met by the double helix whose half-turn corresponds exactly to the component.[38] This relationship between crossings and DNA can be illustrated on the trefoil knot depicted in Figure 2.5(a) with an arbitrary polarity.[89]

Each crossing of the knot is placed into a square to convert portions within the squares into diagonals. The diagonals are meant to divide the square into four regions, with two of these regions being between parallel and two between antiparallel strands. Because the stands of the double helix are antiparallel, pairing has to be designed to run over a half-turn segment (six base pairs) in the regions between antiparallel strands. Thus, the formation of target crossings can be specified by complementary sequences. With two mirror types of crossings existing (positive and negative), two mirror types of DNA are required Figure 2.5(b).

The main DNA form, B-DNA, is a right-handed helix and its crossings can be designated as negative. Although there is no DNA form known that is an exact geometrical mirror image of the B form, topologically the left-handed

Z-form can be used to generate positive crossings,[24,38] Figure 2.5(b). The two DNA forms are qualitatively different and the B → Z transition does not occur spontaneously and must be promoted.[111] In natural systems, the lifetime of Z-DNA is very short but can be induced by specifying sequences and external conditions.[23,111] Although the latter are not necessarily benign and compatible with enzymatic ligation it is possible to assign a Z-forming propensity to short segments by introducing CG repeats or through base modifications.[24] By putting these parameters into a basic design framework Seeman and coworkers successfully constructed a rich repertoire of knotted DNA elements.[6,85]

One example concerns merging two pairing domains containing one double helix turn each into a single molecule. Both domains can switch to Z-DNA but one has a higher transition propensity and does it more readily than the other.[89] The idea is to generate different knotted species by playing on the differences in the sensitivities of the domains towards assembly conditions.[89] It was done by changing ionic strength in DNA solutions,[89] as shown in Figure 2.6.

Briefly, at very low concentrations of dications, none of the domains assembles into the double helix. At increased concentrations both domains fold as B-DNA to give a trefoil knot with negative crossings. Further increases in concentrations create weak Z-inducing conditions at which only one of the domains, the more sensitive, converts into Z-DNA generating a figure-eight knot. Strong Z-favouring conditions force both domains to adopt the Z-form, yielding the trefoil knot with positive crossings,[89,111] Figure 2.6.

An apparent drawback of the design is that all of the conversions are irreversible and the formed topologies are final. In such a setup changing solution conditions cannot facilitate in the transformation of one knot to another without cutting and religating its covalent backbone.

In realising that this is precisely what type I topoisomerase acts upon, the researchers performed the characteristic stepwise interconversions of different knots, both from B to Z and from Z to B forms.[111] The impressive success of the experiment prompted an attempt to find an RNA topoisomerase – a predicted but previously unfound enzyme – with the help of an RNA knot.[112]

Similarly to DNA, RNA molecules can contain hairpins and pseudoknots.[113] But unlike DNA, cellular RNA is a single-stranded molecule and therefore it was long assumed that the formation of functional RNA structures is not hampered by the requirement of solving the RNA molecular topology problem due to its inexistence in the cellular context. The search for an RNA topoisomerase thus did not seem necessary. However, since it was known that DNA topoisomerase III – a recombination-regulating toposimerase – can cleave RNA, the researchers hypothesised that it should be possible to use the enzyme in catalysing the interconversion of RNA topological states. To probe this, a cyclic RNA strand was designed to adopt two topologies – a circular RNA and a trefoil knot.[112] Because the two forms are the final resultant of RNA folding their interconversion can only be achieved by a strand passage event that analogously to DNA knots can only be catalysed enzymatically. Consistent with this, the anticipated RNA strand-passage activity of DNA topoisomerase III accompanied by circle–knot interconversions was observed.

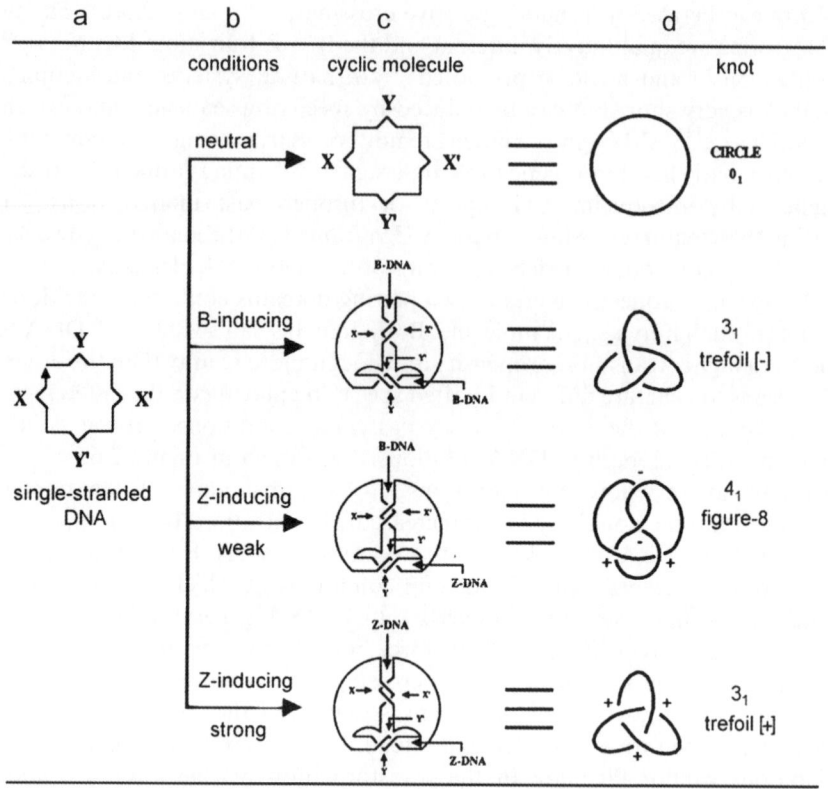

Figure 2.6 Folding of single-stranded DNA knots; (a) A single strand containing four regions, X and Y and their respective complements, X′ and Y′, (b) Four independent conditions of folding and ligation, (c) Generated target products, and (d) Their corresponding molecular topologies, (adapted with permission from Du, S. M., Stollar, B. D. and Seeman, N. C. A synthetic DNA molecule in three knotted topologies. *J. Am. Chem. Soc.*, **117**, 1194–1200. Copyright (1995) American Chemical Society).

The role of topoisomerases in the topological expression of the double helix appears to be unequivocal and is evenly balanced between DNA unknotting and supercoiling. In fact, the enzymes support the formation of more complex structures that can be thought of as other and somehow higher order forms of knots. This is catenanes – the products of catenation.

Catenation is the process of interlocking of DNA rings catalysed by topoisomerases. Catenated states of DNA that result from the compaction of DNA substrates and favoured at high local concentrations[114] can be rationally directed in a fashion similar to that of DNA knots. Indeed, knots and catenanes are closely related and it is relatively straightforward to interconvert knots and catenanes by a simple operation on a crossing.

An example is illustrated in Figure 2.7. One crossing may consist of four polar strands connected in pairs, with two before and two after the crossing.

Figure 2.7   A knot–catenane–knot conversion; a 5_1 knot is converted to a catenane after two strands in a crossing exchange of their outgoing partners, the catenane converts to a trefoil knot with another exchange, (reproduced with permission from Seeman, N. C. DNA Nanotechnology: Novel DNA Constructions. *Annu. Rev. Biophys. Biomol. Struct.*, **27**, 225–248 (1998)).

Knots and catenanes become interconverted every time the crossing is removed through the pairs exchanging outgoing partners. Consequently, DNA that is, as been described, used to construct knots, should be equally effective in constructing catenanes. Gratifyingly, this proves to be the case in both natural and synthetic systems. Catenanes as simple as dimeric[115,116] are common in natural systems and can, for example, be found as intermediates in DNA replication[117] or products of site-specific recombination systems.[118]

Understandably, recombination by integrative systems (*e.g.* bacteriophage λ) is found to provide convenient methods for the synthesis of simple linear catenanes that mainly find use in biochemical studies.[116] However, most such catenanes are substantially more flexible molecules than knots and are not particularly compelling elements in designing geometrically fixed nanoscale scaffolds.

The problem is similar to that of topological engineering based on single-stranded DNA[38] that has proved to be solved, at least to some extent, by "topological protection" techniques involving the temporary complexation of single strands with their complements.[119]

Nonetheless, from the nanodesign standpoint most notable examples are higher-generation catenanes, the topology of which is translated into more complex and ordered forms such as DNA stick polyhedra: DNA cube[83] and a truncated octahedron.[41]

Clearly however, even these structures may be described only on the topological level of linking as geometrically they are not characterised by well-defined coordinates. For this reason, illustrations in Figure 2.8 representing the structures are very schematic.

In reality, in branching points the angles between the double-helical edges are more loosened than they appear. Furthermore, each edge contains an exact number of double-helical turns that makes each of the faces correspond to one cyclic single strand. With the cube composed of six different cyclic strands and the octahedron of fourteen, the objects become hexa- and tetradeca-catenanes respectively, Figure 2.8.

Consistent with the principle assumptions of topology-driven DNA design, the synthesis of such catenated DNA can serve as a proof of topology and *vice versa*. Indeed, linear catenanes can be targeted through cutting restriction sites

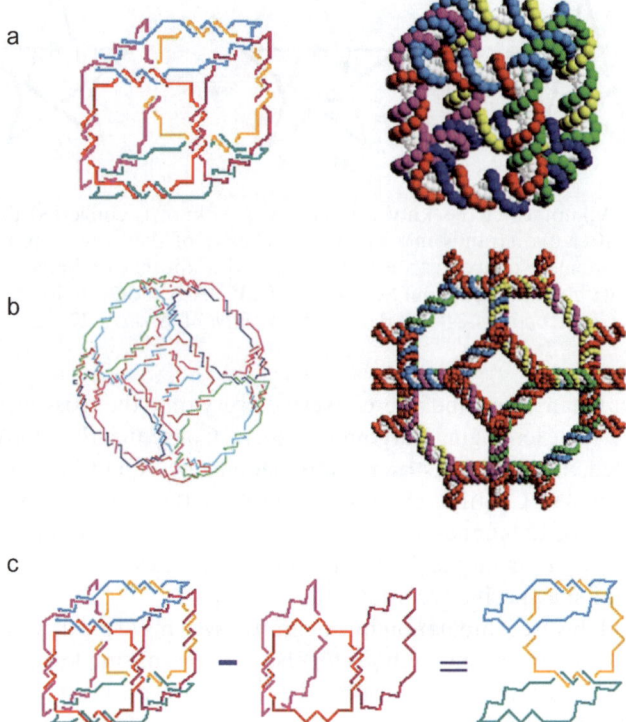

Figure 2.8   DNA catenane designs; (a) and (b) sketch (left) and model (right) representations of (a) hexacatenane with the connectivity of a cube, and (b)
tetradecacatenane with the connectivity of a truncated octahedron, c) the
formation of the cube through linking linear triple catenanes, (reprinted
from Seeman, N. C. DNA engineering and its application to nanotechnology. *Trends Biotechnol.*, **17**, 437–443, Copyright 1999, with permission from Elsevier; and from Seeman, N. C. *et al.* New motifs in DNA
nanotechnology. *Nanotechnology*, **9**, 257–273 (1998), with permission
from IOPP).

incorporated into specific edges of the structures. This breaks the structures
down to individual catenanes that can be subsequently characterised according
to their electrophoretic properties.[120]

   Figure 2.8(c) illustrates the experiment exemplified by the synthesis of the
cube. In the first step, a linear triple catenane serves as a starting material
corresponding to left, front and right faces of the cube. Once the cube is synthesised it is cut on the two front edges in the catenane to give another linear
triple catenane that corresponds to the top, back and bottom faces of the
cube.[83] Conformably, if the synthesis is unsuccessful the starting catenane is
destroyed.[83] The synthesis of the truncated octahedron was proved likewise.[41]
The strands equating to the six square faces were confirmed first. This was
followed by restricting the octacatenane that matches the eight hexagonal faces

down to the tetracatenane flanking each of the squares.[41] As supported by these examples, catenanes and knots present a collection of motifs interesting in itself, and some of their types demonstrate certain versatility and convenience in designing nanoscale objects. However, their use in the construction of more difficult targets, such as secondary motifs and further static 2D arrays or 3D discrete materials, is restricted by the mechanically untranslatable geometrical "floppiness"[7,10,40] of the generated crossings that in their present state makes them practically unsuitable for functional nanodesigns.

Nevertheless, in designing knot-related motifs the DNA expression of one topology stands out as particularly challenging and promising and cannot be omitted.

This is the Borromean rings, a topological link named after the three-ring heraldic design of the Italian Renaissance Borromeo family,[43] Figure 2.9. Considered by many as the most ambitious and somewhat an ultimate test for the designer, the link presents a rich family of topological structures that are very similar to catenanes; similar indeed but only geometrically.

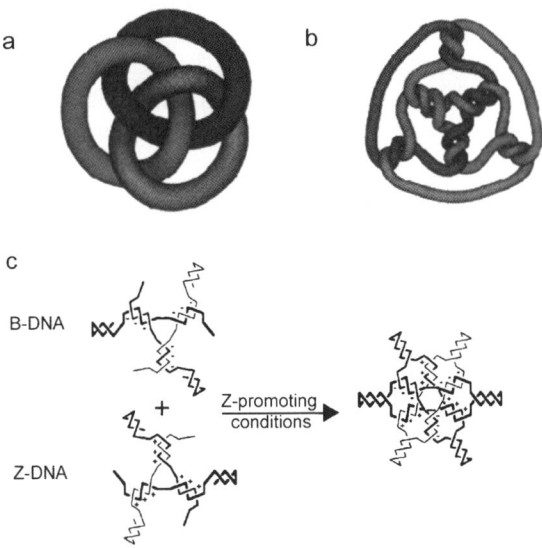

Figure 2.9   Design of DNA Borromean rings; (a) The traditional representation, (b) Borromean rings with each crossing replaced by three to reflect one and a half turns of the double helix. The inner helices are right-handed and correspond to B-DNA, the outer are left-handed and correspond to Z-DNA (a) and (b), reprinted by permission from Macmillan Publishers Ltd: Mao, C. *et al.* Assembly of Borromean rings from DNA. *Nature*, **386**, 137–138, Copyright (1997)), (c) The assembly of the Borromean rings from two 3-arm junctions, B-DNA and Z-DNA, (reproduced with permission from Seeman, N. C. DNA Nanotechnology: Novel DNA Constructions. *Annu. Rev. Biophys. Biomol. Struct.*, **27**, 225–248 (1998)).

The interbonding of rings in catenanes is analogous to that of the links of a chain. As in the chain, if one ring (link) breaks then what is left is the two parts of the chain flanking the broken ring. Borromean rings are topologically totally different. The hallmark of this motif is the inseparable arrangement of three interlocked circles or rings that can be unraveled into three independent pieces in one step with the removal of any of the individual circles. In other words, if one ring is broken the whole structure will fall apart, giving rise to a number of intact but unlinked circles. The critical nature of this key difference is particularly evident from the synthetic point of view. While the syntheses of catenanes and knots are more or less of routine operations the unique makeup of noncatenated, albeit interlocked, Borromean rings in finding a universal synthetic strategy still presents a formidable task.[121]

This may be a subject of future successful attempts and studies, but the sufficient and necessary criteria in tackling the problem always start with establishing the direct relationship between the topology and its molecular adaptation by the chosen type of assembly. In this vein, for DNA-based designs the traditional representation of Borromean rings as planar Venn diagrams, in which the innermost three crossings can be viewed as negative and the outermost three as positive, is modified, Figure 2.9.

For DNA, it can be redrawn to replace all of the crossings by one and a half turns of the double helix as opposed to a single half-turn in knot designs. Unless required by a particular target the use of one and half turns is more convenient.[42] In this conversion each crossing is replaced by three with the innermost segments corresponding to a three-arm branching point of B-DNA, Figure 2.9(b).

Therefore, the three inner helices are right-handed and the three outer helices are left-handed, which preserves the original equal number of positive and negative crossings. Because any topological picture can be deformed an infinite number of times, the generated projection can be adjusted to give a clear retrosynthetic disconnection path involving the ligation of two stable three-arm junctions constructed from B-DNA and Z-DNA, Figure 2.9(c). In such a structure each of the rings contains a different hairpin motif. Arranging hairpins at the equator of the representation is convenient in assigning ligation and restriction sites of the two junctions that can be made of different sizes aiding in the separation of restriction products, Figure 2.9(c). As in both previous catenane-based designs, this feature ensures an unambiguous characterisation of the resultant structure providing the proof of the target topology. Specifically, when one ring is cleaved the other two remain but in a separate, *i.e.* nonbonded, form, which confirms the innate feature of the Borromean rings assembly.

The approach has been experimentally tested in the synthesis of six ssDNAs corresponding to the three strands of each branched junction, B-DNA and Z-DNA, that were annealed individually and then combined and ligated under Z-DNA-promoting conditions into the first DNA Borromean link.[42]

The method represents an excellent example of topological control that in principle may be implemented into other biomolecular systems. Nevertheless, despite such an impressive execution of the topology using DNA the motifs share the same design limitations that are characteristic for knotted and catenated structures. Remaining to be geometrically nebulous, which is largely due to the entropic conformational nature of interconnections merely represented by crossings, these motifs are barely useful as building components for periodically configured structures featured over lengthier scales.

2.2   Fixing Spatial

Invariably, building more complex nanoscale objects demands additional geometrical constraints capable of offering spatially better defined and molecularly more conserved cohesion patterns. Notwithstanding the different complexities the primary motifs exhibit, their architectural frameworks are almost exclusively mediated by sticky-ended associations. Generally, and as shown by the outlined examples, the sequence specificity of sticky ends proves to be a powerful tool in meeting a number of nanoconstruction criteria, but again these are set at the level of local structural predictability that limits the types of possible designs. Although this may be satisfactorily sufficient for engineering even infinitely long objects such as infinite double helices,[122] it is clear that a larger-scale design requires operationally more advanced motifs.

Elements that offer subsidiary means to the sticky-end cohesion and thus meet the requisites for the constructing arrangements repetitive on the nanoscale are termed here "secondary" motifs.

2.2.1   Hinting Geometric: Secondary Motifs

The term "secondary", which might seem to have been taken liberally, is still central to DNA nanodesign. This is not only to reflect a higher order of secondary motifs as compared to primary or individual elements, but also to stress the need for finding more robust nanoscale components capable of withstanding rigorous testing of mechanically sustainable nanofabrication.[9] The practical authenticity of the statement can be illustrated using the following examples.

2.2.1.1   Crossing Double

The first, and in many respects also a typical, representative of secondary motifs is the so-called double-crossover element.

The motif can be thought of as a more complex topological expression of the aforedescribed junction motifs. However, the main drive for its exploitation is functional and is chiefly stimulated by the urgency of developing topologically tighter structures. Initiated by Seeman and colleagues and aimed at geometrically rigid and structurally more independent motifs that can be used as spatially regulated oligomerisation elements, the main attempts were long focused on different classes of junctions that have been explored fairly extensively.[8,80,85,86,90] Previously mentioned elegant works on engineering anti- and mesojunctions[86] as well as site-specifically constraint bulged junctions[80,81] serve as classical examples. However, the apogee of this endeavour was the discovery of the double-crossover motif to have become one of the most exciting and influential DNA elements.[44,45]

The molecule, normally abbreviated DX, is a recombinatorial motif related to the Holliday junction.[44,119] DX consists of two four-arm Holliday junctions joined by two double helical arms at two adjacent arms, Figure 2.10.

There may be five distinct DX isomers. These are designated as DPE, DPOW, DPON, DAE and DAO, Figure 2.10. The isomers differ by (1) having parallel, first three, and antiparallel, the last two, orientation of their helix axes, and (2) the number of double helical half-turns, odd (DAO, DPOW, DPON) or even (DPE, DAE) between their crossover points.[44] DPOW and DPON molecules also contain the extra half-turn corresponding to a major (wide) or minor (narrow) groove spacing. Thus, in the abbreviations given to each of the isomers D, A, P, E, O, W and N stand for double, antiparallel, parallel, even, odd, wide and narrow, respectively.

Importantly, the DX molecules differ also in their behaviour, which defines the preference of their applications in nanodesign.[44,45] The parallel double-crossovers, transient motifs relevant to biological processes, are usually not well behaved in nondenaturing gels,[44] particularly when the separation between their crossover points is short, unless their ends are furnished into hairpin loops.[123] The parallel orientation of the helical domains implies that the minor and major grooves of both engaged helices come together each turn, which ultimately leads to long-range repulsive interactions destabilising the structures. By contrast, antiparallel DX, in which minor grooves of one helix are facing the major grooves of the other, give more stable structures.[45] For this reason, the main focus has been on antiparallel DX molecules – DAE and DAO.[6,22] The molecular topologies of these isomers differ only in the number of double-helical half-turns between crossovers, *i.e.* odd and even. The molecules were shown to be twice as rigid as linear duplexes[67] and were proposed as oligomerising elements in nanofabrication.[45] To facilitate in the interpretation of resultant nanomaterials, an additional third motif, DAE + J, was introduced.[44,124] This element is the DAE motif containing one or more bulged junctions (hence + J) that are placed between the crossovers, Figure 2.10. The bulge, which is composed of two nucleotides, enables the helix that is interrupted to be stacked and so to become an extra double-helical hairpin protruding from the plane of the two DX helices. This then can be used as a topographic marker or reporter in microscopy and ligation experiments.

2.2.1.1.1 Reporting Visible. The principle, for instance, was used in designing DNA tiles reported by Winfree *et al.*[124] DX motifs modified with hairpin junctions were employed to tile a plane giving rise to periodically striped 2D lattices that can be directly visualised using AFM, Figure 2.11.

One of the used motifs (DAE) with 4×16 nm dimensions was shown to self-assemble into single-domain crystals of 2×8 micrometres with uniform thickness of 1–2 nm.

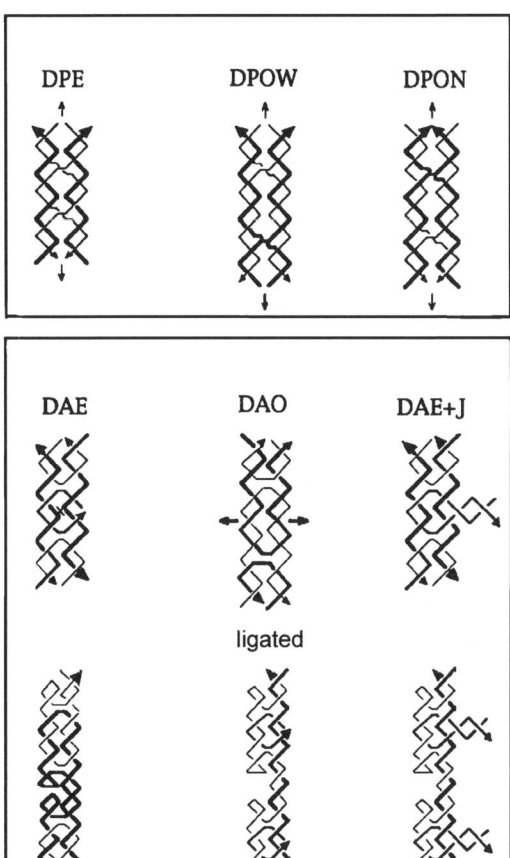

Figure 2.10 Double crossover (DX) motifs; parallel (top panel) and antiparallel (bottom panel) motifs are shown with zigzag helical structures denoting strands. Two consecutive perpendicular lines correspond to a full helical turn for a strand. Line thickness is related by symmetry elements shown as double-headed arrows – two-fold axis lying in the plane of the page; and by 'x' (DAE) – an approximate dyad perpendicular to the plane of the page that for DAE + J is removed by the bulge. Thick lines in the ligation products of antiparallel motifs (bottom) correspond to reporter strands for DAE and DAE + J and catenated rings for DAO, (reprinted with permission from Li, X. *et al.* Antiparallel DNA double crossover molecules as components for nanoconstruction. *J. Am. Chem. Soc.*, **118**, 6131–6140. Copyright (1996) American Chemical Society, and Fu, T.-J. and Seeman, N. C. DNA double crossover structures. *Biochemistry*, **32**, 3211–3220. Copyright (1993) American Chemical Society).

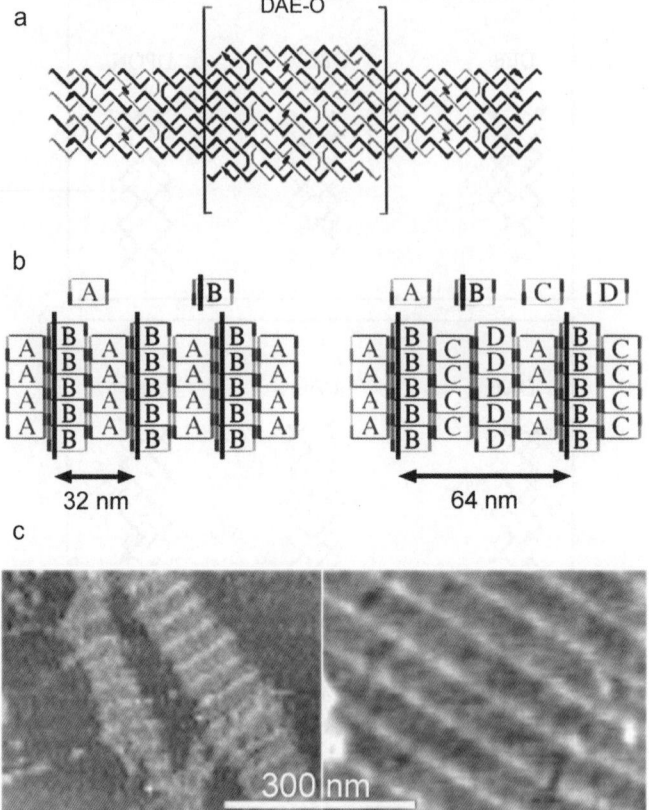

Figure 2.11   DX striped lattices; (a) DAE lattice topology with an odd (O) number
of half-turns between crossings on adjacent units, one unit is bracketed,
(b) Two structural types composed of two and four units, one unit
carrying a hairpin is marked by a thick line in each type, (c) Their
characteristic stripe patterns as seen by AFM, (reprinted by permission
from Macmillan Publishers Ltd: Winfree, E. *et al.* Design and self-
assembly of two-dimensional DNA crystals. *Nature*, **394**, 539–544,
Copyright (1998)).

Because the vertical and horizontal repeats of the assembled arrays are
correspondingly 4 and 32 nm, and because features of 4 nm are beyond the
resolution of AFM, the protruding periodicities could only appear at distances
of 32 nm, Figure 2.11(b). Indeed, this was observed.

The helices derived from hairpin junctions and sticking out of the plane
generated visible stripes above the surface at the intervals,[124] Figure 2.11(c).
The pattern was highly periodic and additional modifications were revealed to
give a larger spacing by using, for instance, different components. Specifically,
when four components (instead of two that gave the 32 nm spacing) were used a
spacing of 64 nm was apparent, Figure 2.11(c).

The approach shows that it is possible to control surface topology and such self-assembling lattices can be used as scaffolding materials for templating the assembly of other systems through chemical or molecular recognition recruitment.[124]

Another type of experimental design in which the introduction of periodic reporter strands can be effectively enhancing is used in ligation-closure assays. These originally biochemical methods are being progressively applied to the characterisation of DNA nanoarchitectures (*e.g.* see the above experimental designs in relation to knotted and Borromean links).[81] In the case of DX elements such assays can be used to assess the rigidity of the resultant structures.

In a typical experiment, DNA strands are ligated and their ligation products are examined to reveal if cyclisation led to one product or many. Functionally, the presence of many cyclic products would indicate that the angles at the junction points of the tested molecule are not fixed.[81] A prominent attribute here is that each of the oligomerised constructs contains an accessible reporter strand whose fate is identical to that of the whole complex. In this way, topological consequences for the ligated DAE molecules containing five strands and the DAO motifs with only four would mean that only DAE molecules could generate a reporter strand, Figure 2.10.

The fifth, central, strand in DAE, however, often cannot be closed off. Therefore, an ideal solution to generating a reporter strand would be to extend it by introducing a bulged junction; that is, converting DAE into DAE + J. Both motifs contain five strands, two of which are continuous and three are crossovers including one central cyclic strand that in DAE + J forms a three-arm bulged junction. Ligating both DAE and DAE + J molecules generates reporter strands and resulted in negligible or no cyclisation.[44] By marked contrast, the DAO motif lacking a reporter strand upon ligation gave a series of catenated molecules, Figure 2.10.

2.2.1.1.2 Translating Symmetrical. The main advantage of the double-crossover motif, as intended and designed,[44,80] is its significant structural rigidity in comparison to single branched junction motifs that fail to adequately support periodic nanoscale configurations.[9,85] As argued by Seeman and coworkers[40] rigidity is precisely the sought criterion that is overwhelmingly conducive to structural integrity in the assembly of periodic matter, notably represented by crystal patterns.[8]

The hypothesis can be extended to the point of specifying particular sticky-ended associations for the fabrication of short-range finitely defined constructs, as described above, in a highly controlled manner. Since the sticky ends were specified or made different they are deemed asymmetric that, as experienced in most such nanodesigns, alludes to the fact that the best way of manipulating finite nanoscale species lies in their symmetry minimisation. Logically, this should be taken adversely in constructing longer-range crystal-alike objects for which symmetry is a major dominant. Indeed, the key distinction of crystalline materials is the translational symmetry of their compiling units or motifs, which in turn implies that the motifs must not be flexible to maintain the same spatial relationship between each other. In this context, cyclisation, knotting and in some cases catenation are the consequence reactions of structural flexibility and can be seen as assembly poisons for periodic structures.[80–81,125] Based on this, it can be concluded that no apparent cyclisation observed for the ligation-closure of DAE/DAE + J motifs confirms the hypothesis.

Further confirmation came from a number of different edge-sharing DAE-based designs. In one, for example, it has been shown that the DAE motif furnished into one side (edge) of a DNA triangle maintains its structure, which allows it to oligomerise into micrometre-long 1D arrays with nanoscale features,[126] Figure 2.12(a).

The main objective of this design was to establish the compatibility of a rigid geometrical form (such as a triangle) with a rigid DNA motif (DX). The relative stiffness of the DX motif within the triangle was chosen as the parameter to judge against in ligation-closure experiments[44,81] that revealed the ability of triangles with incorporated DX motifs along one edge to generate noncyclic constructions. This notion was further strengthened by AFM imaging providing strong support to the drawn relationship between the model of ligation products and the observed zigzag assemblies, Figure 2.12(a).

Intriguingly, it appears that either DAE or DAE + J triangles can be used as building blocks for 2D hexagonal lattices.[44] To realise these, every triangle edge can be made to contain one DAE or one DAE + J molecule giving different types of arrangements. If, in the exclusively DAE-built triangle, one edge is made of DAE + J, an extra helical domain will be produced within each hexagon.[44] Instead, in the purely DAE + J lattice the extra junction is involved in the formation of triangles and is to brace branched junctions to prevent them from bending and thus keeping them straight at their axes.[6] However, experimental feasibility of these arrays proved to be difficult to conform with the DX motifs whose rigidity appeared not to match that inherent to triangles. The need to solve this problem stimulated the development of a new and more efficient motif.

A construction to fill this gap was proposed as a DX triangle,[127] Figure 2.12(b).

The element, the combined derivative of two previously engineered motifs – the double crossover[45] and bulged triangle motifs,[81,128] has two key characteristics.

One rests upon the fact that the DX molecules are twice as stiff as conventional duplexes,[67] and therefore the thicker edges of the formed triangle are expected to be considerably more rigid than those of the simple bulged triangular junction. The second dwells on the innate property of the DX motifs for double intermolecular interactions as opposed to single helical interactions that makes the designed triangle less prone to errors in twist.

These features, the synergetic combination of which all previous motifs lacked, were shown to be crucial in constructing trigonal arrays.[127] Two triangles were designed to coassemble into a continuous DX structure of a pseudohexagonal lattice arrangement containing thirteen double helical turns ($\sim$46 nm), Figure 2.12(b).

The edges of each of the triangles were formed by 65 nucleotide pairs in each of their DX helices (four turns per edge) terminating with six-nucleotide sticky

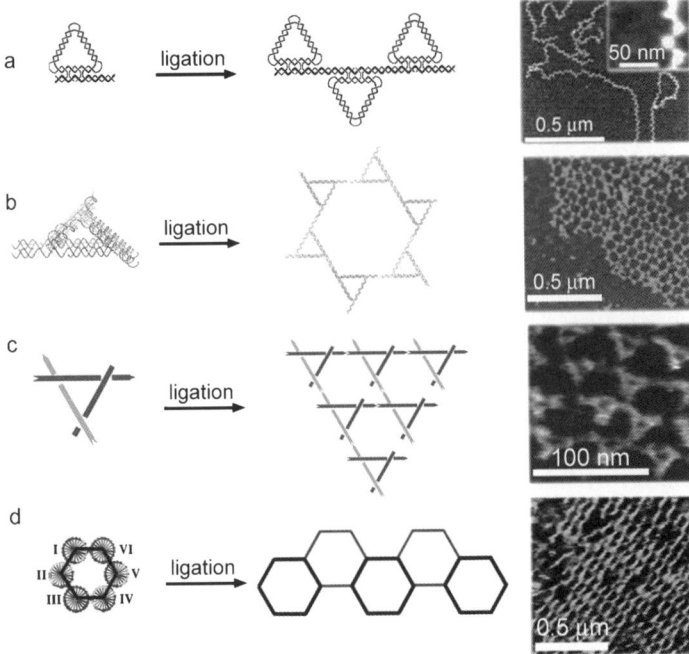

Figure 2.12 DX translational symmetry elements and their assembly into periodic arrays as seen by AFM (far right column); (a) Ligation of a double-stranded triangle with one DAE-edge into a zigzag of 1D arrays, (reprinted with permission from Yang, X. *et al.* Ligation of DNA Triangles Containing Double Crossover Molecules. *J. Am. Chem. Soc.*, **120**, 9779–9786. Copyright (1998) American Chemical Society), (b) DX triangle assembled into a pseudohexagonal trigonal lattice, (reprinted with permission from Ding, B. *et al.* Pseudohexagonal 2D DNA Crystals from Double Crossover Cohesion. *J. Am. Chem. Soc.*, **126**, 10230–10231. Copyright (2004) American Chemical Society), (c) 2D triangle array built from an equilateral tensegrity triangle, (reprinted with permission from Liu, D. *et al.* Tensegrity: Construction of Rigid DNA Triangles with Flexible Four-Arm DNA Junctions. *J. Am. Chem. Soc.*, **126**, 2324–2325. Copyright (2004) American Chemical Society), (d) A ligation of a six-helix bundle (a cross-sectional view with numerically assigned helices) into a 2D hexagonal lattice, the characteristic corrugated arrangement is schematically shown with the black layer being closer to the reader than the grey layer, (reprinted with permission from Mathieu, F. *et al.* Six-Helix Bundles Designed from DNA. *Nano Lett.*, **5**, 661–665. Copyright (2005) American Chemical Society).

ends. Both pair- and sequencewise the trigonal frameworks are identical and only differ in the sticky ends. A hexagon flanked by six triangles, three of each triangle, demonstrates the arrangement, Figure 2.12(b).

Because the edge of the hexagon lacks one triangle and has nine turns (~ 30 nm) the centre-to-centre distance can be estimated as ~ 34 nm. Consistent with the design, a characteristic honeycomb structure was identified by AFM,

Figure 2.12(b). As predicted, each unit of the structure has an hexagonal geometry formed by six triangles, with the centre-to-centre hexagon having an edge of about 38 nm. Importantly, to provide experimental evidence to the engineered structural characteristics of the DX triangle, namely doubly thick sticky ends, the same set of experiments was repeated with the triangles having the sticky ends removed from one of the helices. As expected, no lattices were generated[127] showing that the use of double helical arms is truly critical in promoting robust 2D grid-alike assemblies.

The idea was further tested on other systems having been extended to other interesting concepts that can be exemplified by the construction of stable tensegrity nanostructures, DNA equilateral triangles reported by Mao and co-workers,[129] and six-helix bundle motifs designed by Seeman with colleagues.[130]

The first design relies on the combining tense and integrity of connected assembly units rather than on their rigidity, Figure 2.12(c).

The motif, built up of three four-arm junctions fused in a triangle, with each vertex formed by a four-arm junction and each edge by a duplex, was shown to assemble into 1D and 2D lattices consisting of individual triangles that appeared as uniformly sized triangular particles. The success of the design hinted that it should be possible to use the motif to guide DNA assembly in 3D, with all its three components extending in three different directions making it essentially nonplanar.

The intention of applying the design in 3D stems from the assumption that although 3D arrays can be generated from motifs that are planar and rigid (such as DX), providing that their interhelical angles are flexible enough for them to undergo small and fixable rotations, 3D motifs that are inherently constrained rotations are more capable of setting up 3D structures. To demonstrate this experimentally a DX version of the motifs – a 3D-DX triangle – was produced.[131]

The motif exhibits emphasised tensegrity geometry entailing the arrangement of a 3D triangular construction in which each side of the triangle is placed below one and above another.[131] The design thus has crossover separations with even and mixed even and odd numbers of half-turns between all junctions. The repeat units contain seven and eight double-helix turns roughly corresponding to 25 and 28 nm, respectively. As revealed by AFM, the motif self-assembled into extended and periodic 3D arrays, with repeat distances being in prefect agreement with the estimates.

The second design, 2D arrays assembled from hexagonal six-helix bundle DNA motifs,[130] can serve as an angle-related approach that carries major similarities with both the DX motif[45] and the tensegrity triangle.[129,130]

This motif, dubbed 6HB, was constructed from a series of DX motifs positioned in parallel to each other at a specific angle such that the resulting structure would present a discrete bundle of six aligned helices interconnected in a fashion similar to that of DX motifs. Because the helicity of DNA is 10.5-fold[132,133] it was possible to assign crossover separations that would rotate the helices by 120°,[134] Figure 2.12(d). In In this way, the bundle can be assembled in 1D, 2D or 3D. For instance, if the sticky ends of each helix in the bundle are made complementary to each other 6HB molecules will organise into a 1D array. By contrast, the generation of 2D and 3D arrays from the motif is best achieved by bonding the helices in adjacent pairs.[135]

In short, pairing sticky ends on the front of helices I and VI in Figure 2.12 with the back ends of helices III and IV will give a 3D array (same for front–back pairs of II-III with IV-V and IV-V and II-I), whereas a 2D array of alternating grooves and ridges is constructed when only two of the three pairs contain sticky ends.[131]

2.2.1.2 Extending Cohesive

The experimental robustness of the outlined concepts is likely to transfer to other motifs based on the same principles of geometrical certainty. The dominating factors in their constructions, however, will continue to serve the cause of the sticky-ended cohesion, and additional factors that may emerge are bound to reinstate its versatility.

One such factor, defined here as double cohesion[131] and exhaustively employed in the context of sticky ends appears to have provided a compensatory mechanism for individual variations in connections between assembling components. The fabrication of 2D and 3D nanostructures based on such a structural refinement looks very promising. Nevertheless, any novel objects whose assembly derives from the point-to-point bonding offered by sticky ends will always require geometric constraints to add further dimensions to their assembly.

As has been pointed out by Yan and Seeman[47] different contact modes can be employed to carry this through. In terms of polyhedra, a perfection test in nanodesign, these can be corner (between vertices), edge (between two edges) or face (between all edges of two faces) sharing. With the sticky-end cohesion corresponding to the former and the latter (face sharing) currently seeming practically formidable, edge sharing is prompted by the very architecture of DX motifs, Figure 2.13.

2.2.1.2.1 Sharing Mutual. Yan and Seeman set about extending the sticky-ended assembly to edge sharing by designing two new motifs. The molecules were constructed as DAE-fused modules shaped in an edge-sharing fashion as two and three DNA triangles, E2T and E3T, Figure 2.13(a).

The first, E2T, is of two triangles linked by one DAE construct and is topologically closed. The molecule has four strands: one outside strand, two cyclic inner strands and another smaller cyclic strand involved in the two crossover structures. To render the inner strands triangular three dT_4 bulges were introduced into each cycle to bend the corners, Figure 2.13(a).

E3T motif is an extended version of E2T and was constructed using the same set of parameters, but three triangles were fused together by two DAE molecules giving a composition of six strands; one outside strand, three triangular strands bent by the same type of bulges and two circular strands. The outside edges of each triangle are three-turn long double-helical domains. In this format, both motifs are catenated structures, with E2T adopting a rhombus-like shape and E3T taking a half-hexagonal-like trapezoidal form.[47]

The motifs were tested for their ability to assemble into different periodic arrangements. A DX molecule was incorporated on each edge of a E2T motif to give a construction in which one domain of the molecule is used as an extension for each outside edge of E2T and the second domain is to interact with a similar domain of another copy of the motif. Interactions between the units of the motif were directed by sticky ends. Because the rhombus shape has four edges 1D linear arrays can be readily assembled using two specific sets of sticky ends. Moreover, two distinct types of arrays as defined by different spacings between individual units of two different types could be envisaged.

Specifically, in Figure 2.13(b) the units are indicated as having A/A′ and B/B′ sites. These were designed to give two different repeat patterns of centre-to-centre distances of 20 and 30 nm, respectively, Figure 2.13(c). Accordingly, the distances between centres of the designed units in necklace-like patterns observed by AFM appeared to be consistent with the estimated ones. Encouraged by the results, the authors went further in their pursuit of applying the motifs to the construction of more complex structures. They argued, for example, that in building 2D lattices repeating units need to be positioned in a coplanar arrangement. To test whether this can be implemented into edge sharing, a conjugated version of the E2T was designed. In this molecule two units are covalently linked at one set of B/B′ sites, Figure 2.13(c), and blunt ended at the other inhibiting thus the 1D assembly of B/B′. Thus, the assembly is solely driven by the sticky ends at A/A′ sites. The distance between the two fused rhombi is estimated to be about 30 nm, whereas the self-assembly of two newly produced units is meant to give a double-row array with a repeat distance of about 20 nm, Figure 2.13(c). Consistent with this, AFM experiments confirmed the formation of double-row 1D arrays with the distances between the units and rows matching those predicted, Figure 2.13(c).

The concept successful in building 1D arrays can be so in 2D given that it is planarity amenable in 2D. However, 2D arrays, albeit anticipated,[136] have yet to be demonstrated with the help of more robust motifs. Similarly, to facilitate building 3D structures the edge-sharing concept may be extended to face sharing.[36] For example, mimicking the octahedral geometry of the Buckminster Fuller's sphere could give discrete 3D DNA structures. In its classical form the sphere is a 12-connected octet truss structure – the stick version of cubic-close packing – presented by a series of fused tetrahedra. Within the edge-sharing concept, the structure, the construction of which would otherwise require twelve-arm junctions, can be built as an array of face-sharing motifs analogous to the one described above. However, once again, the feasibility of such an approach is to be demonstrated.

2.2.1.2.2 Multiplying Traversal.
An option to extending edge sharing in the context of hedging the necessity of sticky ends in the construction of latticed structures can be offered by a number of other strategies, leading to the introduction of useful motifs that can be exemplified by paranemic[137] and triple[46] cohesion and square-aspect-ratio tiling.[138]

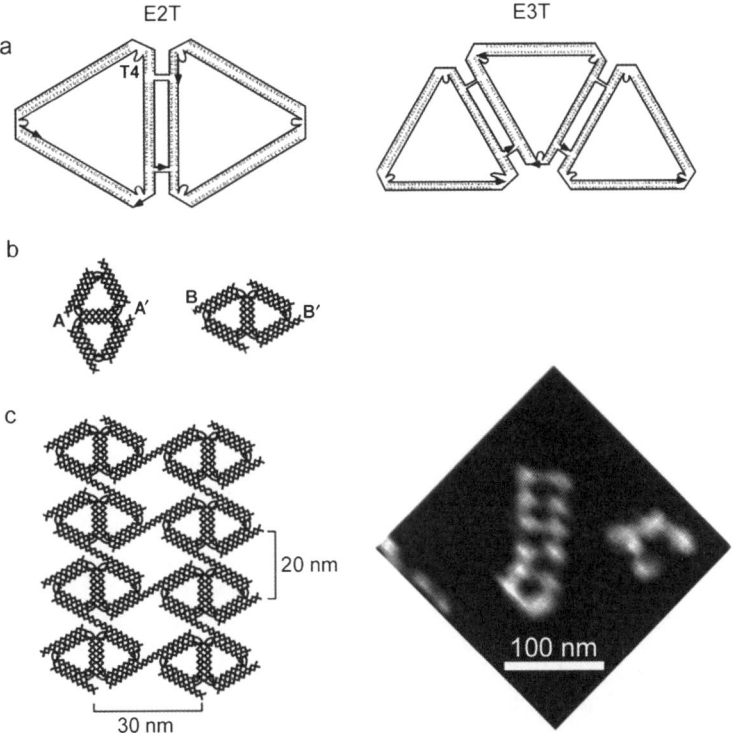

Figure 2.13 Edge-sharing DX motifs; (a) E2T and E3T motifs composed of two and three topologically closed triangles, correspondingly. The triangles are held together by DAE molecules formed by cyclic central and inner triangular strands. The angles of each triangle are fixed by T_4 bulges in the inner strands, (b) Two E2T types with two different sets of sticky ends designated A/A′ and B/B′, (c) A double-row array of a conjugated version of E2T in which two units are covalently linked at one set of B/B′ sites and blunt ended at the other leaving only A/A′ sticky ends (left). The distance between the two fused rhombi is ∼30 nm and a repeat distance in the array is ∼20 nm, as confirmed by AFM (right), (reprinted from Yan, H. and Seeman, N. C. Edge-sharing motifs in structural DNA nanotechnology. *J. Supramol. Chem.*, **1**, 229–237, Copyright 2001, with permission from Elsevier).

Paranemic cohesion is represented by a class of topologically shut complexes – coaxial four-stranded motifs termed paranemic crossovers, PX.[137]

PX motifs are stabilised by every nucleotide paired by Watson–Crick bonding that, in contrast to other known four-stranded DNA, eliminates the need for any other nonhydrogen bonding interactions, Figure 2.14(a).

One PX molecule contains a central dyad axis relating two flanking parallel duplexes that form crossovers wherever possible, with crossover points flanking the axis at every major and minor groove. Within this framework, two other helix axes that flank the crossover axis have a repeat similar to the normal twist

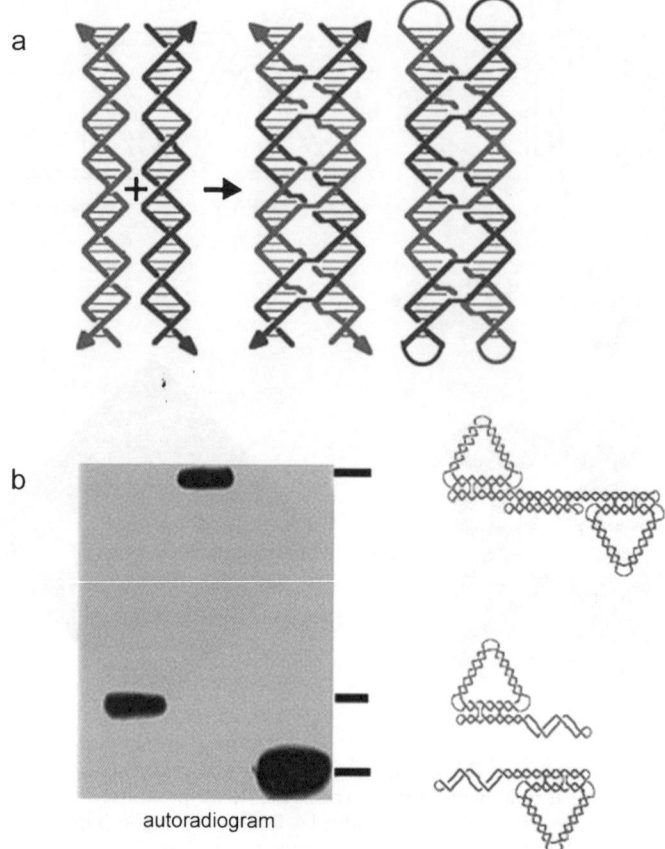

Figure 2.14   Paranemic crossover (PX) motif; (a) The formation of PX from two double helices exchanging strands at every point leading to the inter-digitation of the helices. Capping of the helices in hairpin loops emphasises their paranemic relationship, (b) Test of paranemic cohesion based on two triangles with one DX-edge each. Each tailed nontriangular DX domain produces PX with its complementary partner on the other triangle, (reprinted with permission from Zhang, X. *et al.* Paranemic Cohesion of Topologically Closed DNA Molecules. *J. Am. Chem. Soc.*, **124**, 12940–12941. Copyright (2002) American Chemical Society).

in duplexes so that every strand in PX crosses over every helical repeat, but the structural period of each strand, and hence that of the PX structure, is of two helical repeats. Thus, PX motifs are very similar to and at the same time principally different from DX motifs in that both types are composed of two parallel helices, however, crossovers in PX are formed at every possible point with no backbone juxtapositions, Figure 2.14(a).

Such indefinite pairing of two component duplexes without physical linking between them shows the structure paranemic. The interdigitation of the duplexes can also be referred to as plectonemic as it is analogous to that of the strands in a double helix, and likewise observed in supercoiled structures. Taken together, the conventions give rise to a very efficient form of assembly that offers an alternative to sticky ends.

This has been demonstrated in different static geometrical designs[48,139] as well as in sequence-dependent dynamic nanodevices.[140–142] However, arguably the best illustration of its efficacy and potential is the 1:1 binding of large triangle motifs.[48]

There are two main factors behind this type; first, sticky ends that are normally introduced by restriction enzymes are very short to support complex large and stable structures, and, second, assemblies based on long sticky ends are hampered by twist errors. With this, a PX construct topologically similar to edge-shared[47] and DX triangles[126,127] has been designed, Figure 2.14(b).

Each of two triangles was constructed with one edge presented by a DX molecule. As with edge-shared motifs the two triangles are topologically closed molecules with, however, small central strand left unsealed. As opposed to those of the two other types of triangles, the triangle-linking extensions were made as complementary PX molecules to give the type of cohesion that was meant to go to completion with the formation of a one-to-one structure. This was confirmed in gel-electrophoresis experiments using equimolar mixtures of the two triangles with each containing a tailed nontriangular DX molecule paired with the opposite segment of the other triangle by PX binding.[48]

An equally efficient extension to the point-to-point cohesion, triple cohesion,[46] was introduced in an attempt to accentuate the problem of coplanarity in building 2D lattices. Triple crossover, TX, is a motif consisting of three adjacent helices (parallel and antiparallel) lying in the same plane, Figure 2.15(a).

The helices are held by two or more sets of crossover sites between each of the domains rather than three crossover sites between two domains as the term might suggest. Although the presence of a third helix entails otherwise the motif was designed to have all its axes coplanar,[46] for which domains within each helical pair were made antiparallel.

This is evident from Figure 2.15(a) showing that minor grooves abut major grooves in both interhelical regions. The molecule has four strands shared by three helices and four crossover points, with each vertical pair separated by an odd number of half-turns, three between the left two and five between the right two helices. The central helix is capped with hairpin loops, Figure 2.15(a).

In this arrangement TX appears as a double framework of two DAO components with a common central duplex. In designing such a topology three main points could be identified.

The first assumes that motifs with more contiguous patterns of reporter strands would allow for more accurate acquisition of outputs in DNA-based computation that largely relies on reporter strands.[8,143] The second is stimulated by the encapsulation properties of 2D lattices that are desirable for the accommodation of guest macromolecules in assembling, for example, molecular memory devices.[143,144] In contrast to DX constructs, a TX-based array would be more intrinsic in generating a one-helix gap at a high density with just

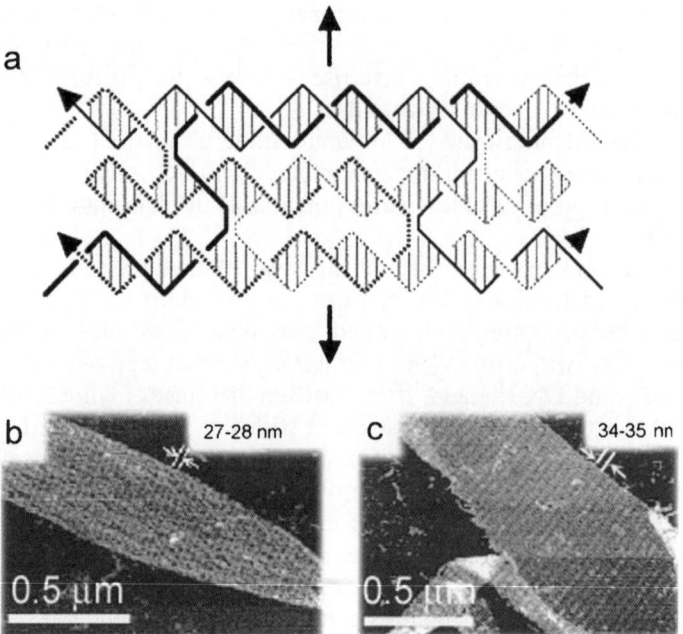

Figure 2.15   Triple crossover (TX) motif; (a) The TX assembly of four strands into
three helices with coplanar axes. The middle helices are capped with
hairpin loops. The two vertical arrows indicate the dyad axis interrelating
two pairs of strands: continuous and dotted, (b) 2D arrays assembled
from two TX units. Visible crosshatched stripes corresponding to bulged
reporters (TX + 2J) are uniformly separated at distances of 27–28 nm, (c)
2D arrays assembled from three TX units. Visible stripes caused by the
capped outer helices of one unit with uniform spacings of 34–35 nm,
(reprinted with permission from LaBean, T. H. *et al.* Construction,
Analysis, Ligation, and Self-Assembly of DNA Triple Crossover Complexes. *J. Am. Chem. Soc.*, **122**, 1848–1860. Copyright (2000) American
Chemical Society).

one species in the array.[8] The third point is for ensuring large displacements in
nanomechanical devices that can exceed 10 nm when TX components are
used.[46,140,145]

Experimental proof for this was provided using two different varieties of TX
motifs for the assembly of 2D arrays, Figures 2.15(b) and (c). Similarly to those
based on DAE designs (Figure 2.11), in one type one TX module coupled with
a TX + 2J motif (TX with two bulged junctions) to give 2D arrays with characteristic topographic features caused by bulged reporters is seen as uniformly
separated crosshatched stripes, Figure 2.15(b). In contrast to DAE designs, in
this design each component in the array is located between two 2-nm gaps,
which can be used as encapsulation or binding sites.

In another array type, distance marking was performed using three TX
components. Two, A and B, are typical TX molecules, but with their central

loops made sticky ended as opposed to capped by hairpins in the conventional counterparts. Conversely, in the third, C, outer helices are converted to hairpins and sticky ends complementary to those of the first two are incorporated to its central domain. Because it is physically impossible to fit the third between A and B in the same orientation, the C component is rotated about 103° relative to the place of A and B motifs. In such an orientation capped helices of C serve as topographic labels protruding from the plane generating a uniform and predicted spacing.

The extended ligation of hairpin reporters in itself, achieved at least within the 20 units range, in the formed arrays permits computational applications of TX-based arrays that have also been tested.[46] However, the main advantage in providing progress points in nanodesign PX and TX that motifs give – the notion of multiple cohesion elements that in some terms challenge and in others contribute to the seemingly exhaustive role of sticky-ended cohesion – stands its ground strengthened by the emergence of novel applications.[138,140–143] These also offer specific concepts that are related to this mission either reciprocally through introducing mixed topologies[139] or in an originally independent manner.[146]

Regarding the former, it has been shown that PX edge sharing expressed in a triangular form with four strands per edge can be used synergistically with the TX motif.[139]

In this design two sides of one edge of the triangle were modified with two different sticky ends to bind one TX molecule, which acted as a mediator of the triangle assembly into a zigzag 1D array. Conceptually similar designs can include skewed TX triangle and double PX cohesion elements.[131]

Regarding the latter an intriguing study suggests that nanostructures can also be cloned.[37] The idea is inspired by the occurrence of single-stranded DNA in certain viruses, *e.g.* M13 bacteriophage, which in principle makes it perfectly compatible with DNA nanodesign. The first test on this route was the replication of PX DNA using rolling circle amplification (RCA).[146] Templating RCA on the PX motif was dictated by the topology of the crossover as this, first, can be generated from two closed dumbbells or a cyclic DNA, and, second, requires a DNA polymerase to "read through" the molecule to untangle all the cross-overs before catalysing the synthesis. The obtained results principally confirmed the idea of replicating artificial nanostructures by having shown the moderate amplification of the motif. This positively shortens the distance towards the desired target of nanostructure cloning also prompting the discovery of possible implications of complex DNA motifs in natural systems.[37,146]

The described motifs constitute a powerful toolkit in nanodesign. However, as emphasised in their topologies, the motifs of the series are largely based on dyad symmetry, which is characterised by several intrinsic shortcomings.

Relatively serious with regards to controlling intended assemblies is branch migration – an isomerisation event accompanied by the relocation of branch points.

The event appears to be more evident in a DX tile when the structure is viewed as an approximation to the collection of two four-arm junctions linked by two adjacent arms.[44] Although this is partly inhibited by branch migration having to occur as a coherent process of both crossover points its possibility increases with the increased complexity of motifs and target architectures.[45]

On the other hand, an alternative to the motifs – immobile low-branched junctions more stable but also excessively flexible at their branch points – initiated the very search for geometrically more constraint crossover elements.

As a consensus to this dilemma Wang and Seeman[79] proposed junction motifs with the number of double-helical arms greater than eight. The geometry of the motifs enables the total exclusion of dyad symmetry and hence branch migratory formations. The key principle of the concept is relatively straightforward: since there are only four base pairs in conventional DNA each of them can be used only once to flank a junction. With this, junction motifs are to be limited to four-arm junctions. But if two identical pairs are placed adjacent to each other their pairing in the middle would be impossible. As a consequence, junctions containing up to eight arms can be formed.[79] Thus the basic principle of the concept focuses on creating an entropic barrier for branch migration at the expense of the four base pairs arranged symmetrically about one junction centre. If applied to a higher-arm junction containing, for example, eight branches, the sequence of base pairs around the junction for arms 1–8 can be arbitrary and could be made T–A, T–A, G–C, G–C, T–A, T–A, G–C, G–C. Repeating the same dinucleotide sequence every four arms was shown to be most effective in generating a 12-arm junction with the sequence of G–C, T–A, C–G, A–T, G–C, T–A, C–G, A–T, G–C, T–A, C–G, A–T. Correspondingly, using such sequence symmetry in the construction of 8- and 12-branched junctions revealed no branch migration.[79]

2.2.1.2.3 Tiling Square.

A promising solution to finding the best combination of the strong points of both element types has been formulated by Yan *et al.*[138] in a high square-aspect ratio concept.

The concept addresses key points in designing 2D lattices with controllable and definable encapsulation properties.[147] It is underpinned by a four-by-four tiling approach represented by the design of four four-arm junctions generated by nine strands, with one participating in all four junctions, Figure 2.16(a).

When assembled into a tile the junctions become fixed to points in four different directions in the plane. To enable this, bulged loops were introduced into the shared strand to fold at each of the four corners inside the tile core. This was also meant to decrease stacking interactions between adjacent junctions, the inhibitory effect of which on the assembly was observed in other aforedescribed designs. However, the main constraint to each individual branch junction is provided by three others within the tile and a further one in an adjacent tile. Altogether this makes the resulting structure a very rigid molecule, which can be and was used as a building block in larger superstructures.

Different assembly strategies were applied to control the periodicity of square cavities in the set lattice formation. In one, the distances between neighbouring tile centres were made as even numbers of helical half-turns (four full turns), which allowed for the same face of each tile to point towards the same lattice

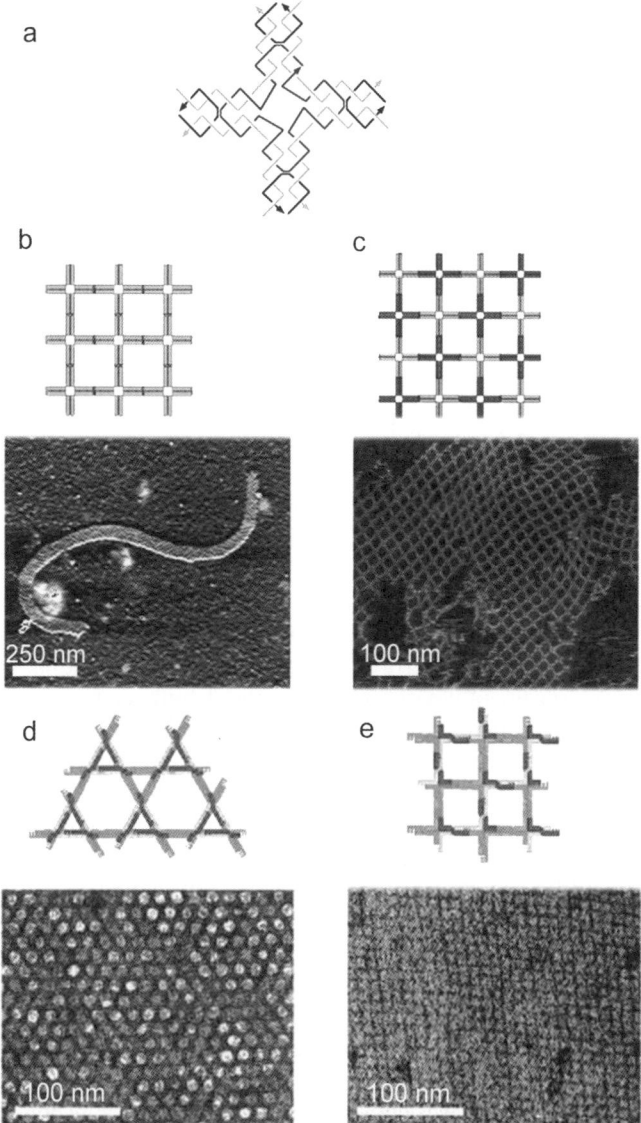

Figure 2.16 Design of high aspect ratio lattices; (a)–(c) 4 × 4 motif, (a) Self-assembly of nine strands into one four-arm tile. One strand (dark grey) contains T_4 loops connecting adjacent junctions, 2D arrangements of the tile into nanoribbons (b) and nanogrids with corrugated design (c) and as seen by AFM, (c) The gridlike structure is formed by tiles with two surfaces: one facing out of the plane (in light grey) and the other into the plane (in dark grey), (from Yan, H. *et al.* DNA–Templated Self-Assembly of Protein Arrays and Highly Conductive Nanowires. *Science*, **301**, 1882–1884 (2003). Reprinted with permission from AAAS), (d), (e) DNA crystals, schematic representations and TEM images of a Kagome lattice formed by assembly of χ-stacked junctions (d) and a square lattice assembled from square-planar junctions held by protein RuvA (e), (Malo, J. *et al.* Engineering a 2D Protein-DNA Crystal. *Angew. Chem. Int. Ed.*, **44**, 3057–3061 (2005). Copyright Wiley-VCH Verlag GmbH and Co. KGaA. Reproduced with permission).

face. Self-assembled structures derived from this design led to the dominating formation of micrometre-long ribbon-like structures with uniform widths of about 60 nm, Figure 2.16(b). The structures, however, were found to be double layered as compared to a flat lattice with edges being higher than in the middle. The latter may indicate a finite curvature of a tube-like structure. The formation of such tubular lattices was hypothesised to be a consequence of each component tile being oriented in the same direction in the lattice plane with possible or incidental locally seeded curvature being extended over the entire lattice. This was tested and found to be reproducible in the second design, referred to as "corrugated".

This design aimed at (1) eliminating the surface curvature inherent in each tile within the assembly; and (2) producing larger pieces of 2D grid with square aspect ratio, Figure 2.16(c). The former was done by interassociating adjacent tiles with the same face of each tile being oriented alternately up and down in adjacent tiles. The latter was ensured by making distances between the centres of adjacent tiles four and a half helical turns, which combined with two double-helical diameters gives approximately 19 nm, the distance consistently measured in AFM images. Consequently, large lattice pieces of up to hundreds of nanometres on each edge were observed with square-shaped cavities.

The most remarkable property of the assembled tiles is the flanking of the four-by-four arrangement cavities by segments of four separate tiles. This is in marked contrast to other similar designs, notably different types of parallelograms,[131,147,148] that contained cavities defined by a single tile.

Such a multitile format logically prompts a technological extension of the concept to combinatorial approaches allowing setting binding cavities within the tiles and consequently leading to specific patterns, the functionalities of which can be addressed by other molecules or biopolymers.[138]

In resonance with this, Turberfield and coworkers[149] reported an approach of synthesising DNA crystals with the help of a bacterial protein, RuvA. The researchers proposed that control over the construction of DNA nanostructures, additional to that of DNA hybridisation and different cohesion modes, can be offered by proteins that are intrinsically manipulative of DNA.

RuvA, one such protein was chosen to demonstrate the concept. The protein is a part of resolvasome, a protein complex possessing Holliday junctions during homologous recombination.[150] The role of RuvA is architectural and concerns holding the junctions in a square-planar configuration that aids in promoting and resolving branch migration.

Set off from this, the researchers assumed that Holliday junctions can be used for building an array and RuvA capable of interacting with the junctions would bind to them during the assembly to determine both the structure and the symmetry of the array. Indeed, this was observed when comparing two types of crystals produced, one made of DNA only and the other in the presence of RuvA, Figure 2.16.

Although both types were built from the same building blocks, immobile junctions of four sticky-ended arms assembled from four strands each, the addition of RuvA changed the lattice symmetry and connectivity dramatically. The effect is truly remarkable particularly from the design perspective.

As discussed above, immobile four-arm junctions are very flexible molecules with unfixed angles at branching points. As a result, they cannot assemble orthogonally unless the angle is fixed, and can form only distorted or skewed structures such as Kagome lattices, Figure 2.16(d). This is precisely what was obtained when DNA-only junctions were assembled. In drastic contrast, in the other lattice type RuvA insertions proved to mediate the fixation of branches to roughly right angles, promoting square-lattice arrangements, Figure 2.16(e).

The design–fabrication path of this approach and the simplicity with which its rationale is realised imply an anticipation of nontrivial approaches in the development of more robust means for constructing novel DNA nanostructures.

Yet, this clearly is viewed from the cognitive perspective of DNA nanodesign as initiated and brought about by all the motifs described thus far.

Serving as a concluding example within this section the design somehow sums up what has been achieved and points to possible progress directions, *e.g.* devising hybrid biomolecular motifs.

It is clear that the time and place for expanding the portfolio of DNA motifs is very favourable to catch the momentum of moving to the next phases towards the challenge of constructing more complex and functional DNA materials. In other words, with all the progress made so far in guiding DNA assembly an obvious question to ask is not "why", as it is the inherent part of the very notion of design to be defined by curiosity and chance, but "where next" all the developed motifs could take DNA and the designer to. A reference to this question in the context of programmable DNA hierarchies is elaborated in the following section.

2.3 Scaffolding Algorithmic

Within the repertoire of the outlined motifs, most, if not all, assembled structures are derivatives of multiple binding interactions between many short strands. In design terms the indefinite set of such interactions is inherently dispensed by robust hybridisation and complementarity offered by the DNA structure. However, the main invariability of this strategy is the strong dependence of the resultant structures on the full completion of all assigned interactions. This renders nanostructures with increasing complexity less probable.

Clearly, to go any further along this route more versatile methods are needed to address this first. Encouragingly, strongly supportive of this is another invariability that remains the key in nanoscale expressions using DNA however complex they may be. Once again, this is the double helix – a molecular code whereby any DNA supramolecular hierarchy can be promoted. The code has been attested via discretionary algorithmic assemblies that allow programming DNA virtually into any shapes, sizes and morphologies, which will be exemplified herein by various hierarchical DNA scaffolds reported to date.

2.3.1   Pursuing Autonomous

2.3.1.1   *Lengthening to Shorten*

A typical protocol in constructing relatively simple hierarchies such as knotted, catenated or tiled structures involves multiple reaction and purification steps making target constructs prone to assembly errors and structural variations ultimately affecting their monodispersity and geometrical integrity. A logical and efficient path around this problem would be to limit the number of steps.

The first example of this strategy was reported by Shih *et al.*[151] As opposed to approaches based on annealing several short strands in assembling DNA nanostructures, the key idea behind this design was to use a considerably longer single-stranded DNA that by undergoing folding adopts a well-defined nanoscale DNA nanostructure. Concurrent to this assumption, which permits a specific assembly mode, another was made to ensure geometrically defined rigidity. This builds upon previously developed architectural paradigms of branched junctions[36–37,85] suggesting that the arrangement of double-helical branches into a continuously triangulated frame as a resistance means to longitudinal deformation and compression should allow for discretely rigid structures. Confirmatively, tetrahedral forms that are of triangulated cages present a mechanically resistant architecture,[152] whereas nontriangulated forms such as previously described DNA cubes[83] and truncated octahedra[41] are easily deformed structures.

Employed in conjunction the two assumptions led to the report of the first triangulated octahedron, the feasibility of which, although predicted,[44] has not been proved before. In designing a single-stranded octahedron whose structure inherently requires components possessing substantial rigidity, immobile junctions were not selected and a combination of crossover motifs was used instead. The octahedron was built from twelve struts constituting the edges of the octahedron connected by six flexible joints representing its vertices, Figure 2.17.

Designed to fold in two stages, a single strand of 1669 nucleotides, dubbed "heavy chain", first binds five "light chains" of 40 nucleotides each to collapse into a branched structure, Figure 2.17(a). During this event five DX struts are formed to give an intermediate structure with terminal half-strut branches – fourteen unique loops with each being complementary to only one other loop. The second stage is the assembly of the loops to the remaining seven struts via intramolecular paranemic cohesion. Thus, the assembled octahedron contains five DX and seven PX components joined as six four-arm junctions, Figure 2.17(b). The structure was confirmed by cryo-electron microscopy and single-particle reconstruction to reveal octahedral objects of the target shape and predicted size, Figure 2.17(c).

Importantly, the designed folding was not accompanied by the formation of knotted or catenated regions, suggesting that no covalent bonding was required for the generation of the octahedron. This, by contrast with the previously reported truncated octahedral and cubic structures, strongly supports the attainment of a single-step construction protocol which by being based on a

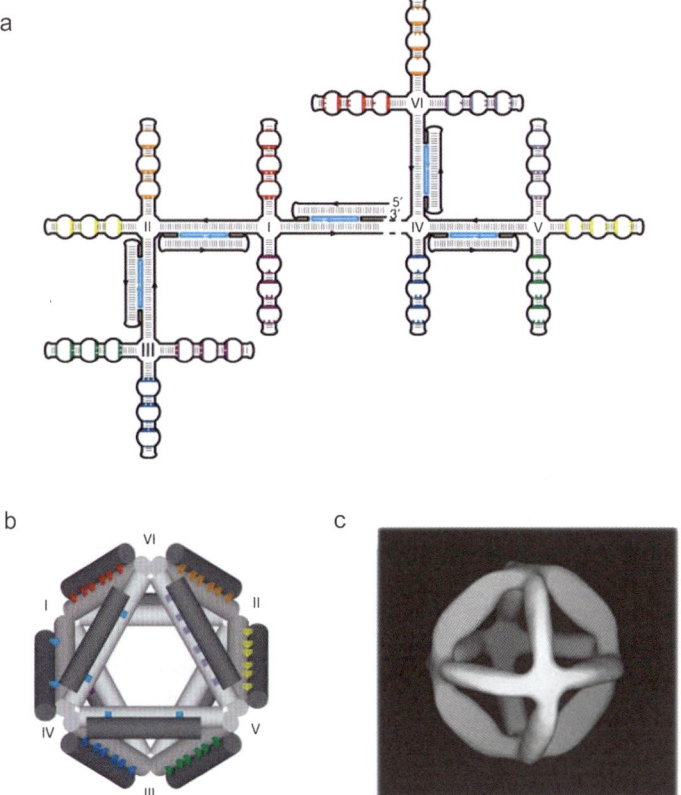

Figure 2.17 Design of DNA octahedron; (a) Branched intermediate assembled from a single heavy chain (black) and five unique light chains seeding five DX struts (cyan). Complementary half-PX loops associating into seven other struts are coloured alike. Roman numerals indicate six four-arm junctions linking the core-layer double helices of each strut. (b) 3D folding of the intermediate into an octahedron. Twelve correspondingly coloured struts make up the edges connected by the six joints forming vertices, (c) 3D map generated from single-particle reconstruction of the octahedron, (reprinted by permission from Macmillan Publishers Ltd: Shih, W. M. *et al.* A 1.7-kilobase single-stranded DNA that folds into a nanoscale octahedron. *Nature*, **427**, 618–621, Copyright (2004)).

single folding event could permit the reversible strategy in the assembly of 3D DNA structures. The design also explicitly shows that a triangulated architecture provides necessary deformation resistance to discrete 3D objects. However, proportionally high ratios of misfolded and distorted octahedra observed in the assembly also indicate that rigidity may not be the sole factor in supporting well-defined monodisperse 3D structures. Additionally, it does not relate to other possible design complications such as relative stability and steric purity of target structures.

2.3.1.2   Gathering to Limit

Although fundamentally predefined by the type of designed hybridisation interactions these, in addition, require enantiomeric selection or stereoselective discrimination of possible diastereomeric forms of a given structure, which to this point has been largely ignored.

An approach allowing a rapid and chirally specific fabrication of rigid DNA geometric constructs proposed by Goodman *et al.*[152,153] may address this. The team reported a synthesis of a family of DNA tetrahedra, each formed from multiple strands in one assembly step.[153] The synthesis completes within a few seconds and is nearly stereospecific, with yields of a single diastereomer as high as 95%. The tetrahedron design consists of rigid triangles built from four single strands, Figure 2.18(a).

When annealed, the strands assemble into one tetrahedral face each forming double-helical edges of the triangles covalently linked at their vertices. Four edges were designed to have single-stranded breaks, nicks, and each vertex was closed through single unpaired "hinge" bases. When combined in equimolar ratio under hybridisation conditions the sequence-specific strands assemble into tetrahedra with nearly quantitative yields. This can be followed by enzymatic ligation of the nick sites to close the structure into a catenane, Figure 2.18(a).

Thus, the hierarchy of cooperative folding assigned to the strands ensures both the one-step character of the synthesis itself and its high efficiency. The assembled tetrahedra can then be used as building blocks for more complex 3D structures. This was demonstrated by bracing different tetrahedra at their gapped edges with programmable single-stranded DNA linkers. In short, linking strands with two identical or different subsequences could be designed to connect preassembled tetrahedra into homo- and heterodimers, respectively. The dimer formation can then be used to investigate the stereoselectivity of the synthesis, Figure 2.18(c). Namely, the position of a gapped edge of one tetrahedron determines the azimuthal orientation of the free end of a linker and by this whether it can reach and hybridise with another tetrahedron. This enables the design of two gap positions, one for each of the two strands forming the edge, and hence two configurations.

In one, the linking strand would project away from its centre, but in the other into the centre, with the latter expected to be inaccessible, Figure 2.18(c). The results of linking experiments proved to be consistent with the presence of large excesses of the diastereomer having the major groove of each helix facing inward at each vertex. This directly states a significant difference between the stabilities of the two possible diastereomers. Furthermore, the rigidity of tetrahedra was assessed as a response to axial compression that was found to be linear and reversible for forces up to 100 pN and undergoing double-helix buckling under higher loads.[152] With such a combination of properties – stereoselective synthesis and structural rigidity – the design conveys a facile route to defining the relative coordinates of any part of a more complex 3D geometry with near-atomic precision.

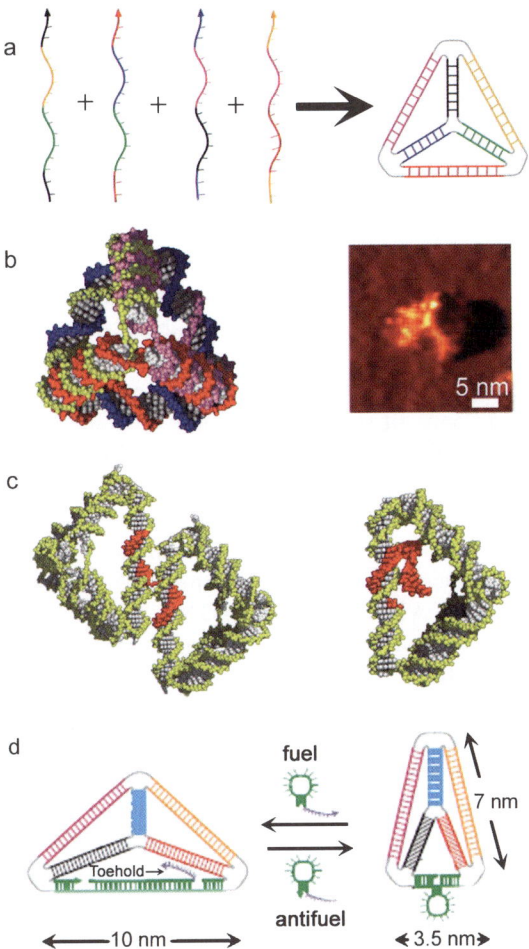

Figure 2.18 Design of DNA tetrahedron; (a) Self-assembly of four single strands containing complementary subsequences (coloured alike) forming the edges of a tetrahedron, (b) Space-filling representation of the tetrahedron, with each strand individually coloured, (left). An AFM image of the tetrahedron with three upper edges resolved (right), (c) Schematic representation of linking experiments. Two possible diastereomers of the tetrahedron can be distinguished by dimerisation via a single-stranded linker (red). In one, the linker is on the outside and is accessible for hybridisation with another tetrahedron (left panel), in the other, the linker is on the inside and hybridisation is prevented (right). A dimer is formed only for the diastereomer with the major groove of each helix facing inward at the vertices, (a)–(c), (from Goodman, R. P. *et al.* Rapid Chiral Assembly of Rigid DNA Building Blocks for Molecular Nanofabrication. *Science*, **310**, 1661–1665 (2005). Reprinted with permission from AAAS), (d) the "closed" tetrahedron (right) is opened (left) upon the addition of a "fuel" hairpin strand and is closed again by adding the "antifuel" hairpin displacing the "fuel" strand by hybridising to the toehold region first, (reprinted with permission from Macmillan Publishers Ltd: Goodman, R. P. *et al.* Reconfigurable, braced, three-dimensional DNA nanostructures. *Nature Nanotech.*, **3**, 93–96, Copyright (2008)).

The practical likelihood of this has been emphasised by two follow-up designs from the same laboratory. In one, a target for a more complex and larger but less symmetric polyhedral nanostructure was set to be demonstrated in the construction of a trigonal bipyramid.[154]

In this design the same single-step annealing procedure was employed to assemble a target trigonal bipyramid from six DNA strands. Each of the strands by running around one of the faces in a way analogous to the original tetrahedral design hybridises to an adjacent strand forming nine edges of approximately 7 nm each (two turns of the double helix).

Again, nicks were introduced into some of the strands, while single unpaired nucleotides were used as joining hinges. Accordingly, the bipyramid is a braced set of triangles formed by three four-arm and two three-arm junctions. Upon ligation of the preassembled structures nine pairs of strands designed to share one edge and the six pairs of strands designed not to share an edge convert to linked and unlinked circles, respectively. The formation of catenated structures found in fifteen different bipyramids confirmed the topological relationship between each pair of strands as designed. However, ligation of more than two strands can give rise to more complex catenanes and can be associated with the formation of by-products. This prompted another concept presented by the group,[155] the essence of which is to make DNA tetrahedra reconfigurable in response to specific molecular signals. To manipulate the shape of a nanostructure in a controllable manner is extremely attractive as it may lead to the development of nanoscale machinery operational on a single-molecule level.

To address this, a tetrahedron with an additional structural module, reconfigurable edge, was designed. Similarly to the two previous designs, the object was assembled from four short strands each running round one side of the tetrahedron,[153] Figure 2.18(a). Six pairs of complementary regions form six edges with each strand containing three of these domains. However, in this design the tetrahedron was made uneven in all its edges: five of the edges were about 7 nm in length and the sixth about 3.5 nm, Figure 2.18(d).

This underpins the key distinction of the design – the insertion of a hairpin loop into the centre of the strand that is opposite the nick of its complementary strand. The loop is meant to act as a reconfigurable motif such that the tetrahedron would be able to adopt two states – "closed" and "open" as imposed by the response to the configurational change of the loop. More specifically, when the motif folds as a hairpin the edge it is inserted to is at its shortest and the tetrahedron is closed. The extension of the edge by a "fuel" strand (an oligonucleotide that opens the hairpin by hybridising to both halves of the stem and to the loop) gives a stretched or "open" form of the tetrahedron. The opened edge thus becomes longer and is of a continuous 10-nm duplex with two nicks. The fuel strand contains a 3.5-nm extension that within the formed tetrahedron remains unhybridised, giving rise to a toehold,[156] Figure 2.18(d). This can then be used to promote the displacement of the fuel strand by an "antifuel" strand. Because the latter is complementary to the entire fuel strand including the extension, its hybridisation propensity is higher and can be used to reverse the hairpin to its original closed state. Thus, sequential addition of the strands can change the length of the reconfigurable edge by a factor of three, repeatedly. Encouraged by the specificity of the reaction, the researchers went further and designed a four-state tetrahedron incorporating two hairpin loops in opposite edges. The loops are topologically similar but assembled from different sequences. Engaging the two edges in an independent manner was shown to promote the switching of the tetrahedron between four different states with all possible combinations of open and closed edges. It was possible, for instance, to convert the tetrahedron from an entirely closed to a partially or an entirely open state by opening either hairpin or both sequentially.[155]

The design represents a DNA geometry dynamically controlled on the nanoscale. This makes it potentially useful for actuating single molecules or discrete assemblies on mesoscopically structured templates or for rendering conformationally defined DNA shapes robotic, *i.e.* reversibly operational in 3D.

The challenge of their technical realisation, however, brings one back to the problem of scalable DNA manipulation and ultimately recaps on the notion of single-step scaffolding or unrestricted shaping using DNA.

2.3.2 Assigning Arbitrary

Whichever geometrical shape and specific function entices DNA nanodesign, in all its existing concepts and constructional forms it makes one, certainly disguised in some cases, but explicitly stated in many others, reference.

The reference is made to the possibility of developing a general method for autonomous construction of artificial DNA patterns. This, what appears to be a well-accredited goal,[6,12] can be broadly defined as a "universal constructor", the von Newmann's postulate[157] of self-reproducing automata capable of converting an algorithmic input into commands specifying the construction of an arbitrary object.

2.3.2.1 Synchronising Local

Considering an algorithmic path towards instructive DNA assembly it would be only fair to say that every DNA design is directly contributory to this. Notably, if not the idea than at least the theoretical foundation for DNA automata has been seeded by pioneering concepts of DNA computing[70] and tiling.[12,158] Subsequently, the introduction of DNA nanotechnology[8] that comes from recognising the empirical arbitrariness of the processing of complex information using DNA inspired the concept of algorithmic assembly formulated by Winfree.[12]

The concept originated as a theoretical attempt of providing a link between DNA self-assembly and computation. In more specific terms, it sets out to demonstrate the compatibility of DNA self-assembly with Turing-type computers; finite-state machines that by moving back and forth on an infinite tape possess memory sufficient to perform any computation otherwise performed by other mechanical devices of any complexity. This gives the computers a claim for universality and the relation with comparably simple DNA structure becomes more tempting to both hone the theory and apply it to prescriptive DNA assembly.

In this vein, Rothemund *et al.*[159] hypothesised that a natural Turing computation model that can be implemented by 2D algorithmic self-assembly would be the class of 1D or local cellular automata. An update rule for local automation was chosen as the binary exclusive disjunction or exclusive-or (XOR) function that allows each cell to be computed as the XOR of its two neighbours at each time step, and as it grows a fractal pattern, namely a Sierpinski triangle, is generated.

This implies that to perform a cellular automation using molecular self-assembly a similar pattern with an intrinsic binary function should be employed. The formation of tiles (triangular pattern) by sticky-ended cohesion (typical binary function) was deemed an obvious choice. Thus, the assigned program is not a sequence of instructions but a 2D aperiodic framework of tiles assembled from different pieces via sticky ends.

However, because, unlike the synchronous cellular automation, molecular self-assembly is asynchronous and prone to errors the necessity of solving four main challenges was identified as preceding to the successful experimental implementation of the chosen cellular automata.

The first challenge, defined as the translation of abstract tiles into molecular tiles capable of forming 2D crystals, is solved with the help of previous designs of highly periodic tiled crystals.[46,124,147] Two other challenges – finding specific binding domains from which molecular tiles can be assembled cooperatively to minimise errors and according to the logic of the chosen type of abstract tiles – is addressed by sequence-specific hybridisation patterns previously introduced to support addressable intertile interactions.[124,160] And for the final challenge – specifically nucleated or input assembly of molecular tiles with the exclusion of spurious nucleation – the answer is the design of single-stranded templates capable of directed nucleation maintaining complex lattice-type assemblies (*e.g.* barcode patterning).[46,161]

The designed strategy was confirmed experimentally in the translation of an abstract pattern into DNA tiles derived from DX motifs and long single strands used as a tile-nucleating input. The produced patterns contained 100–200 correct tiles assembled with low error rates. This explicitly indicates that all four features, stipulated as innate for Turing-type universal computation by crystallisation, can be achieved experimentally with high yields and accuracy. From an input template controlling nucleation of crystal growth via selective and cooperative sticky-end-directing tile associations to the formation of extended crystals this altogether shows that DNA can be expressed into any desired algorithm for computation and, what in the context of this chapter carries higher importance, for any construction target. The latter has been proved in an approach reported by Rothemund[13] that opens up fundamentally new horizons in arbitrary DNA nanodesign.

2.3.2.2   Prescribing General

The approach, introduced as a "scaffolded DNA origami", generalises all preceding concepts of DNA nanoconstruction in an elegant and ingeniously simple one-pot method of using many ssDNA molecules to direct folding of one long single-DNA strand into practically any 2D shape with a diameter of 100 nm and spatial resolution of 6 nm.

In general terms, in assembling DNA origami a computationally assigned layout of a scaffold strand, a long single-stranded DNA, was used to fill a specific pattern row by row. To fix and rigidify the strand within the pattern, staples, short ssDNA molecules, were introduced to cross over from one row to another hybridising with the corresponding complementary regions of the scaffold strand.

In specific terms, the design required the sequence of five steps.

In the first, a geometric model of a DNA structure approximating a desired shape of specified sizes is built. The shape is then filled from top to bottom by an even number of parallel helical duplexes that are cut to fit the shape in sequential pairs. These are constrained by being an integer number of turns in length thus requiring the fixing of the helices together by a periodic array of crossovers to mark locations at which strands running along one helix switch to an adjacent helix and continue from those points.

In the second step, which concerns designing a folding path, the scaffold strand composed of about 900 nucleotides is introduced to fold back and forth in a cellular pattern such that it provides one of the two strands in every helix. This introduces a raster progression from one helix to another, thus generating a set of crossovers, termed "scaffold crossovers", with the scaffold able to form a crossover only at those positions where the DNA twist places it at a tangent point between helices. This, in turn, implies that to transfer the progression from one helix to another and onto a third the separation between successive scaffold crossovers must be an odd number of half-turns. However, where the direction is reversed vertically returning to a previously visited helix, the separation must be an even number of half-turns.

In the third and fourth steps, the designed model and path are represented as lists of DNA lengths and offsets in units of half-turns that together with the sequences of the scaffold and staple strands are input to a computer program. The final two, fourth and fifth steps, are rectifying. The fourth is in that it introduces minimisation parameters to balance the twist strain between crossovers imposed by the asymmetric nature of the helix (*e.g.* major and minor grooves, a non-integer number of base pairs per half-turn). Namely, periodic crossovers are arranged with glide symmetry allowing minor grooves to face alternating directions in alternating columns of periodic crossovers, whereas to balance scaffold crossovers their twist is recalculated and their position is changed (typically by one base pair), with staple sequences recomputed accordingly. The fifth is in that it introduces an additional constraint parameter to strengthen the binding of the staples with the scaffold strand. This is stimulated by the fact that the staple strands at their meeting positions leave nicks in the backbone. By merging adjacent staples across nicks fewer but longer staples with larger binding domains were generated.

Additionally, as the designed folding path compatible with a circular scaffold leaves a "seam" contour that the path does not cross, another pattern of breaks

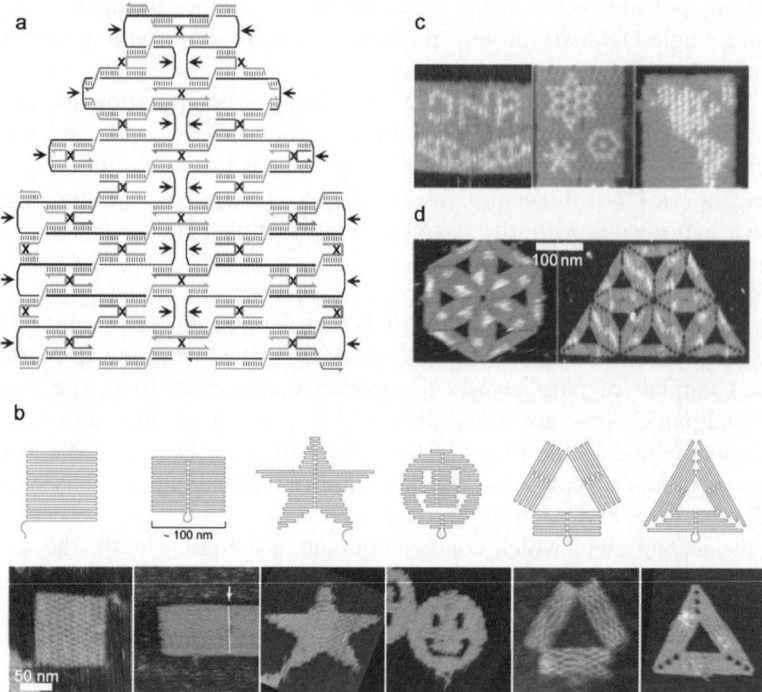

Figure 2.19   Design of DNA origami; (a) A shape created by parallel double helices
assembled from a scaffold strand (black) running through every helix and
joined with staple strands by periodic (x) and scaffold (black arrows)
crossovers, (b) Different folding paths and their corresponding AFM
images; square, rectangle, star, disk with three holes, triangle with rectangular domains, and sharp triangle with trapezoidal domains (from left
to right). Hanging tails and loops denote unfolded sequences, the white
line and arrow indicate blunt-ended stacking, and white features (sharp
triangles) indicate hairpins, (c) AFM images of assembled patterns; a
pattern representing DNA with a pixelated turn of ~ 100 nm, an hexagonal pattern reflecting a nearly hexagonal pixel lattice, map of the
western hemisphere with the scale of $1:200 \times 10^{12}$, (d) Assembly of sharp
triangles with extended staples into hexagons (left) and lattices (right),
(reprinted by permission from Macmillan Publishers Ltd: Rothemund,
P. W. K. Folding DNA to create nanoscale shapes and patterns. *Nature*,
440, 297–302, Copyright (2006)).

and merges may be created to cement the seam alike. In this case, a seam crossed
or spanned by merged staples would be "bridged". Thus, when used in combination the steps can give rise to a single algorithm allowing one-pot synthesis
of an intended DNA shape. Figure 2.19(a) sums up all the design steps.

To provide empirical evidence to the algorithm, a modified version of the 7-kb
circular genomic DNA of a M13 phage was chosen as the scaffold strand.

Not all segments of the obtained strand could be used for folding in most
designs and therefore "remainder strands" were added to complement unused

sequences. A typical experiment involves mixing staple and remainder strands (200–250 strands in a 100-fold excess) with the scaffold strand followed by the annealing of the complex and imaging by AFM.

Six different folds were reported to demonstrate the method, Figure 2.19(b).

In constructing one, the simplest and most flexible, a 26-helix square with no vertical reversals, a linear scaffold with 2.5-turn crossover spacing together with 26-mer staples that bound each of two adjacent helices were used. Consistent with the design, more than ten per cent of observed structures were well-formed squares, with the remaining being deformed rectangular fragments.

In testing the formation of a bridged seam, another design, a rectangle was produced using 1.5-turn crossover spacing, 32-mer staples and a circular scaffold. The central seam in the rectangle assembled with nearly quantitative yields was indeed readily formed and could be easily visualised.

To probe the applicability of the algorithm to the construction of arbitrary shapes, a five-pointed star was designed with 1.5-turn spacing, 32-mer staples and a linear scaffold. Again, as in the case of the square, only ten per cent of the structures were well formed. In both cases, the low yields could be attributed to strand breakage during syntheses that involves enzymatic reactions and subsequent AFM processing. Indeed, when a circular scaffold was folded into stars yields improved to 63%.

The aim of a third form was to show that scaffolds can be routed arbitrarily through different shapes, *i.e.* without the seeming requirement for topological disks. With this, a three-hole disk, a smiley face, was constructed assuming a highly asymmetric folding path characterised by five distinct seams. The design also revealed a tendency to deformations, yet still giving about 70% of well-shaped structures.

In an attempt to combine distinct raster-fill domains in a nonparallel arrangement, a triangular shape consisting of three separate 2.5-turn spacing rectangles held together by only single covalent bonds along the scaffold was designed. The flexibility of this type of single bonding at the vertices may have been the reason behind the rare observation of the target equiangular triangles that also tended to undergo stacking through their rectangular domains.

A solution to this imperfection was found in a sixth form, "sharp triangles", built from trapezoidal domains with 1.5-turn spacing. The domains were designed with slanted edges that by allowing the trapezoids to meet at the triangle vertices generated more extended interfaces to be bridged by additional staples. Considerable increases in observed yields of well-formed triangles; more than 80% for triangles bridged at the vertices and 50% for unbridged against 1% for the rectangular triangles, strongly confirmed the design.

By applying the principles attested in fixing the DNA scaffold in a specific shape, it was hypothesised that staple strands can mediate the decoration of shapes with arbitrary patterns of binary pixels, which can be done through labeling DNA with small molecules such as biotin or fluorophores.[13]

In this regard, if the original set of staples is viewed as binary "0"s, then a new set of labelled staples (one staple for each original) can give binary "1"s. Thus, by mixing certain subsets of these strands any pattern can be created.

To show this, a set of hairpins was designed to decorate the middle of staples at the position of merges. The resulting pixel patterns were found to depend on merge patterns and to be rectilinear, *i.e.* with adjacent columns of hairpins on alternate faces of the shape, or staggered, *i.e.* geometrically packed with all hairpins on the same face. When imaged by AFM, labelled staples expectedly gave greater height contrast than unlabelled staples, 3 nm against 1.5 nm. This resulted in a pattern of light "1" and "0" pixels. Some patterns, each with 200 pixels, constructed using these conventions are shown in Figure 2.19(c).

Similarly, combinations of shapes were proposed to be constructed using "extended staples" that can be designed to connect shapes along their edges, which is done by merging and cutting normal staples along the edges. Starting with the sharp triangles, Figure 2.19(b), this approach was used to create finite (hexagon) and periodic (triangular lattice) structures, Figure 2.19(d).

Designwise, the method proposed by Rothemund is unique in that it deliberately ignores the core of accepted practices in DNA nanotechnology involving careful sequence optimisation and synthesis; its uniqueness, however, is mainly due to the nature of scaffolded DNA self-assembly, the hierarchical success of which is dictated by a single parameter – long single-stranded DNA. Indeed, it may not be equally possible to construct smaller scaffolds based on scaffold strands of considerably shorter hybridisation segments as these would encourage assembly variations. In this vein, the arbitrary one-pot synthesis offered by the concept may be exclusive for building large 2D DNA scaffolds whose construction principles though may provide very robust means for technological applications are to remain less amenable to small-scale systems.

Methodologically, DNA origami is an extremely attractive approach for assembling discrete extended arrays capable of mediating various macromolecular arrangements. As proposed by Rothemund,[13] an application of the method would be the creation of a "nanobreadboard", which can serve as a template for seeding different bio- and physicochemical components in a highly controlled manner. The idea of using nanodefined periodic arrays for nucleating protein crystallisation, modelling protein assemblies, building enzyme factories or multilayered matrices, is very appealing and has already given rise to some interesting designs.

For instance, Liu *et al.*[162] demonstrated that selectively designed DNA aptamers can be used to link proteins to periodic sites of a self-assembled DNA array. These three components compile a simple system, in which a linear TX-built DNA lattice has aptamer-containing docking sites that are regularly separated as loops being exposed upon the lattice assembly to bind thrombin. In another report, Park *et al.*[163] presented an approach to programming the self-assembly of protein arrays with controlled density. The method is based on a previously described family of four-by-four tiles[138] that by exhibiting an array of periodic square cavities can template the construction of periodic streptavidin arrays.[138] With further optimisation of the general four-by-four framework needed to achieve controlled spacing between individual proteins, two types of junction units instead of one in the original design (Figure 2.16) were designed to assemble alternately. This allowed the introduction of the varied periodic spacing between protein molecules on the resulting lattice as the units can be modified selectively with protein-binding biotin groups.

2.3.3 Adding up to Third

Conceptually, DNA origami is likely to expand onto 3D with a variety of more complex systems testing one's ability to manipulate DNA. The idea is not new and its viability is being tried by several laboratories. However, most designs do not expand beyond a few forms, of which the dominant one remains the tubular morphology first proposed by Seeman[9] and experimentally followed by others.

2.3.3.1 Wrapping to Shut

The impression of the special preference made for DNA nanotubes in the exploration of the third dimension may be misleading, but the choice is yet logical. Strictly speaking, nanotubes are not 3D objects but nonplanar 2D arrangements with the characteristics of the third dimension. In In this way, nanotubes[130] may serve as an intermediate type of DNA scaffolding for future systems that are expressed in and employ the dimension in their architectures.[6,9]

All examples of nanotubes designed and characterised to date can be loosely viewed as closed or wrapped tiles. In fact, the first nanotubes were accidentally encountered in tile-based assemblies[124,130] when it was suggested that the tiles are intrinsically inclined to form narrow micrometre-long ribbons by curling and closing on themselves.[164–166] The proposal appears to be consistent with the fact that the tubes exhibit helical order.[164,167] Yet, with several concepts proposed for the construction of tubular DNA structures, a general framework for designing and classifying nanotubes has not been established. Nevertheless, however natural and expected it is the utilisation of secondary DNA motifs in conjunction with the 2D tiling provides a very promising strategy.

The main question here is how to extract the set of rules able to link the design of tiles with their tubular assembly in a predictable manner.

The first on-purpose tiling design[138] that was reported to lead to regular nanotubes relied on the apparent consistency of tile symmetry with the tube-type assembly and morphology. Although the design rationale is pronounced in the systematic observation of uniformly wide ribbon or tubular structures, the relationship between the tile curvature and its geometry and spatial selection in the tube formation was not entirely clear. As a consequence, a variety of tube sizes mixed with fragmentary lattices observed for the design ultimately incites the lack of control over the desired forms and within specific morphologies.

A following design from the same group concentrated on solving this through the introduction of a covalent constraint that also served as an orientation marker.[164] The design aimed at the conversion of flat tiles into uniform nanotubes with the help of thiols introduced to dsDNA loops projecting out of the tile plane. This was meant to render the systems switchable at the expense of reversible disulfide-bond formation. As judged by a series of microscopy techniques, nanotube structures were indeed the only observed products resulting from annealing thiol-modified strands.

The design was featured with two tile building blocks. Both were based on the TX motif[46] and were named accordingly as TAO (tile A) and TAO + 2J (tile B with two bulged loops); that is, as formed by three coplanar double-helical domains (T) with strands reversing direction at crossover points, hence antiparallel (A), having an odd number of helical half-turns between the points (O), Figure 2.20(a).

The two bulged loops in tile B protrudes one into and one out of the tile plane. All B tiles were modified by replacing one loop with two thiols coupled one on each of the termini. Sticky ends were placed at the corners of the tiles to enable two types of tiling, Figure 2.20(b). Nanotubes are formed as a result of alternating assembly of B and A tiles. In this mode, dsDNA helix axes are parallel with the tube axis, and thiols are inside the tubes. The resulting structures were micrometres in length with uniform widths of about 25 nm, Figure 2.20(c).

Importantly, striations with periodicity of 27 nm (predicted length of AB tile units) across nanotubes clearly seen on electron micrographs correspond to outwardly protruding loops and indicate closed-ring structures, rather than spiral structures otherwise characterised by stripes with a diagonal slant. The hypothesised burial of disulfides within the tubes as the indication of the disulfide formation between adjacent B tiles – the force that causes the formed lattice to curve into a tube – was confirmed using thiol-reactive colloidal gold that failed to detect any thiols on the outer surface. This provides sensible evidence for the fact that the structure is formed periodically by an even number of tiles per ring layer, namely eight tiles per layer, and that the rings are flat. One potential drawback of this design is the lack of generality in constructing DNA-only systems without the need for additional modifications.

With this in mind, Rothemund et al.[167] set out to devise general design principles enabling DNA-nanotube assembly by introducing a DNA-only nanotube motif. A proposed motif is composed of two parts. One is a core presented by DX motifs[45] and contains five pseudoknotted strands that form two double-helical domains fixed by a pair of crossover points. Two different cores assigned by two different sequences, were named RE and SE. The second part is the two pairs of ssDNA sticky ends flanking the core allowing it thus to bind other cores, Figure 2.21(a).

A single core yields a single tile able to assemble into a lattice independently or in combination with another tile type, which is gratifyingly done by assigning specific sticky-end codes, Figure 2.21(a).

Lattices may be constructed with stripes running either diagonal to or perpendicular to the long axis of the tiles. This can be accomplished, for example, by using reporter hairpin structures formed on the outside of nanotubes in a way similar to the previously described approaches. A collection of tiles can thus be generated as a program for creating a particular structure using a particular pattern that in turn is set by a particular choice of sticky ends. These were designated by lowercase letters as *s*, *d* and *p* for single-tile, diagonally striped and perpendicularly striped, respectively. In order to drive the assembly of such patterns into tubes it was argued that a lattice should exhibit an appropriate symmetry. The one that the designed DX tiles can satisfy was chosen as a rotational symmetry compatible with an intrinsically curved geometry for a patch of tiles.

The symmetry confers necessary nonzero curvature and promotes tubular formation that is also mediated by the orientation and spacing of the crossover points. These identify factors primarily contributive to the tube formation, with most of them met by an antiparallel DAE type of DX motifs having an even number of half-turns between intermolecular crossovers, *i.e.* DAE-E. Other antiparallel DX types that have spacings with odd numbers of half-turns were therefore considered incompatible with tube formation. Basic features of the resulting model were confirmed by AFM on nanotubes derived from different tile sets, Figure 2.21(b).

For instance, it was shown that REs or SEs alone can be programmed to give tubes similar to those formed from combinations of other tiles. Diagonally striped tubes were assembled from REd and SEd tiles, whilst REp + SEp pairs gave nanotubes with a pattern of stripes perpendicular to the long axis of the tubes, Figure 2.21(b).

In each case, the tubes were very similar and in most cases the nanotubes were closed single-walled tubular structures able to open into sheets. The model assignment of the inside and outside of the tubes was tested by assuming that large decorations on the inside of a given tube would lead to the disruption of tubular formation. Examined by AFM, in general it was confirmed. Of different variants of tiles (mainly of SEs type) with hairpins predicted to lie on the outside and inside of tubes or between helices, the formation of morphologically normal tubes was observed only for tiles with outside hairpin insertions. In the other two the tubular formation was disrupted, leading to variable morphologies from short open to flipped tubes and their combinations. Consistent with this, only outside-located hairpins modified with biotin molecules were found to be accessible for biotin-binding protein streptavidin. The sum of the experiments provided good correlation for the proposed close-packed model of a heptagonal tube.

Other groups thought along the same lines and designed similar symmetry-simplified nanotubes using different motifs.

For example, Wei and Mi[168] reported a new motif based on the TX motif that they referred to as triple-crossover triangle or TXT. The motif is composed of eight ssDNA strands hybridised to three double-helical strands assuming the shape of a triangular prism through strand exchange at six immobile crossover points.[46] Given the shape it was deemed relevant to consider the motif more rigidified as compared to planar motifs and adaptable to algorithmic assembly of DX motifs,[124] but in 3D as, unlike DX, TXT has an extra double helix that can provide arrays assembled in 2D with an extra layer thus addressing the third dimension. The formation of 1D, 2D and 3D arrays was envisaged using this strategy. However, only 1D filamentous arrays designed as longitudinally polymerised TXT were proved experimentally.

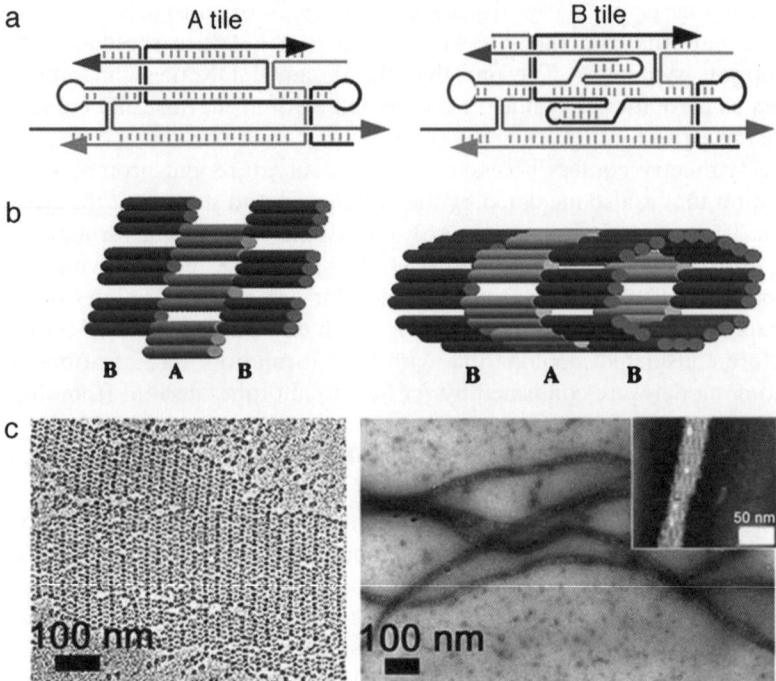

Figure 2.20   Design of DNA nanotubes; (a) Two TAO tiles, (b) Models of TAO lattice
(left) and nanotube (right) with cylinders representing double helices,
(c) TEM images of the lattice (left) and nanotubes (right) with alternating
stripes separated at ~28 nm, Inset is an AFM image of a nanotube,
(reprinted with permission from Liu, D., Park, S. H., Reif, J. H.
and LaBean, T. H. DNA nanotubes self-assembled from triple-crossover
tiles as templates for conductive nanowires. *Proc. Natl. Acad. Sci.*, **101**,
717–722. Copyright (2004) National Academy of Sciences, USA).

In a similar but independent study by Park *et al.*,[169] within which such an
arrangement of double helices was termed a three-helix bundle or 3HB, the
formation of 1D filaments was also observed. These were uniform in size but
thinner, within the diameter of a single tile (~4 nm). They also were shown to
be closed as metallisation experiments aiming at the fabrication of metal
nanowires using the filaments as templates were considerably larger (20–50 nm
wide) suggesting their deposition on the outside of the tubes.

Mitchell *et al.*[166] took advantage of the chiral nature of the double
helix and designed a DX-based tiling system that formed chiral ribbons
with straight edges. The chirality of the nanotubes results from the pattern
of connections between two types of tiles that were designed to tessellate
by assembling as alternating rings or nested helices, Figure 2.21(c). By TEM,
the ring and helical order of the tiles was confirmed by biotin-streptavidin
labeling revealing straight and spiral stripes across the assembled tubules,
Figure 2.21(d).

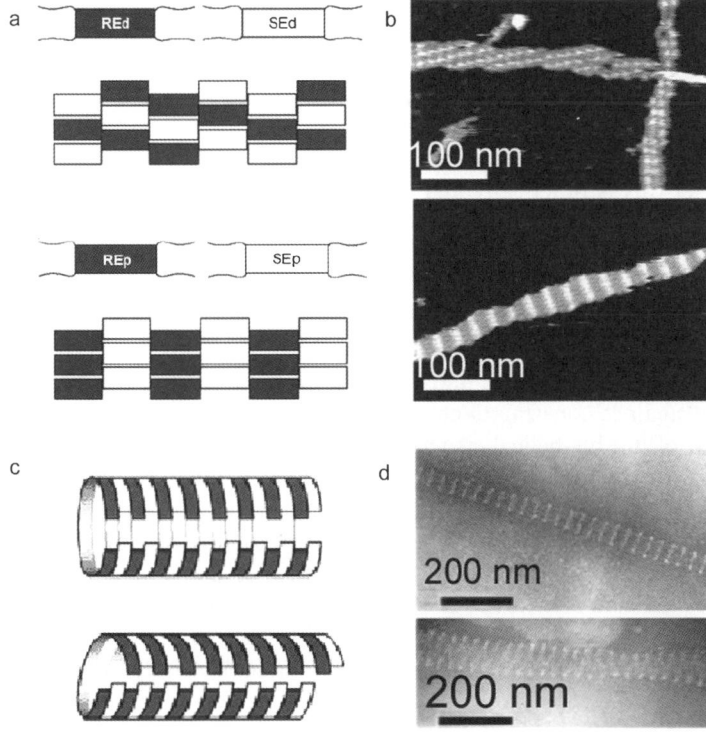

Figure 2.21 Design of programmable DNA nanotubes; (a) Schematic representation of DNA tiles with two different DAE-E cores, RE and SE. In combination the tiles can assemble into lattices with diagonal, REd + SEd, or perpendicular, REp + SEp, stripes generated by reporter hairpins (not shown), (b) AFM images of the diagonally (top) and perpendicularly (bottom) striped open nanotubes, (reprinted with permission from Mitchell, J. C. *et al.* Self-Assembly of Chiral DNA Nanotubes. *J. Am. Chem. Soc.*, **126**, 16342–16343. Copyright (2004) American Chemical Society), (c) Folding of 2D arrays assembled from two types of tiles (indicated in grey and white) into tubes with alternating rings (top) and nested helices (bottom) of the tiles, (d) TEM images of correspondingly stripped nanotubes. Electron transparent stripes indicate protein-labelled tiles, (reprinted with permission from Rothemund, P. W. K. *et al.* Design and Characterisation of Programmable DNA Nanotubes. *J. Am. Chem. Soc.*, **126**, 16344–16352. Copyright (2004) American Chemical Society).

2.3.3.2   Framing to Classify

The designs present the current state-of-the-art and give the set of structural parameters successfully applied to designing nanotubes primarily stressing on rigidity provided by crossover motives and curvature-promoting symmetry in tiling. However, the high occurrence of morphologically polydisperse tubes in all of the examples prompted the development of more robust theoretical

frameworks. In pursuing this goal, Sherman and Seeman[165] set forth a classification approach to designing minimally strained nucleic acid nanotubes.

It was noted that all nanotube designs were disposed to the idea of simpli-
fying design processes by adapting certain types of symmetry. This invariably
led to most nanotubes being of rigid regular polygons, which can be regarded as
a limitation to the problem of nanotube design.

Whether this is practically unavoidable or is justified by the lack of experimental foundation for establishing interrelation between local and global
geometries, that would otherwise allow a much broader variety of tubes
including those with concave cross sections, will probably continue being an
open question.[9,165] Nevertheless, the proposal by Sherman and Seeman aims to
determine parameters in response to which nanotubes can be classified.
Approximating this to the most stable and minimally strained structures
modelled against constraints characteristic of DX motifs was suggested as a
perfect fit of double-helical structures to a set of duplex parameters.

With a number of further approximations allowed by such parameters,
classes of low-stress DNA nanotubes deferring by a number of helices and
cavities of varied shapes were predicted. Although the approach does not
consider strained forms it may be used as a model for secondary and tertiary
structures, particularly in the context of primary and secondary motifs.

Subsequently, Kuzuya *et al.*[170] reported direct experimental evidence for the
model in designing tubular sheaths with predicted cavity shapes and sizes,
which can also give an adaptation model for the earlier introduced six-helix
bundle motif.[130]

The main objective of this design was to provide a hollow system that can
partially cover and then sheath its content as opposed to the more common and
recognised approach of delivering target entities into the preformed cavity. The
solution to this was found as the design of DNA half-tubes that cover one side of
an encapsulated species and then by forming full nanotubes completely cover it.

The structures were assembled from newly designed six- and eight-helix
bundles, 6HB and 8HB. A typical six-helix-based tube is composed of two
different TX molecules with a bend of $120°$ between their two DX components.
Two molecules of the resulting motif join face-to-face to produce the tube. The
eight-helix bundle derives from a single arched four-helix that links with
another copy of itself to give rise to the eight-helix tube. As revealed by
AFM,[170] such half-tube components formed 6HB and 8HB nanotube arrays of
regular thicknesses.

With this example, all designed tubes based on crossover motifs could be
roughly put into two main design categories.

In one, nanotubes are built from monomers that polymerise with their
terminal sticky ends under a nonplanar angle leading to 2D arrays that by
closing on themselves give tubes.[166,167] This type is prone to size and shape
variations due to the axial rotation (about the tube axis) it is subject to, thereby
slanted or skewed tubes can be produced. In the second, nanotubes are strongly
correlated with the exact number of helices assembling into a building
block with a strict requirement for a bend angle between crossover components.

The latter is generated by strand switching that in principle can be designed at specific positions to lead to specific structures.[165] Both cases, however, assume certain promiscuity within the key parameters that is likely to be attributed to the multistrand nature of the DX motifs. An expected question in this regard would concern how many strands are required to produce nanotubes.

In their "approaching the limit" paper, Liu *et al.*[171] answered the question by designing nanotubes that can be assembled from a minimum number of strands. In pursuing the limit, the researchers emphasised sequence symmetry in building a DX motif to arise from a single strand.

The strand was designed to have four palindromic (self-complementary) segments that by associating with other copies of themselves formed a two-stranded complex. The motif contains two duplex domains and four single-stranded overhangs, thereby it assembles into 2D arrays. Because the following folding of the arrays into tubes is an intracomplex process, as opposed to intercomplex growth of extended 2D lattices, it was presumed that at low DNA concentrations nanotube assemblies would be kinetically more favourable. Low concentrations could also help to avoid the formation of unstable amorphous and sporadic structures whose assembly may result from the tendency of the designed palindromes to associate into complexes other than the target. An exclusive morphology represented by long nanotubes uniform in height and width clearly seen by AFM confirmed these assumptions.[171]

Characteristically of all nanotube designs, these nanotubes were rigid structures and their collections could be well-organised arrays. Although expected, the problem of organising nanotubes into specific arrays presents certain technological interest, but evidently requires nanotubes being periodically aligned, which can be hampered by their physical properties such as the increased persistence length and rigidity. As demonstrated by Lin *et al.*[172] an effective strategy to aligning nanotubes into well-defined arrays is to combine top-down and bottom-up methods. In the proof-of-principle study it was shown that nanotubes can be assembled by simply being annealing from a pool of single-stranded DNA[171] and then randomly deposited on solid substrates. This is followed by the alignment of the tubes into arrays by a surface-patterning technique, *e.g.* soft lithography, which allows the versatile manipulation of the nanotubes within the limits of their morphological integrity.[172] The latter can be enhanced to make nanotubes more robust to practical applications, for example, by ligation of multiple nick sites between tiles in assembled tubes yielding sturdier assemblies.[173]

2.4 Outlook

One of the main aims of prescriptive self-assembly is to program structures necessarily functional on the nanoscale. This makes nanodesign a multistep process whose progress can be monitored at each individual stage. The progress outlined in this chapter is that of initial stages that can be viewed as constructing "craft materials"[174] or indeed structural precursors for operational devices. The transition may not be straightforward and may require more robust models to underpin stimuli-adjustable designs whose structural responses can be translated into meaningful mechanical work. Concentrating on devising more universal models presents a systematic approach and is not always the right choice as it restrains intuitive and imaginary input into nanodesign. Practically, however, the generalisation of DNA self-assembly can directly lead to dynamic models performing in 3D that are of the most value but also are more challenging to construct.[175,176] This certainly does not prevent us from designing spatially "less sophisticated" nanoscale systems, with a number of dynamic designs with mechanical,[140,156,177] catalytic,[178,179] switching,[180,181] optoelectronic,[182,183] plasmonic[184] and inductive[185] properties being reported. The designs emerge rapidly, which essentially indicates that the anticipated transition has already begun. Restating the potential of DNA to template the construction of nanoscale machinery their success discloses that merging the attested rationales behind structurally finite constructs and static motifs with functional integrity of dynamic designs serves as a branching point in creating higher-generation 3D architectures – the scaffolding of autonomous robotic systems.

References

1. J. D. Watson and F. H. C. Crick, Molecular Structure of Nucleic Acids: A Structure for Deoxyribose Nucleic Acid, *Nature*, 1953, **171**, 737–738.
2. M. H. F. Wilkins, A. R. Stokes and H. R. Wilson, Molecular Structure of Nucleic Acids: Molecular Structure of Deoxypentose Nucleic Acids, *Nature*, 1953, **171**, 738–740.
3. J. D. Watson and F. H. C. Crick, Genetical Implications of the Structure of Deoxyribonucleic Acid, *Nature*, 1953, **171**, 964–967.
4. R. E. Franklin and R. G. Gosling, Molecular Configuration in Sodium Thymonucleate, *Nature*, 1953, **171**, 740–741.
5. R. E. Franklin and R. G. Gosling, Evidence for 2-Chain Helix in Crystalline Structure of Sodium Deoxyribonucleate, *Nature*, 1953, **172**, 156–157.
6. N. C. Seeman, DNA in a material world, *Nature*, 2003, **421**, 427–431.
7. U. Feldkamp and C. M. Niemeyer, Rational Design of DNA Nanoarchitectures, *Angew. Chem. Int. Ed.*, 2006, **45**, 1856–1876.
8. N. C. Seeman, Nucleic Acid Junctions and Lattices, *J. Theor. Biol.*, 1982, **99**, 237–247.
9. N. C. Seeman, Macromolecular design, nucleic acid junctions, and crystal formation, *J. Biomol. Struct. Dyn.*, 1985, **3**, 11–33.
10. N. C. Seeman, DNA Nanotechnology: Novel DNA Constructions, *Annu. Rev. Biophys. Biomol. Struct.*, 1998, **27**, 225–248.
11. C. Mao, The Emergence of Complexity: Lessons from DNA, *PLoS Biol.*, 2004, **2**, 2036–2038.
12. E. Winfree, On the computational power of DNA annealing and ligation. DIMACS Ser, *Discrete Math. Theoret. Comput. Sci.*, 1996, **27**, 199–219.

13. P. W. K. Rothemund, Folding DNA to create nanoscale shapes and patterns, *Nature*, 2006, **440**, 297–302.

14. F. H. C. Crick, L. Barnett, S. Brenner and R. J. Watts-Tobin, General nature of the genetic code for proteins, *Nature*, 1961, **192**, 1227–1232.

15. F. H. C. Crick, The origin of the genetic code, *J. Mol. Biol.*, 1968, **38**, 367–379.

16. S. Itzkovitz and U. Alon, The genetic code is nearly optimal for allowing additional information within protein-coding sequences, *Genome Res.*, 2007, **17**, 405–412.

17. W. Saenger, *Principles of Nucleic Acid Structure*, Springer Verlag, New York, 1988.

18. E. S. Balakirev and F. J. Ayala, Pseudogenes: are they "junk" or functional DNA?, *Ann. Rev. Genet.*, 2003, **37**, 123–151.

19. E. F. Vanin, Processed Pseudogenes: Characteristics and Evolution, *Ann. Rev. of Genet.*, 1985, **19**, 253–272.

20. P. M. Harrison, H. Hegyi, S. Balasubramanian, N. M. Luscombe, P. Bertone, N. Echols, T. Johnson and M. Gerstein, Molecular Fossils in the Human Genome: Identification and Analysis of the Pseudogenes in Chromosomes 21 and 22, *Genome Res.*, 2002, **12**, 272–280.

21. S. Hirotsune, N. Yoshida, A. Chen, L. Garrett, F. Sugiyama, S. Takahashi, K. Yagami, A. Wynshaw-Boris and A. Yoshiki, An expressed pseudogene regulates the messenger-RNA stability of its homologous coding gene, *Nature*, 2003, **423**, 91–96.

22. A. Condon, Designed DNA molecules: principles and applications of molecular nanotechnology, *Nat. Rev. Genet.*, 2006, **7**, 565–575.

23. A. H. Wang, G. J. Quigley, F. J. Kolpak, J. L. Crawford, J. H. van Boom, G. van der Marel and A. Rich, Molecular structure of a left-handed double helical DNA fragment at atomic resolution, *Nature*, 1979, **282**, 680–686.

24. A. Rich, A. Nordheim and A. H.-J. Wang, The chemistry and biology of left-handed Z-DNA, *Ann. Rev. Biochem.*, 1984, **53**, 791–846.

25. Q. Du, C. Smith, N. Shiffeldrim, M. Vologodskaia and A. Vologodskii, Cyclisation of short DNA fragments and bending fluctuations of the double helix, *Proc. Natl. Acad. Sci.*, 2005, **102**, 5397–5402.

26. S. Cocco, R. Monasson and J. F. Marko, Force and kinetic barriers to initiation of DNA unzipping, *Phys, Rev.*, 2002, **E65**, 041907.

27. I. M. Kulic and H. Schiessel, DNA Spools under Tension, *Phys. Rev. Lett.*, 2004, **92**, 228101.

28. J. Bednar, P. Furrer, V. Katritch, A. Z. Stasiak, J. Dubochet and A. Stasiak, Determination of DNA Persistence Length by Cryo-electron Microscopy. Separation of the Static and Dynamic Contributions to the Apparent Persistence Length of DNA, *J. Mol. Biol.*, 1995, **254**, 579–594.

29. P. J. Hagerman, Flexibility of DNA, *Annu. Rev. Biophys. Biophys. Chem.*, 1988, **17**, 265–286.

30. V. V. Demidov and M. D. Frank-Kamenetskii, Two sides of the coin: affinity and specificity of nucleic acid interactions, *Trends Biochem. Sci.*, 2004, **29**, 62–71.
31. N. C. Seeman, De Novo Design of Sequences for Nucleic Acid Structural Engineering, *J. Biomol. Struct. Dyn.*, 1990, **8**, 573–581.
32. Y. Yang, T. P. Westcott, S. C. Pedersen, I. Tobias and W. K. Olson, Effects of localized bending on DNA supercoiling, *Trends Biochem. Sci.*, 1995, **20**, 313–319.
33. A. V. Vologodskii, A. V. Lukashin, V. V. Anshelevich and M. D. Frank-Kamenetskii, Fluctuations in superhelical DNA, *Nucl. Acids Res.*, 1979, **6**, 967–982.
34. G. Chirico and J. Langowski, Brownian dynamics simulations of super-coiled DNA with bent sequences, *Biophys. J.*, 1996, **71**, 955–971.
35. A. V. Vologodskii and N. R. Cozzarelli, Conformational and Thermo-dynamic Properties of Supercoiled DNA, *Ann. Rev. Biophys. Biomol. Str.*, 1994, **23**, 609–643.
36. N. C. Seeman, The Use of Branched DNA for Nanoscale Fabrication, *Nanotechnology*, 1991, **2**, 149–159.
37. N. C. Seeman, Construction of three-dimensional stick figures from branched DNA, *DNA Cell Biol.*, 1991, **10**, 475–486.
38. N. C. Seeman, The design of single-stranded nucleic acid knots, *Mol. Eng.*, 1992, **2**, 297–307.
39. N. C. Seeman, DNA engineering and its application to nanotechnology, *Trends Biotechnol.*, 1999, **17**, 437–443.
40. N. C. Seeman, H. Wang, X. Yang, F. Liu, C. Mao, W. Sun, L. A. Wenzler, Z. Shen, R. Sha, H. Yan, M. H. Wong, P. Sa-Ardyen, B. Liu, H. Qiu, X. Li, J. Qi, S. M. Du, Y. Zhang, J. E. Mueller, T.-J. Fu, Y. Wang and J. Chen, New motifs in DNA nanotechnology, *Nanotechnology*, 1998, **9**, 257–273.
41. Y. Zhang and N. C. Seeman, Construction of a DNA–Truncated Octa-hedron, *J. Am. Chem. Soc.*, 1994, **116**, 1661–1669.
42. C. Mao, W. Sun and N. C. Seeman, Assembly of Borromean rings from DNA, *Nature*, 1997, **386**, 137–138.
43. C. Liang and K. Mislow, On Borromean links, *J. Math. Chem.*, 1994, **16**, 27–35.
44. X. Li, X. Yang, J. Qi and N. C. Seeman, Antiparallel DNA double crossover molecules as components for nanoconstruction, *J. Am. Chem. Soc.*, 1996, **118**, 6131–6140.
45. T.-J. Fu and N. C. Seeman, DNA double crossover structures, *Bio-chemistry*, 1993, **32**, 3211–3220.
46. T. H. LaBean, H. Yan, J. Kopatsch, F. Liu, E. Winfree, J. H. Reif and N. C. Seeman, Construction, Analysis, Ligation, and Self-Assembly of DNA Triple Crossover Complexes, *J. Am. Chem. Soc.*, 2000, **122**, 1848–1860.
47. H. Yan and N. C. Seeman, Edge-sharing motifs in structural DNA nanotechnology, *J. Supramol. Chem.*, 2001, **1**, 229–237.

48. X. Zhang, H. Yan, Z. Shen and N. C. Seeman, Paranemic Cohesion of Topologically Closed DNA Molecules, *J. Am. Chem. Soc.*, 2002, **124**, 12940–12941.

49. U. Feldkamp, R. Wacker, W. Banzhaf and C. M. Niemeyer, Microarray-Based *in vitro* Evaluation of DNA Oligomer Libraries Designed in silico, *ChemPhysChem*, 2004, **5**, 367–372.

50. M. Zuker, Mfold web server for nucleic acid folding and hybridisation prediction, *Nucl. Acids Res.*, 2003, **31**, 3406–3415.

51. N. R. Markham and M. Zuker, DINAMelt web server for nucleic acid melting prediction, *Nucl. Acids Res.*, 2005, **33**, W577–581.

52. N. Le Novere, computing the melting temperature of nucleic acid duplex, *Bioinformatics*, 2001, **17**, 1226–1227.

53. K. J. Breslauer, R. Frank, H. Blocker and L. A. Marky, Predicting DNA duplex stability from the base sequence, *Proc. Natl. Acad. Sci.*, 1986, **83**, 3746–3750.

54. M. R. Shortreed, S. B. Chang, D. Hong, M. Phillips, B. Campion, D. C. Tulpan, A. Condon, H. H. Hoos and L. M. Smith, A thermodynamic approach to designing structure-free combinatorial DNA word sets, *Nucleic Acids Research*, 2005, **22**, 4965–4977.

55. M. R. Shortreed, S. B. Chang, D. Hong, M. Phillips, B. Campion, D. C. Tulpan, A. Condon, H. H. Hoos and L. M. Smith, A thermodynamic approach to designing structure-free combinatorial DNA word sets, *Nucl. Acids Res.*, 2005, **33**, 4965–4977.

56. R. M. Dirks, M. Lin, E. Winfree and N. A. Pierce, Paradigms for computational nucleic acid design, *Nucl. Acids Res.*, 2004, **32**, 1392–1403.

57. J. Sager and D. Stefanovic, DNA Computing, 2006, 275–289.

58. I. L. Hofacker, Vienna RNA secondary structure server, *Nucl. Acids Res.*, 2003, **31**, 3429–3431.

59. J. H. Reif, T. H. LaBean, S. Sahu, H. Yan and P. Yin, in Unconventional Programming Paradigms, 2005, 173–187.

60. A. A. Podtelezhnikov, N. R. Cozzarelli and A. V. Vologodskii, Equilibrium distributions of topological states in circular DNA: Interplay of supercoiling and knotting, *Proc. Natl. Acad. Sci.*, 1999, **96**, 12974–12979.

61. A. A. Podtelezhnikov, C. Mao, N. C. Seeman and A. Vologodskii, Multimerisation-Cyclisation of DNA Fragments as a Method of Conformational Analysis, *Biophys. J.*, 2000, **79**, 2692–2704.

62. A. Vologodskii, Exploiting circular DNA, *Proc. Natl. Acad. Sci.*, 1998, **95**, 4092–4093.

63. Q. Du, A. Kotlyar and A. Vologodskii, Kinking the double helix by bending deformation, *Nucl. Acids Res.*, 2008, **36**, 1120–1128.

64. H. Jacobson and W. H. Stockmayer, Intramolecular Reaction in Polycondensations. I. The Theory of Linear Systems, *J. Chem. Phys.*, 1950, **18**, 1600–1606.

65. D. Shore, J. Langowski and R. L. Baldwin, DNA flexibility studied by covalent closure of short fragments into circles, *Proc. Natl. Acad. Sci. USA*, 1981, **78**, 4833–4837.

66. D. Shore, J. Langowski and R. L. Baldwin, DNA flexibility studied by covalent closure of short fragments into circles, *Proc. Natl. Acad. Sci.*, 1981, **78**, 4833–4837.

67. P. Sa-Ardyen, A. V. Vologodskii and N. C. Seeman, The Flexibility of DNA Double Crossover Molecules, *Biophys, J.*, 2003, **84**, 3829–3837.

68. J. SantaLucia and D. Hicks, The thermodynamics of DNA structural motifs, *Ann. Rev. Biophys. Biomol. Str.*, 2004, **33**, 415–440.

69. A. Brenneman and A. Condon, Strand design for bio-molecular computation, *Theoret. Comput. Sci.*, 2002, **287**, 39.

70. L. M. Adleman, Molecular computation of solutions to combinatorial problems, *Science*, 1994, **266**, 1021–1024.

71. K. J. Baeyens, H. J. De Bondt, A. Pardi and H. R. Holbrook, A curved RNA helix incorporating an internal loop with GA and AA non-Watson–Crick base pairing, *Proc. Natl. Acad. Sci.*, 1996, **93**, 12851–12855.

72. S.-H. Ke and R. M. Wartell, Influence of Neighboring Base Pairs on the Stability of Single Base Bulges and Base Pairs in a DNA Fragment, *Biochemistry*, 1995, **34**, 4593–4600.

73. I. L. Kuznetsova, M. A. Zenkova, H. J. Gross and V. V. Vlassov, Enhanced RNA cleavage within bulge-loops by an artificial ribonuclease, *Nucl. Acids Res.*, 2005, **33**, 1201–1212.

74. J.-Y. Wang and K. Drlica, Modelling hybridisation kinetics, *Math. Biosci.*, 2003, **183**, 37–47.

75. A.-C. Déclais and D. M. J. Lilley, New insight into the recognition of branched DNA structure by junction-resolving enzymes, *Curr. Opin. Struct. Biol.*, 2008, **18**, 86–95.

76. N. D. F. Grindley, K. L. Whiteson and P. A. Rice, Mechanisms of Site-Specific Recombination, *Annual Review of Biochemistry*, 2006, **75**, 567–605.

77. R. A. Holliday, mechanism for gene conversion in fungi, *Genet. Res.*, 1964, **5**, 282–304.

78. Y. Wang, J. E. Mueller, B. Kemper and N. C. Seeman, Assembly and characterisation of five-arm and six-arm DNA branched junctions, *Biochemistry*, 1991, **30**, 5667–5674.

79. X. Wang and N. C. Seeman, Assembly and Characterisation of 8-Arm and 12-Arm DNA Branched Junctions, *J. Am. Chem. Soc.*, 2007, **129**, 8169–8176.

80. B. Liu, N. B. Leontis and N. C. Seeman, Bulged 3-arm DNA branched junctions as components for nanoconstruction, *Nanobiology*, 1994, **3**, 177–188.

81. J. Qi, X. Li, X. Yang and N. C. Seeman, The ligation of triangles built from bulged three-arm DNA branched junctions, *J. Am. Chem. Soc.*, 1996, **118**, 6121–6130.

82. J. Chen, N. R. Kallenbach and N. C. Seeman, A Specific Quadrilateral Synthesized from DNA Branched Junctions, *J. Am. Chem. Soc.*, 1989, **111**, 6402–6407.

83. J. Chen and N. C. Seeman, Synthesis from DNA of a molecule with the connectivity of a cube, *Nature*, 1991, **350**, 631–633.

84. J. E. Mueller, S. M. Du and N. C. Seeman, The design and synthesis of a knot from single-stranded DNA, *J. Am. Chem. Soc.*, 1991, **113**, 6306–6308.

85. N. C. Seeman, DNA Nicks and Nodes and Nanotechnology, *Nano Lett.*, 2001, **1**, 22–26.

86. S. M. Du, S. Zhang and N. C. Seeman, DNA junctions, antijunctions, and mesojunctions, *Biochemistry*, 1992, **31**, 10955–10963.

87. H. Wang and N. C. Seeman, Structural Domains of DNA Mesojunctions, *Biochemistry*, 1995, **34**, 920–929.

88. M. E. Churchill, T. D. Tullius, N. R. Kallenbach and N. C. Seeman, A Holliday recombination intermediate is twofold symmetric, *Proc. Natl. Acad. Sci.*, 1988, **85**, 4653–4656.

89. S. M. Du, B. D. Stollar and N. C. Seeman, A synthetic DNA molecule in three knotted topologies, *J. Am. Chem. Soc.*, 1995, **117**, 1194–1200.

90. N. C. Seeman, S. Zhang and J. Chen, DNA Nanoconstructions, *J. Vac. Sci. Technol. A*, 1994, **12**, 1895–1903.

91. Y. Li, Y. D. Tseng, S. Y. Kwon, L. d'Espaux, J. S. Bunch, P. L. McEuen and D. Luo, Controlled assembly of dendrimer-like DNA, *Nature Mater.*, 2004, **3**, 38–42.

92. T. F. Roth and D. R. Helinski, Evidence for circular DNA forms of a bacterial plasmid, *Proc. Natl. Acad. Sci.*, 1967, **58**, 650–657.

93. T. Cremer and C. Cremer, Chromosome territories, nuclear architecture and gene regulation in mammalian cells. *Nat. Rev. Genet.*, 2001, **2**, 292–301.

94. K. Timmis, F. Cabello and S. N. Cohen, Covalently closed circular DNA molecules of low superhelix density as intermediate forms in plasmid replication, *Nature*, 1976, **261**, 512–516.

95. J. Vinograd, J. Lebowitz, R. Radloff, R. Watson and P. Laipis, The twisted circular form of polyoma viral DNA, *Proc. Natl. Acad. Sci.*, 1965, **53**, 1104–1111.

96. S. Y. Shaw and J. C. Wang, Knotting of a DNA chain during ring closure, *Science*, 1993, **260**, 533–536.

97. A. V. Vologodskii and N. R. Cozzarelli, Effect of supercoiling on the juxtaposition and relative orientation of DNA sites, *Biophys. J.*, 1996, **70**, 2548–2556.

98. A. V. Vologodskii and N. R. Cozzarelli, Supercoiling, knotting, looping, and other large-scale conformational properties of DNA, *Curr. Opin. Str. Biol.*, 1994, **4**, 372–375.

99. F. B. Fuller, The Writhing Number of a Space Curve, *Proc. Natl. Acad. Sci.*, 1971, **68**, 815–819.

100. J. H. White, Self-Linking and the Gauss Integral in Higher Dimensions, *Am. J. Math.*, 1969, **91**, 693–728.

101. S. A. Wasserman and N. R. Cozzarelli, Biochemical topology: Applications to DNA recombination and replication, *Science*, 1986, **232**, 951–960.

102. C. C. Adams, *The Knot Book: An Elementary Introduction to the Mathematical Theory of Knots*, W. H. Freeman and Company, New York, 2004.

103. A. V. Vologodskii, N. J. Crisona, B. Laurie, P. Pieranski, V. Katritch, J. Dubochet and A. Stasiak, Sedimentation and electrophoretic migration of DNA knots and catenanes, *J. Mol. Biol.*, 1998, **278**, 1–3.

104. J. Hoste, M. Thistlethwaite and J. Weeks, The first 1701936 knots, *Math. Intell.*, 1998, **20**, 33–48.

105. S. A. Wasserman, J. M. Dungan and N. R. Cozzarelli, Discovery of a predicted DNA knot substantiates a model for site-specific recombination, *Science*, 1985, **229**, 171–174.

106. S. A. Wasserman and N. R. Cozzarelli, Supercoiled DNA-directed knotting by T4 topoisomerase, *J. Biol. Chem.*, 1991, **266**, 20567–20573.

107. P. O. Brown and N. R. Cozzarelli, A sign inversion mechanism for enzymatic supercoiling of DNA, *Science*, 1979, **206**, 1081–1083.

108. V. V. Rybenkov, C. Ullsperger, A. V. Vologodskii and N. R. Cozzarelli, Simplification of DNA topology below equilibrium values by type II topoisomerases, *Science*, 1997, **277**, 690–693.

109. F. B. Dean, A. Stasiak, T. Koller and N. R. Cozzarelli, Duplex DNA knots produced by Escherichia coli topoisomerase I. Structure and requirements for formation, *J. Biol. Chem.*, 1985, **260**, 4975–4983.

110. L. F. Liu, C. C. Liu and B. M. Alberts, Type II DNA topoisomerases: enzymes that can unknot a topologically knotted DNA molecule via a reversible double-strand break, *Cell*, 1980, 697–707.

111. S. M. Du, H. Wang, Y.-C. Tse-Dinh and N. C. Seeman, Topological transformations of synthetic DNA knots, *Biochemistry*, 1995, **34**, 673–682.

112. H. Wang, R. J. Di Gate and N. C. Seeman, An RNA topoisomerase, *Proc. Natl. Acad. Sci.*, 1996, **93**, 9477–9482.

113. A. Xayaphoummine, T. Bucher and H. Isambert, Kinefold web server for RNA/DNA folding path and structure prediction including pseudoknots and knots, *Nucl. Acids Res.*, 2005, **33**, W605–610.

114. M. A. Krasnow and N. R. Cozzarelli, Catenation of DNA rings by topoisomerases, *J. Biol. Chem.*, 1982, **257**, 2687–2693.

115. S. D. Levene, C. Donahue, T. C. Boles and Cozzarelli, Analysis of the structure of dimeric DNA catenanes by electron microscopy, *Biophys. J.*, 1995, **69**, 1036–1045.

116. A. Bucka and A. Stasiak, Construction and electrophoretic migration of single-stranded DNA knots and catenanes, *Nucl. Acids Res.*, 2002, **30**, e24.

117. D. E. Adams, E. M. Shekhtman, E. L. Zechiedrich, M. B. Scmid and N. R. Cozzarelli, The role of topoisomerase IV in partitioning bacterial replicons and the structure of catenated intermediates in DNA repliation, *Cell*, 1992, **71**, 277–288.

118. J. B. Bliska and N. R. Cozzarelli, Use of site-specific recombination as a probe of DNA structure and metabolism *in vivo*, *J. Mol. Biol.*, 1987, **194**, 205–218.

119. T.-J. Fu, Y. C. Tse-Dinh and N. C. Seeman, Holliday junction crossover topology, *J. Mol. Biol.*, 1994, **236**, 91–105.

120. J. Chen and N. C. Seeman, The electrophoretic properties of a DNA cube and its substructure catenanes, *Electrophoresis*, 1991, **12**, 607–611.

121. S. J. Cantrill, K. S. Chichak, A. J. Peters and J. F. Stoddart, Nanoscale Borromean Rings, *Acc. Chem. Res.*, 2005, **38**, 1–9.

122. H. Qiu, J. C. Dewan and N. C. Seeman, A DNA decamer with a sticky end: the crystal structure of d-CGACGATCGT, *J. Mol. Biol.*, 1997, **267**, 881–898.

123. T.-J. Fu, Y.-C. Tse-Dinh and N. C. Seeman, Holliday junction crossover topology, *J. Mol. Biol.*, 1994, **236**, 91–105.

124. E. Winfree, F. Liu, L. A. Wenzler and N. C. Seeman, Design and self-assembly of two-dimensional DNA crystals, *Nature*, 1998, **394**, 539–544.

125. N. B. Leontis, M. T. Hills, M. Piotto, A. Malhotra, J. Nussbaum and D. G. Gorenstein, A model for the solution structure of a branched, three-strand DNA complex, *J. Biomol. Struct. Dyn.*, 1993, **11**, 215–223.

126. X. Yang, L. A. Wenzler, J. Qi, X. Li and N. C. Seeman, Ligation of DNA Triangles Containing Double Crossover Molecules, *J. Am. Chem. Soc.*, 1998, **120**, 9779–9786.

127. B. Ding, R. Sha and N. C. Seeman, Pseudohexagonal 2D DNA Crystals from Double Crossover Cohesion, *J. Am. Chem. Soc.*, 2004, **126**, 10230–10231.

128. R.-I. Ma, N. R. Kallenbach, R. D. Sheardy, M. L. Petrillo and N. C. Seeman, Three-arm nucleic acid junctions are flexible, *Nucl. Acids Res.*, 1986, **14**, 9745–9753.

129. D. Liu, M. Wang, Z. Deng, R. Walulu and C. Mao, Tensegrity: Construction of Rigid DNA Triangles with Flexible Four-Arm DNA Junctions, *J. Am. Chem. Soc.*, 2004, **126**, 2324–2325.

130. F. Mathieu, S. Liao, J. Kopatsch, T. Wang, C. Mao and N. C. Seeman, Six-Helix Bundles Designed from DNA, *Nano Lett.*, 2005, **5**, 661–665.

131. P. E. Constantinou, T. Wang, J. Kopatsch, L. B. Israel, X. Zhang, B. Ding, W. B. Sherman, X. Wang, J. Zheng, R. Sha and N. C. Seeman, Double cohesion in structural DNA nanotechnology, *Org. Biomol. Chem.*, 2006, **4**, 3414–3419.

132. J. C. Wang, Helical Repeat of DNA in Solution, *Proc. Natl. Acad. Sci.*, 1979, **76**, 200–203.

133. D. Rhodes and A. Klug, Helical periodicity of DNA determined by enzyme digestion, *Nature*, 1980, **286**, 573–578.

134. R. Sha, H. Iwasaki, F. Liu, H. Shinagawa and N. C. Seeman, Cleavage of Symmetric Immobile DNA Junctions by Escherichia coli RuvC, *Biochemistry*, 2000, **39**, 11982–11988.

135. R. Sha, F. Liu and N. C. Seeman, Direct Evidence for Spontaneous Branch Migration in Antiparallel DNA Holliday Junctions, *Biochemistry*, 2000, **39**, 11514–11522.

136. N. C. Seeman, Y. Zhang, T.-J. Fu, S. Zhang, Y. Wang and J. Chen, Chemical Synthesis of Nanostructures, *Mater. Res. Soc. Symp. Proc.*, 1994, **330**, 45–56.

137. Z. Shen, H. Yan, T. Wang and N. C. Seeman, Paranemic Crossover DNA: A Generalized Holliday Structure with Applications in Nano-technology, *J. Am. Chem. Soc.*, 2004, **126**, 1666–1674.

138. H. Yan, S. H. Park, G. Finkelstein, J. H. Reif and T. H. LaBean, DNA–Templated Self-Assembly of Protein Arrays and Highly Conductive Nanowires, *Science*, 2003, **301**, 1882–1884.

139. W. Liu, X. Wang, T. Wang, R. Sha and N. C. Seeman, PX DNA Triangle Oligomerized Using a Novel Three-Domain Motif, *Nano Lett.*, 2008, **8**, 317–322.

140. H. Yan, X. Zhang, Z. Shen and N. C. Seeman, A robust DNA mecha-nical device controlled by hybridisation topology, *Nature*, 2002, **415**, 62–65.

141. S. Liao and N. C. Seeman, Translation of DNA Signals into Polymer Assembly Instructions, *Science*, 2004, **306**, 2072–2074.

142. B. Ding and N. C. Seeman, Operation of a DNA Robot Arm Inserted into a 2D DNA Crystalline Substrate, *Science*, 2006, **314**, 1583–1585.

143. A. V. Garibotti, S. Liao and N. C. Seeman, A Simple DNA-Based Translation System, *Nano Lett.*, 2007, **7**, 480–483.

144. B. H. Robinson and N. C. Seeman, The design of a biochip: a self-assembling molecular-scale memory device, *Prot. Eng.*, 1987, **1**, 295–300.

145. C. Mao, W. Sun, Z. Shen and N. C. Seeman, A nanomechanical device based on the B-Z transition of DNA, *Nature*, 1999, **397**, 144–146.

146. C. Lin, X. Wang, Y. Liu, N. C. Seeman and H. Yan, Rolling Circle Enzymatic Replication of a Complex Multi-Crossover DNA Nano-structure, *J. Am. Chem. Soc.*, 2007, **129**, 14475–14481.

147. C. Mao, W. Sun and N. C. Seeman, Designed Two-Dimensional DNA Holliday Junction Arrays Visualized by Atomic Force Microscopy, *J. Am. Chem. Soc.*, 1999, **121**, 5437–5443.

148. R. Sha, F. Liu, D. P. Millar and N. C. Seeman, Atomic force microscopy of parallel DNA branched junction arrays, *Chem. Biol.*, 2000, **7**, 743–751.

149. J. Malo, J. C. Mitchell, C. Vénien-Bryan, J. R. Harris, H. Wille, D. J. Sherratt and A. J. Turberfield, Engineering a 2D Protein-DNA Crystal, *Angew. Chem. Int. Ed.*, 2005, **44**, 3057–3061.

150. D. Zerbib, C. Mézard, H. George and S. C. West, Coordinated actions of RuvABC in Holliday junction processing, *J. Mol. Biol.*, 1998, **281**, 621–630.

151. W. M. Shih, J. D. Quispe and G. F. Joyce, A 1.7-kilobase single-stranded DNA that folds into a nanoscale octahedron, *Nature*, 2004, **427**, 618–621.

152. R. P. Goodman, I. A. T. Schaap, C. F. Tardin, C. M. Erben, R. M. Berry, C. F. Schmidt and A. J. Turberfield, Rapid Chiral Assembly of Rigid DNA Building Blocks for Molecular Nanofabrication, *Science*, 2005, **310**, 1661–1665.

153. R. P. Goodman, R. M. Berry and A. J. Turberfield, The single-step synthesis of a DNA tetrahedron, *Chem. Commun.*, 2004, **1**, 1372–1373.

154. C. M. Erben, R. P. Goodman and A. J. Turberfield, A Self-Assembled DNA Bipyramid, *J. Am. Chem. Soc.*, 2007, **129**, 6992–6993.

155. R. P. Goodman, M. Heilemann, S. Doose, C. M. Erben, A. N. Kapanidis and A. J. Turberfield, Reconfigurable, braced, three-dimensional DNA nanostructures, *Nat. Nanotechnol*, 2008, **3**, 93–96.

156. B. Yurke, A. J. Turberfield, A. P. Mills, F. C. Simmel and J. L. Neumann, A DNA-fuelled molecular machine made of DNA, *Nature*, 2000, **406**, 605–608.

157. J. von Neumann, *Theory of self-reproducing automata*, ed. A. Burks, University of Illinois Press., Urbana Illinois, 1966.

158. H. Wang, *Dominoes and the AEA Case of the Decision Problem.* ed. Fox, J. Polytechnic Press, Brooklyn, N.Y, 1963.

159. P. W. K. Rothemund, N. Papadakis and E. Winfree, Algorithmic Self-Assembly of DNA Sierpinski Triangles, *PLoS Biol.*, 2004, **2**, 2041–2053.

160. C. Mao, T. H. LaBean, J. H. Reif and N. C. Seeman, Logical computation using algorithmic self-assembly of DNA triple-crossover molecules, *Nature*, 2000, **407**, 493–496.

161. H. Yan, T. H. LaBean, L. Feng and J. H. Reif, Directed nucleation assembly of DNA tile compelxes for barcode-patterned lattices, *Proc. Natl. Acad. Sci.*, 2003, **100**, 8103–8108.

162. Y. Liu, C. Lin and H. L. H. Yan, Aptamer-Directed Self-Assembly of Protein Arrays on a DNA Nanostructure, *Angew. Chem. Int. Ed.*, 2005, **44**, 4333–4338.

163. S. H. Park, P. Yin, Y. Liu, J. H. Reif, T. H. LaBean and H. Yan, Programmable DNA Self-Assemblies for Nanoscale Organisation of Ligands and Proteins, *Nano Lett.*, 2005, **5**, 729–733.

164. D. Liu, S. H. Park, J. H. Reif and T. H. LaBean, DNA nanotubes self-assembled from triple-crossover tiles as templates for conductive nano-wires, *Proc. Natl. Acad. Sci.*, 2004, **101**, 717–722.

165. W. B. Sherman and N. C. Seeman, Design of Minimally Strained Nucleic Acid Nanotubes, *Biophys. J.*, 2006, **90**, 4546–4557.

166. J. C. Mitchell, J. R. Harris, J. Malo, J. Bath and A. J. Turberfield, Self-Assembly of Chiral DNA Nanotubes, *J. Am. Chem. Soc.*, 2004, **126**, 16342–16343.

167. P. W. K. Rothemund, A. Ekani-Nkodo, N. Papadakis, A. Kumar, D. K. Fygenson and E. Winfree, Design and Characterisation of Programmable DNA Nanotubes, *J. Am. Chem. Soc.*, 2004, **126**, 16344–16352.

168. B. Wei and Y. Mi, A New Triple Crossover Triangle (TXT) Motif for DNA Self-Assembly, *Biomacromolecules*, 2005, **6**, 2528–2532.

169. S. H. Park, R. Barish, H. Li, J. H. Reif, G. Finkelstein, H. Yan and T. H. LaBean, Three-Helix Bundle DNA Tiles Self-Assemble into 2D Lattice or 1(d) Templates for Silver Nanowires, *Nano Lett.*, 2005, **5**, 693–696.

170. A. Kuzuya, R. Wang, R. Sha and N. C. Seeman, Six-Helix and Eight-Helix DNA Nanotubes Assembled from Half-Tubes, *Nano Lett.*, 2007, **7**, 1757–1763.

171. H. Liu, Y. Chen, Y. He, A. E. Ribbe and C. Mao, Approaching The Limit: Can One DNA Oligonucleotide Assemble into Large Nanostructures?, *Angew. Chem. Int. Ed.*, 2006, **45**, 1942–1945.

172. C. Lin, Y. Ke, Y. Liu, M. Mertig, J. Gu and H. Yan, Functional DNA Nanotube Arrays: Bottom-Up Meets Top-Down, *Angew. Chem. Int. Ed.*, 2007, **46**, 6089–6092.

173. P. O'Neill, P. W. K. Rothemund, A. Kumar and D. K. Fygenson, Sturdier DNA Nanotubes via Ligation, *Nano Lett.*, 2006, **6**, 1379–1383.

174. N. C. Seeman, Molecular craftwork with DNA, *Chem. Intell.*, 1995, **1**, 38–47.

175. C. M. Niemeyer and M. Adler, Nanomechanical devices based on DNA, *Angew. Chem. Int. Ed.*, 2002, **41**, 3779–3783.

176. U. Feldkamp and C. M. Niemeyer, Rational Engineering of Dynamic DNA Systems, *Angew. Chem. Int. Ed.*, 2008, **47**, 3871–3873.

177. P. Yin, H. M. T. Choi, C. R. Calvert and N. A. Pierce, Programming biomolecular self-assembly pathways, *Nature*, 2008, **451**, 318–322.

178. D. Y. Zhang, A. J. Turberfield, B. Yurke and E. Winfree, Engineering Entropy-Driven Reactions and Networks Catalysed by DNA, *Science*, 2007, **318**, 1121–1125.

179. G. Seelig, B. Yurke and E. Winfree, Catalysed Relaxation of a Metastable DNA Fuel, *J. Am. Chem. Soc.*, 2006, **128**, 12211–12220.

180. S. J. Hurst, H. D. Hill and C. A. Mirkin, "Three-Dimensional Hybridisation" with Polyvalent DNA-Gold Nanoparticle Conjugates, *J. Am. Chem. Soc.*, 2008, **130**, 12192–12200.

181. C. M. Niemeyer, M. Adler, S. Lenhert, S. Gao, H. Fuchs and L. Chi, Nucleic Acid Supercoiling as a Means for Ionic Switching of DNA-Nanoparticle Networks, *ChemBioChem.*, 2001, **2**, 260–264.

182. C. A. Mirkin, R. L. Letsinger, R. C. Mucic and J. J. Storhoff, A DNA-based method for rationally assembling nanoparticles into macroscopic materials, *Nature*, 1996, **382**, 607–609.

183. S.-J. Park, T. A. Taton and C. A. Mirkin, Array-Based Electrical Detection of DNA with Nanoparticle Probes, *Science*, 2002, **295**, 1503–1506.

184. C. Sonnichsen, B. M. Reinhard, J. Liphardt and A. P. Alivisatos, A molecular ruler based on plasmon coupling of single gold and silver nanoparticles, *Nature Biotech.*, 2005, **23**, 741–745.

185. K. Hamad-Schifferli, J. J. Schwartz, A. T. Santos, S. Zhang and J. M. Jacobson, Remote electronic control of DNA hybridisation through inductive coupling to an attached metal nanocrystal antenna, *Nature*, 2002, **415**, 152–155.

CHAPTER 3
Recaging Within

Creating enclosed environments on the nanoscale is the universal application of a widespread natural phenomenon – molecular encapsulation.[1] This is defined as an ability to form and sustain cavities capable of hosting guest elements.[2,3] Various size-constrained molecular systems with concave surfaces capitalise on the effect to act as caged receptors for different (bio)chemical conversions.[3,4]

Both covalently and noncovalently built encapsulants are known and can be emulated by design.[2,4] The former typically concerns specific chemical transformations and often employs trapping mechanisms, in which reactants are captured by a complex catalyst that simultaneously serves as a protecting capsule.[5] Starting materials as well as their probable intermediates are "protected" until the required transformation occurs; this is marked by the release of reaction products. Strictly speaking, the strategy is of a reversible protecting scheme and necessitates very tight interactions between the trapping and the entrapped. The latter type – noncovalent – relies more on the three-dimensional surrounding of attracted guests, subsequent covalent rearrangements of which, though not necessarily, can also be fostered.[2,4] The final release of guests is not a compulsory event, which allows for sequential transformations and storage. In this notation, the functioning of noncovalent encapsulants is not kinetically limited and does not require high precision and directionality of guest–host interactions that are important for covalent protecting capsules. Noncovalent encapsulants are thus more versatile and can be applied to a greater number of reactions or guests. They can be less specific and are more thermodynamically flexible, which is definitively attributed to the nature of their structural organisation. Noncovalent encapsulants are self-assembled and, as many other self-assembled objects, are multistep supramolecular couplings of complementary subunits whose associations can lead to discrete nano-to-microscopic architectures. Such length scales are practically unachievable for covalent

RSC Nanoscience and Nanotechnology No. 7
Bionanodesign
By M Ryadnov
© Maxim Ryadnov 2009
Published by the Royal Society of Chemistry, www.rsc.org

encapsulants because the synthesis of monodisperse and spatially intralocked molecules of increased complexity poses very costly demands.[4] An emergence of "encapsulation chemistry" using natural self-assembling elements is not therefore surprising.

There is being reported an impressive and steadily growing number of self-assembled encapsulants that exhibit various functions and utilise different chemistries. The amount of information is indeed enormous and can hardly be covered within the scope of a single chapter. Excellent reviews published to date can give the reader a more comprehensive and in some ways debatable representation of the field.[2,4] This chapter will continue building upon the immediate interest of the book to focus on challenges and attainments in designing artificial enclosed self-assemblies inspired by and derived from natural self-assembling motifs.

3.1   Enclosing to Deliver

Isolating molecules and materials as well as performing chemical reactions in encapsulating assemblies is becoming a mainstream focus of nanodesign. Encouragingly, there exists a rich repertoire of natural systems that can offer a variety of approaches to advance this. However, any given one, be it applicable to a single molecule or to a complex material, is determined by a guest type and the function it introduces. This constitutes the essence of nanoscale encapsulation, realisations of which are extensively found in nature.

3.1.1   Transporting Foreign

Viruses commonly viewed as deadly poisons are little more than nucleic acids encased and conserved in protein sheaths. These sheaths, referred to as capsids, form nanoscale objects with various sizes ranging from just several nanometres to several hundreds of nanometres, Figure 3.1.

Capsids loaded with genetic materials (DNA or RNA) become infective virions or stealths that protect and transport genes from cell to cell. Encapsidated nucleic acids, their nature, length and information they carry, encode the very shape and morphology of protein capsids to give a truly fascinating diversity of viruses, Figure 3.1. However, from the structural point of view or that of a designer the wealth of viral forms is apparent. All viruses share common features and follow similar principles in their organisation.[6] Virions can be linear or helical (tobacco mosaic virus), spherical, adopting isometric symmetry (adenovirus) or of either type enveloped in a membrane (helical influenza or icosahedral herpes simplex).[7]

Irrespective of their shape and appearance viral capsids furnish perfect and ready-to-go templates for materials synthesis and design.[8] Firstly, very few viruses respond to external stimuli, which is partly the reason for them not being considered as living organisms. Secondly, their assembly is symmetry driven and is reproduced from generation to generation. And, finally, once

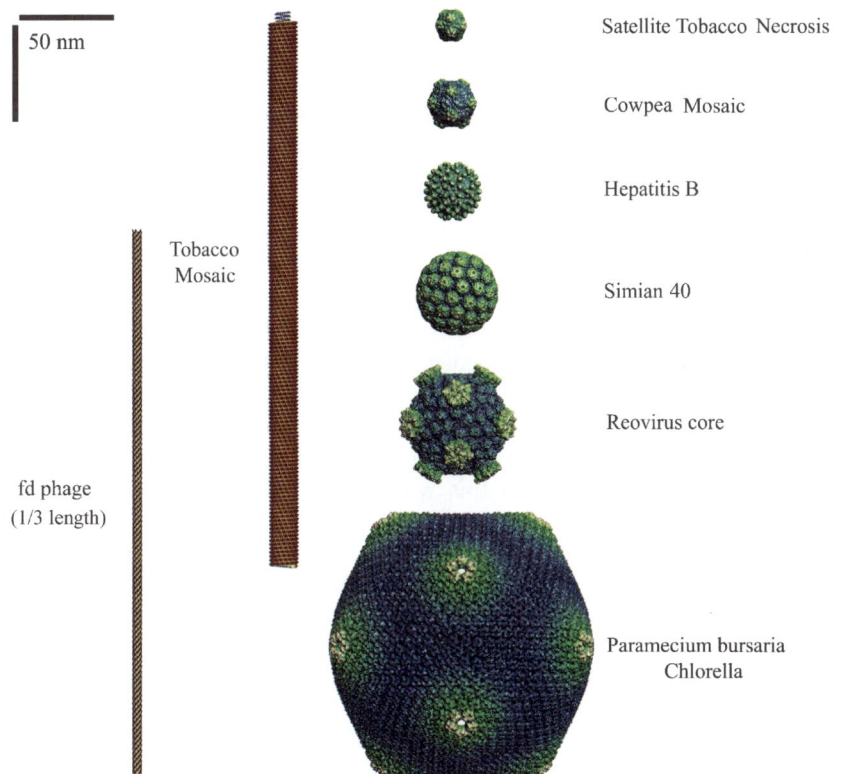

Figure 3.1 Viruses; space-filling image reconstructions of different viral forms, (reproduced with permission from VIPER (http://viperdb.scripps.edu)[15] and http://virology.wisc.edu/virusworld/).

formed, a virus can be reloaded with another material without compromising its native morphology.

Architectural principles underlying discrete viral assemblies are relatively well understood and thoroughly described.[9] The same main rationale that applies to all self-assembled structures is true for viruses. Moreover, this was in fact specifically introduced to explain virus architecture back in the 1950s by Crick and Watson as a theory of identical subunits,[10] and arguably presents a seminal point for the technological spreading of molecular self-assembly in general.

3.1.1.1 Fitting Flat and Straight

The logic behind the theory lies in the presumption that small genomes of viruses may be sufficient to code only a limited number of proteins insufficiently sized to build up an entire capsid. The theory thus offers an approach according to which proteins encoded by viral genomes are used as building blocks multiplied in a regular fashion into a geometrically configured assembly. This in turn implies that the blocks in the capsid are packed symmetrically; that is, each block is surrounded by the identical number of identical subunits. And finally, an ideal solution to pack protein units in equivalent environments has to be a cubic symmetry, which can be rotated around its several axes to produce multiple even projections. This is a minimum free-energy structure. The hypothesis universally pertinent to spherical viruses has been experimentally supported and is now widely accepted. One particularly important consequence from the theory is the implementation of general geometric rules into the description and prediction of viral constructions.

For example, the abundance of icosahedral symmetry in the structure of protein shells has been confirmed in many genetically unrelated viruses. Human rhinoviruses or adenoviruses, plant rice yellow mottle virus or cowpea mosaic virus, animal porcine parvovirus, and bacterial tailless phages may belong to different taxonomic groups, but are all proteinaceous icosahedra.[11] An obvious question to this would be – why such a preference for building viral shells?

An icosahedron is a geometric object with 20 flat faces of equilateral triangles arranged around a sphere and is characterised by 5/2 3 symmetry, Figure 3.2(a).

Thus, the entire surface of an icosahedron being applied to a protein shell would require 60 asymmetric protein blocks or subunits to cover (the subunits are arranged around icosahedron vertices giving three subunits per triangle), Figure 3.2(a). There are six 5-, ten 3- and fifteen 2-fold symmetry axes in an icosahedron, none of which is to be occupied by blocks. Providing this requirement each of 60 protein subunits would be identical to its neighbours, Figure 3.2(a). By being symmetry related the subunits can also be compiled by the smallest possible number of different proteins. This may explain the minimalistic approach viruses take in hosting small genomes, which stays in good agreement with the theory of identical subunits. However, in many cases viral genomes are nonsegmented that tend to produce polycistronic or overlapped genes, genes that by taking the same space as a given single gene translated into one protein encode a larger number of proteins. Concomitantly, protein blocks may contain more than one chain, which in turn may arise from a single superprotein being cleaved into several segments, each of which takes an assigned position necessarily directed by and away from the symmetry axes. As a result, assembled capsids can be more complex and formed from more than 60 subunits. Indeed, with very few exceptions most viruses are composed of more subunits: mosaic viruses contain 100–150 blocks and adenoviruses can have as many as 252 subunits.[12–15] However, unless compromised in equality the number of subunits an icosahedron may accommodate is restricted to 60, Figure 3.2(b). This paradox required finding a consensus or structural adjustment. One is supported by observations that the subunits of capsids are (1) nonidentical, (2) symmetrical, and (3) likely placed on symmetry axes.[16,17] Although these did not appear to be consistent with what was proposed by Crick and Watson, the 5/3 2 icosahedral symmetry remained preserved due to

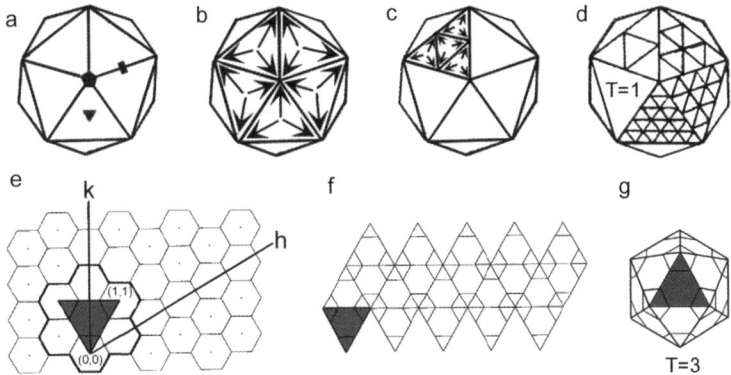

Figure 3.2 Geometric principles of icosahedral symmetry; (a) Schematic representation of an equilateral icosahedron containing 20 flat triangular faces and 12 vertices arranged around a sphere and related by 5-(diamond), 3-(triangle) and 2-(rectangle)fold axes of rotational symmetry, (b) An icosahedron accommodating 60 *equivalent* subunits. Proteins (arrows) are arranged around the icosahedron vertices giving three subunits per triangular face (shown only in five triangles), (c) An icosahedron accommodating more than 60 subunits. Each face is divided into four smaller facets. Proteins (small arrows) are placed into each corner of the facets (shown only in one face) giving in total 240 *quasiequivalent* subunits, with 180 arranged into 30 hexamers and 60 into 12 pentamers, (d) Triangulation numbers with $P = 1$ clockwise: 1, 4, 9, 16, and 25, (e) A hexagonal net lattice relating icosahedral and quasiequivalent symmetries. An icosahedral face for a $T = 3$ surface lattice is defined in the net by the grey triangle with bold lines ($h = 1$, $k = 1$), (f) A net of 20 identical faces, with each planar hexamer composed of six triangles. Convex pentamers are generated when one of the triangles is removed. This ensures the connection of the two free edges of one pentagon in 3D with subsequent folding into a $T = 3$ icosahedron (g) ((e)–(g), reprinted from Johnson, J. E. and Speir, J. A. Quasiequivalent viruses: a paradigm for protein assemblies. *J. Mol. Biol.*, **269**, 665–675, Copyright 1997, with permission from Elsevier).

the saving assumption that the observed subunits, termed capsomers,[18] may themselves be composed of asymmetric subunits.

Based on this recognition Caspar and Klug (CK) developed a concept of quasiequivalent positions or a quasiequivalence theory.[19]

Generally, the concept has several regulations: (1) each protein subunit is positioned in similar but not identical environments (quasiequivalent), (2) protein subunits are clustered in capsomers, and (3) each of 20 triangular faces of icosahedron is subdivided into smaller triangles, facets that are conveniently described by a triangulation number T reflecting the number of facets,[9,12] Figures 3.2(c) and (d).

T can be decided graphically from an equilateral triangular net, Figures 3.2(e)–(g), or using a simple equation: $T = f^2 P$, where f is any integer and $P = (k^2 + k*h + h^2)$, with k and h being any integer with no common factor.

P thus presents a class of icosahedra that can be any number of values 1, 3, 4, 7, *etc.*[20] The way the theory works is relatively straightforward.

A *T* = 1 icosahedron will have 20 triangular faces, with each being assigned three identical facets. Given one facet takes one protein a total of 60 identical proteins will be generated, Figure 3.2(b). In a *T* = 4 each triangle is subdivided into four smaller triangles to place one protein subunit at each corner. This gives 12 blocks per face, which when multiplied by a total of 20 faces will give 240 blocks, Figure 3.2(c). The blocks are organised as penta (those that are at vertices of the faces; 12 in total), and hexa (those located at the verteces of facets; 30 in total) capsomers. Thus, nonequivalent 60 pentamers and 180 hexamers can be only bonded in a similar or quasiequivalent fashion.

Using the same logic, most *T* = 3 viruses will have 180 nonidentical subunits (3 proteins multiplied by 60 triangles) that may follow various arrangement patterns, Figure 3.3. As a rule of thumb, in *T* > 1 viruses proteins will always have different numbers of neighbours due to the fold differences of the triangular vertices of both faces and facets and none of the proteins in a given pair will be in a strictly equivalent environment.

In this context, the most critical point in designing virus-alike shells would be to ensure that all proteins within a capsomer are sequence and folding identical. Nonidentical proteins would still adopt an icosahedral architecture as they do, for example, in polio[21] or rhinoviruses.[22] Yet, predicting protein stoichiometry in relation to bonding structures of icosahedra would become a formidable task.

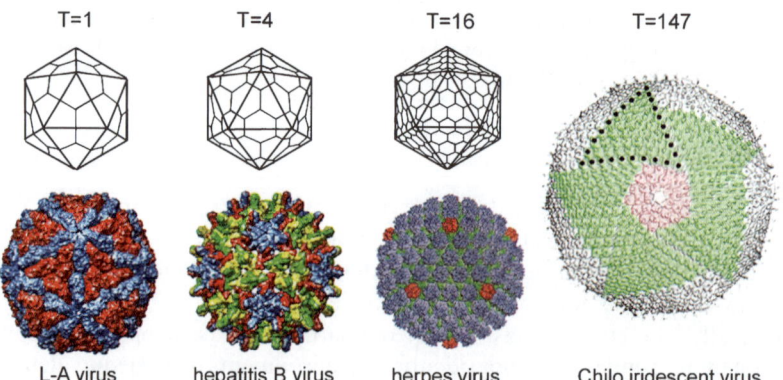

T=1   T=4   T=16   T=147

L-A virus   hepatitis B virus   herpes virus   Chilo iridescent virus

Figure 3.3   Classification of viruses according to triangulation numbers, (reproduced form VIPER (http://viperdb.scripps.edu);[15] reprinted with permission from Chiu, W. and Rixon, F. J. High resolution structural studies of complex icosahedral viruses: a brief overview. *Virus Res.*, **82**, 9–17. Copyright 2001, with permission from Elsevier; by permission from Macmillan Publishers Ltd: Yan, X. *et al.* Structure and assembly of large lipid-containing dsDNA viruses. *Nature Struct. Mol. Biol.*, **7**, 101–103, Copyright (2000)).

The CK concept partly adapts to this by employing pseudoequivalence relations, in which mathematical triangulation numbers of nonquasiequivalent viruses are adjusted to their structural resemblances with quasiequivalent viruses.[20,23] However, this does not address other probable complications that occur in viral icosahedra. For instance, right- or left-handed icosahedrons can still be modelled by CK theory, but mixed or defective would not. The CK theory does not take into consideration inter-, and intracapsomer interactions that exhibit different bonding strengths.[24] Assembly "mutants" that may result in distorted icosahedra often yielding tubular structures (*e.g.* human papillomavirus)[25] also do not obey the theory. Nevertheless, the CK concept provides best mathematical fits for virus structures and forms a firm foundation for modern taxonomy of viruses.[9] A number of databases of virus structures derived from the theory is also available, with ICTVdB (International Committee on Taxonomy of Viruses database)[9] and Viper (Virus Particle Explorer, Scripps Institute)[12,15] being amongst the most popular.

Icosahedral viruses have been of steady interest for researches from different scientific fields and in particular for those engaged at the interface between biology and mathematics. The structure of viruses continues challenging new prediction approaches, one major ambition, however, aims at elucidating assembly pathways of viral capsids. A number of computational models stimulated by new experimental observations are being developed. To name a few, internal "switching" concept;[26] topographic potential surfaces,[27] minimal thermodynamics-kinetics[28] and stochastic growth[29] models, classical nucleation,[30] group[31] and tiling[32] theories all add to advance our understanding of virus structure and organisation. Ultimately, the approaches may lend themselves to general or at least guiding design criteria.[33] This is obviously crucial from a designer's perspective as emulating principles of viral assemblies experimentally remains in its infancy. Nevertheless, there is no doubt that spherical morphology of viruses will keep drawing primary attention in engineering nanoscale encapsulants. Partly this is due to the outlined and truly advanced understanding as well as to the rich abundance of icosahedral viruses in Nature, but partly to similarities that are reached by synthetic examples.

There is emerging a new class of nanoscale designs that are liberally put under the heading of artificial viruses.[34] These structures are dominated by self-assembled particulate systems and are indeed reminiscent of viruses, however, only by their final sphere-like appearances. Most such structures are densely packed vesicles or micelles rather than delineated convex deltahedra. The field with its pros and cons is, however, burgeoning and will be covered in Section 3.3.

Although most viruses adopt icosahedral morphology, other viral geometries applicable to nanodesign also exist. An interesting class is represented by filamentous virions[9,35] including filoviruses,[36] plant mosaic viruses[37] and bacteriophages.[38] Structures of all of these viruses conform to strict symmetry parameters analogously to those of icosahedral viruses.

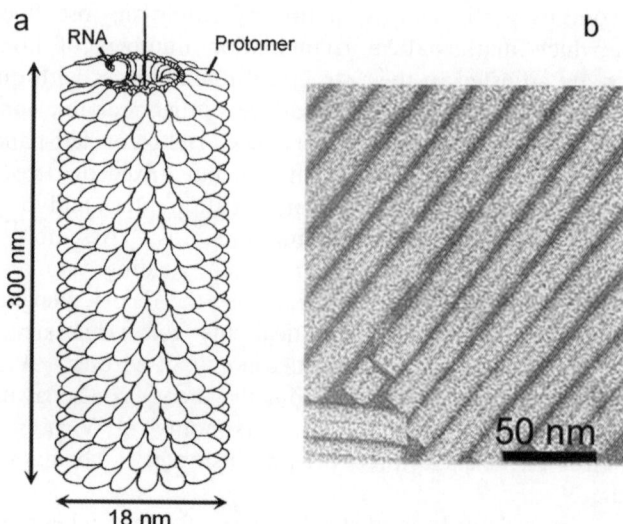

Figure 3.4 Tobacco mosaic virus; (a) Schematic representation of a TMV rod assembled from protomers arranged around RNA, (adapted with permission from Caspar, D. L. D. Assembly and stability of the tobacco mosaic virus particle. *Adv. Protein Chem.*, **18**, 37–121, Copyright 1963, with permission from Elsevier), (b) An electron micrograph of TMV, (reprinted from Hibi, T. *et al.* High resolution electron microscopy of tobacco mosaic virus. *J. Electron Microsc.* (*Tokyo*), **22**, 243–253 (1973) by permission of Oxford University Press).

3.1.1.2 Spiralling Along

Filamentous viruses are best thought of as linear and tubular by morphology and helical by assembly.[39] Tobacco mosaic virus (TMV) was the first to be shown to self-assemble *in vitro* from purified components.[40] Since then, TMV has been a standard material for structural and biochemical studies of viruses, Figure 3.4.

TMV is composed of about 2130 identical protein subunits, called protomers. By winding around a single-stranded RNA protomers self-assemble into a rigid and hollow rod.[41–43] The assembly is initiated by a specific lollipop-shaped region found in the viral RNA, which interacts with protomers to nucleate the formation of a disk.[42,44] Elongation of nucleated rods proceeds through the stacking of further disks.[44] To accommodate the elongation the RNA is believed to dislocate to a protohelix that allows for its complete burial inside the rod.[45] Such a helical arrangement of protomers proves to be energetically optimal as (1) nucleic acids are prone to adopt helical conformations that ensures (2) identical bondings between the encased RNA and the subunits, which in turn leads to (3) cooperative assembly of disks. Therefore, the physical dimensions of TMV, and filamentous viruses in general, predominantly depend on those of the encapsidated nucleic acid. A helix espouses a simple geometry and is characterised by its diameter and the width of one complete turn to be for TMV 8 and 2.3 nm, respectively. The repeating of these parameters is confined to a 280–300 × 18 nm tube of the fully assembled TMV.[45–47]

The tube is featured by a central pore or channel with a uniform diameter of 4 nm that is sustained by the RNA associated with the protomers near their inner surfaces.[48] Another important feature characteristic for TMV is that the two ends of its tube are unlike. One end is capable of end-to-end complexation and hence is often referred to as "active" as opposed to the other, inactive end,[46] Figure 3.4. This gives TMV a strong tendency to form dimers but no higher aggregates.[39,41] The difference consists in a directional structure of the encased RNA. The fully enclosed RNA helix lies in the interstices of adjacent turns of the helix formed by protomers. Thus, the rod can be regarded as a single linear structure of two coincident and mutually supporting helices.[49]

Another filamentous virus, equally influential in nanoscience, is M13 bacteriophage. M13 is a bacteria virus and unlike TMV contains a double-stranded DNA. The virus capsid is assembled from three different types of proteins, designated pVIII, pIII and pIX. 2700 copies of pVIII – the major coat protein – are arranged in a repeating helical array around the DNA to shape a flexuous cylinder, which is longer (860 nm) and thinner (6.3 nm) than TMV. The role of the other two proteins is to cap the cylinder at both ends – another distinction from TMV. The number of pVIII copies in a virion is strictly regulated by the size of its DNA. In a microphage variant of M13[50] an assembled cylinder was shortened to 50 nm. The DNA reduced by approximately thirty times was coated with just 95 copies of pVIII. This property, common to other bacteriophages, offers a rationale for control over the size and shape of virus-derived nanostructures.

TMV and M13 are the most comprehensively studied but surely not the only filamentous viruses known. Others including M13-related phages, for example almost identical Enterobacteria fd phage,[38] Figure 3.1, plant viruses (Flexiviridae)[51] and pleomorphic primate filoviruses[52] are also frequently encountered in research laboratories. Many have unique and potentially useful properties and some are being considered applicable to nanotechnological applications.[53] Nevertheless, TMV and MS13 remain the templates of choice for the synthesis of a variety of nanostructured materials that result from many inorganic transformations, the rationale of which is the subject of Section 3.2.

3.1.2 Packing Out and In

Self-instructive and reproducible assembly of virion components into fully functional viruses suggests that the spherical and helical forms are the most energy favoured. This economical use of space at the expense of self-assembly and geometrical constraining of building blocks closely relates viruses to another important class of encapsulants – eukaryotic gene packages, *i.e.* nucleosomes, discovered 30 years ago.[54,55]

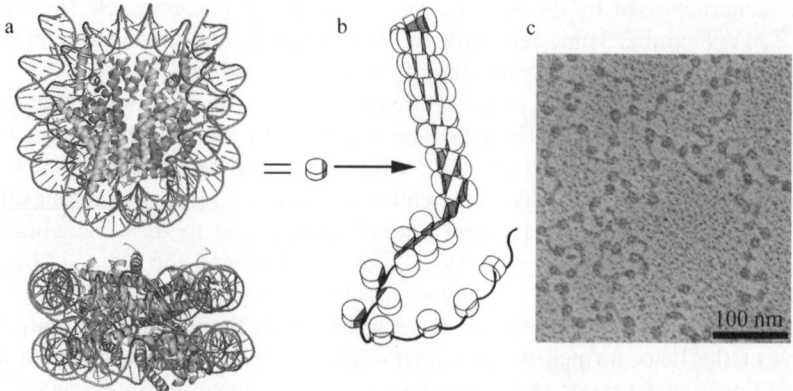

Figure 3.5   Nucleosome; (a) Structure of the nucleosome core particle, down and side views with respect to the DNA superhelix axis (PDB 1P3(a)), (b) Schematic of the chromatin fibre assembly, and (c) An electron micrograph of a nucleosomal beads-on-string arrays, (reproduced with permission from Thoma, F. J. 1979. Originally published in *J. Cell Biol.*, **83**, 403–427).

3.1.2.1   Spooling Around

Eukaryotic DNA is much larger than viral genomes and is not as space compromising. For example, if unwound, an entire human genome would span over two metres of DNA.[56] However, this is packed inside the cell nucleus with a size of just several cubic micrometres. Naturally, this demands another mechanism of protecting encapsulation, more efficient for the packing of large and continuous double-stranded DNA. Nucleosomes offer an ingenious solution to that.[55] Instead of building a complex protein coat, encoding of which would require a considerable space of the genome, eukaryotic DNA winds on multiple copies of symmetrically arranged protein spools called histones.[54] DNA is wrapped around and by such bobbins that subsequently become caged in spiral DNA nanotubes,[57] Figure 3.5.

In other words, histones embrace themselves with multiple rings along the double-stranded DNA that are held from inside and are stacked in a spiral fashion.[58] Thus, an assembly approach structurally reversed to that of viruses is employed. Here, it is proteins that get encapsulated by nucleic acids. Taken liberally, nucleosomes can be thought of as inside-out viruses. Indeed, like viruses they are regularly built capsules, like in viruses their encapsulated components guide the assembly and morphology of the capsule, and together with viruses they exemplify the very definition of gene packaging. Yet, unlike viruses nucleosomes are not standalone assemblies. On the contrary, nucleosomes constitute the initial phase of the organisation of more complex structures – chromatins.[59–61] The multifunctional purpose of nucleosomes is evident from the next phase in chromatin assembly.

Briefly, 11-nm thick DNA nanotubes folded by histones coil further to form 30-nm wide filaments.[62] Directly involved in cellular processes these filaments are required to be able to fold reversibly without causing structural tensions to mechanically rigid DNA structures. Rotating mobility ensured by the piled and flexibly linked rings of nucleosomes thus produces an efficient solution. Decompacted fibres are seen by electron microscopy as intermittently knotted threads, with the structure being referred to as "beads on a string",[54,63] Figure 3.5(c). Needless to say that chromatin assembly is a dynamic process that is flawlessly choreographed. For example, genes that are coiled or spooled into nucleosomes are locked or switched off,[64] but can be activated as nucleosomes are site-specifically dismantled[65] and reassembled,[66] the movement of nucleosomes along the DNA string can be induced,[67] so can be their distribution.[68]

3.1.1.2 Tunnelling Through

Amongst innumerable purposes biological encapsulation serves, two have particular importance for proteins. These stand out as the extreme poles of protein processing: their synthesis and degradation. The former is performed by ribonucleoprotein subunits that clap together to form a tunnel, through which a newly made protein is released into the cell; whereas another tunnel of several heaped protein rings functions as a "death trap" for defected or mis-folded proteins, Figure 3.6. These complex assemblies are ribosomes and proteasomes, respectively.

The ribosome is a cellular organelle that translates a messenger RNA (mRNA) to polypeptide chains that subsequently mature to proteins. It is an RNA-based molecular machine that serves as a messenger site of protein synthesis.[69] The two subunits of ribosome are different in size and morphology but perfectly fit together to reside on a single strand of mRNA by forming a joined 20-nm bead pierced by the mRNA.[70] The mRNA strand is 3 nm thick and is placed between the subunits to occupy a 5-nm tunnel.[71] The width, which is constant along the tunnel, is also sufficient to accept two perpendicularly oriented 3–5 nm molecules of aminoacyl-tRNA at a time to allow for stepwise protein synthesis.[72] This seemingly exiguous offer of space is crucial for ensuring the necessary proximity between the carried (by the t-RNAs) growing polypeptide chain and the next amino acid residue to form a nascent peptide bond.[72,73]

Ribosomes have been extensively studied over several decades.[74] Constant attempts to establish the routes of ribosomal self-assembly are modelled and tried experimentally *in vitro*.[75–78] Relative progress in these experiments, however, suggests that the task is far from trivial and a ribosomal type of assembly is formidably complex for design. As shown in Figure 3.6 even one of the subunits assembles from twelve different proteins and two RNA molecules, each of which is designated to a certain assembly step involving specific partnering. A subunit, as equally as the whole ribosome, that lacks any of the components is not functional, neither is an incorrectly assembled ribosome.[79] The assembly of fully functional ribosomes is poorly tolerant to changes caused

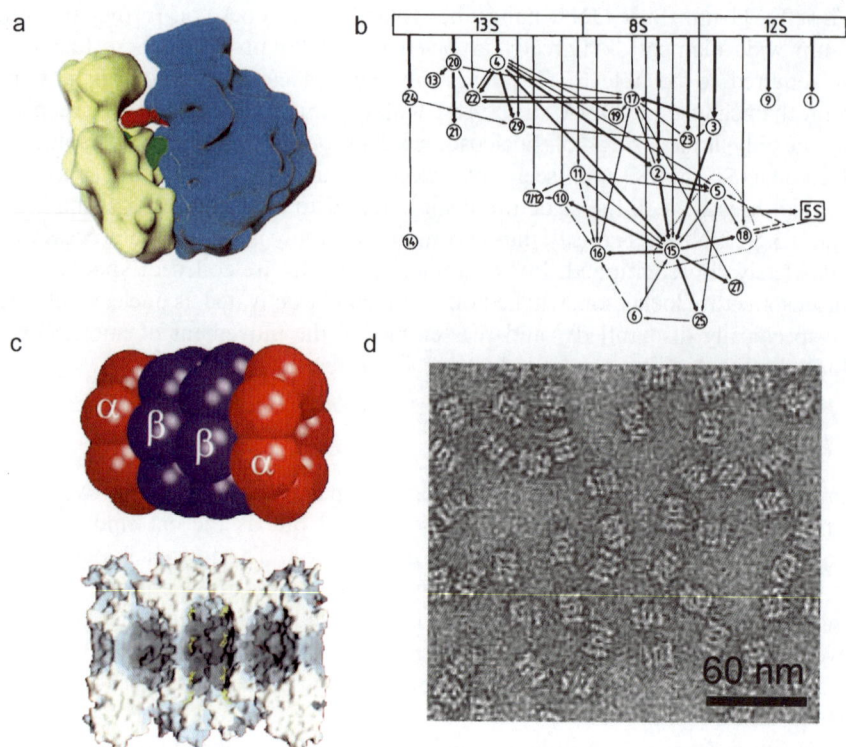

Figure 3.6   Ribosome and proteosome; (a) 70S ribosome assembled from 30S (white)
and 50S (blue) subunits. tRNAs occupying the A and P sites are shown in
green and red, respectively, (reproduced with permission from Nierhaus,
K. H. *et al.* Structure of the elongating ribosome: arrangement of the two
tRNAs before and after translocation. *Proc. Natl. Acad. Sci. USA*, **95**,
945–950. Copyright (1998) National Academy of Sciences, USA), (b)
Assembly map of the 50S subunit. The main fragments of 23S RNA (13S,
8S and 12S) are shown (top). Proteins (numbered circles) are arranged
according to their binding regions on 23S RNA. Arrows indicate the
binding dependence of protein–protein and protein–RNA interactions;
the stronger the dependence the thicker the arrow. Proteins circled with a
dotted line are essential for binding 5S RNA to 23S RNA, (reproduced
from Rohl, R. and Nierhaus, K. H. Assembly map of the large subunit
(50S) of Escherichia coli ribosomes. *Proc. Natl. Acad. Sci. USA*, **79**,
729–733 (1982), with permission from the authors), (c) Sphere and surface
models of a proteosome showing α and β subunits (top) and inner
chambers (bottom), (reproduced with permission from Bochtler, M., *et al.*
The proteasome. *Ann. Rev. Biophys. Biomol. Struct.*, **28**, 295–317 (1999)),
(d) An electron micrograph of bacterial proteosomes, (from Grziwa, A.
et al. Dissociation and reconstitution of the *Thermoplasma* proteasome. *Eur. J. Biochem.*, **223**, 1061–1067 (1994). Copyright Wiley-VCH
Verlag GmbH and Co. KGaA. Reprinted with permission from
Wiley-Blackwell Ltd).

by either the breach of its algorithm or mutagenesis.[80] This is justified by the exceptional task the ribosome is assigned to that demands an execution of extreme precision, which is exemplary in prescriptive self-assembly. Therefore, the full reconstitution of the ribosome may be seen as a first step in reaching equal precedents in designed systems.

Nature masters building enclosed nanoscale spaces at different length scales, even at levels as low as a few nanometres. Proteasomes offer good examples to demonstrate this.[81,82] The tunnel of a proteasome has three chambers formed by combinations of heterogeneous protein rings, α and β, Figures 3.6(c) and (d). A central chamber, which is assembled from two 10-nm β rings, is the largest and connected with neighbouring rings through pores of only 2 nm diameter.[83] These serve as gates to and out of the central chamber. The central chamber has a dozen or so active proteolytic sites spaced 3 nm apart. The figure is not just fortuitous. The proteolytic sites form one continuous substrate-binding surface that was suggested to force different folds of proteins into fully stretched β strands.[81,84] Given that the length of one extended β strand of n residues is $\sim n \times 0.34$ nm, which gives approximately 3 nm, this distribution appears to be ideal to yield chains of eight to ten residues.[85] Remarkably, this is precisely what is observed: proteins entrapped in the central chamber are cleaved by the proteasome down to octa- and decapeptides.

3.1.3 Escaping Walled

3.1.3.1 Capturing On and Off

This chapter would not be complete without outlining another fundamental encapsulation process whose prime purpose is focused on intracellular delivery. This is endocytosis, the process that allows the cell to ingest outside materials such as proteins and solutes.[86] There are several types of endocytosis assigned by the cell that are being extensively studied and very well covered in the literature. However, from the standpoint of this chapter self-assembling formations that constitute and are directly involved in the process are of the most instant interest. In this vein, perhaps the most captivating and arguably major type is the one that is mediated by a protein clathrin,[86] Figure 3.7(a).

Clathrin is indeed a fascinating molecule.[87] It is famously known for its shape, a spiderlike triskelion with three bent legs that form an easily recognisable cross, Figure 3.7(b). The cross is recruited on to membranes by adaptor proteins, which also promote the assembly of clathrin into polyhedral lattices or cages, Figure 3.7(c). The adaptor proteins select membrane-bound cargo to be included into bilayered membrane vesicles. The lattices coat a new vesicle that entraps the cargo for transportation.[88]

The main characteristic of coats is a regular arrangement of open pentagonal and hexagonal facets, Figures 3.7(a)–(c). The structure and symmetry of coats are variable and are specified by a process they are engaged in. Typically, an outside coat has a diameter of 70 to 200 nm, with the vertices of its lattice separated at 16 nm, Figure 3.7(d). The hubs of clathrin triskelions are located at

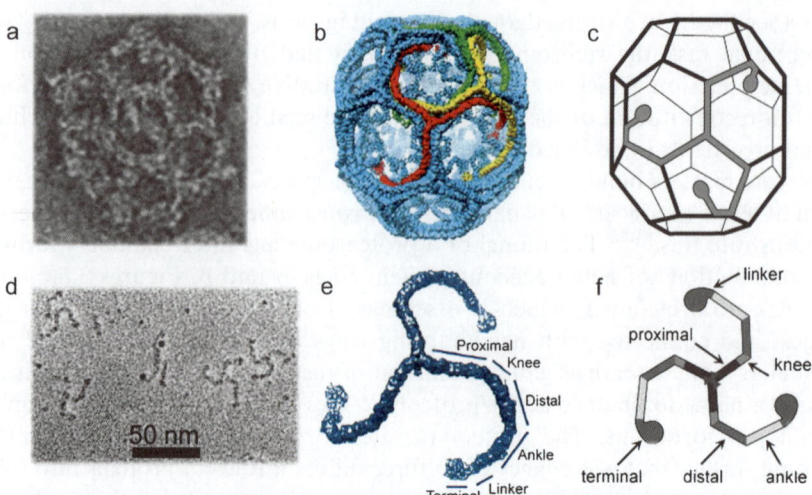

Figure 3.7 Clathrin; (a) Electron micrograph of a clathrin coat, (b) Image reconstruction of 36 clathrin triskelions assembled into an hexagonal barrel. Three overlapping triskelions are highlighted in different colours, (c) Schematic representations of a D6 lattice with a position of a triskelion highlighted in grey, (d)–(f) Clathrin triskelion; an electron micrograph of triskelions (d), image reconstruction (e) and schematic representation (f) of the triskelion labeled with domain names, (reproduced by permission from Kanaseki, T. and Kadota, K. 1969. Originally published in *J. Cell Biol.*, **42**, 202–220; from Macmillan Publishers Ltd: Edeling, M. A. *et al.* Life of a clathrin coat: insights from clathrin and AP structures. *Nature Rev. Mol. Cell Biol.*, **7**, 32–44, Copyright (2006), and Fotin, A. *et al.* Molecular model for a complete clathrin lattice from electron cryomicroscopy. *Nature*, **432**, 573–579, Copyright (2004); and from Jin, A. J. and Nossal, R. Rigidity of triskelion arms and clathrin nets. *Biophys. J.*, **78**, 1183–1194. Copyright (2000). Biophysical Society).

the centres of vertices, Figures 3.7(b) and (c). Each leg of a triskelion radiating from a vertex has bending sites that allow it to extend around two edges of pentagonal and hexagonal cells of a lattice to then protrude inward, Figures 3.7(b) and (c). The sites are introduced by a knee and a linker. The knee creates a strong twist near the middle of a leg to divide one into proximal and distal segments. These segments are relatively straight and fit together with legs of other triskelions to form the edges of facets.[88] The segments can interact in different combinations; distal–distal, proximal–proximal and distal–proximal. Generally, three proximal segments of a triskelion extend towards three adjacent vertices and the distal segments run further to the next vertices, Figures 3.7(e) and (f). For example, a proximal segment of one triskelion extends along that of a neighbouring triskelion. At the kinked (knee) site the leg of the first triskelion curves for its distal segment to lie just beneath the proximal segment of a second leg of the adjacent triskelion, Figures 3.7(b) and (e). The knee of a triskelion is thus placed at an adjacent vertex, where no interactions occur.[89] This imparts a gentle curvature to the leg. The main role of the knee is therefore to bend and twist, which ensures a simple mechanism in adapting to pentagonal and hexagonal facets and coats of various diameters and curvatures,[89] Figure 3.7. Indeed, the differences in leg conformations in a clathrin barrel are mainly attributed to the flexion regions near the vertex, *i.e.* knees.[90,91] Furthermore, the universal design of knees implies that the introduced flexions may be responsible for additional capacities inherent to clathrin cages.[89] For example, the legs can relax local stresses and strains associated with entrapping asymmetric or large cargo without the reassembly of coats.[92]

As already stated, the assembly of triskelions into lattices is promoted by adaptor proteins.[92] The proteins encompass an interface between an outside coat and an enclosed membrane by binding the membrane and N-terminal domains of each leg. In order to accommodate such an arrangement a terminal domain should be preceded by a flexible segment capable of bending. A linker between the distal and terminal domains is the other bending site of the leg. The linker gives a smooth turn to the leg to position a foot-like terminal domain roughly perpendicular to the main axis of the leg,[87] Figures 3.7(e) and (f). The conformation of the linker is thought to be readily adjusted by the terminal domain itself upon binding adaptor proteins.[92] Altogether this helps the triskelion acquire a three-dimensional tripod shape with a height of 20 nm.[88] The latter is a major contribution to the distance between the network of distal and proximal segments and an enclosed membrane.

Analogously to viral capsids, the assembly of clathrin triskelions obeys strict symmetry parameters.[93,94] The basic design of clathrin coats is based on dihedral topology assuming 6-fold space group symmetry (D6), Figures 3.7(b) and (c). This is an hexagonal lattice that to be fully closed requires an exact number of pentagon facets, 12.[88,89] Building on this pentagon–hexagon ratio clathrin can assemble into a range of shapes. Nevertheless, *in vitro* reconstitution of purified coats with adaptor proteins revealed a dominant proportion of D6 barrels.[89,95] Hexagonal barrels are indeed abundant clathrin structures observed in all nucleated cells.[88,93,95] As icosahedral viruses, symmetrical barrels prove to ensue from independent asymmetric units. Each unit is composed of three triskelions or nine legs. The legs are perfectly aligned in the proximal segments and diverge at the knees with different degrees of bending.[89,92] The latter is emphasised to lead to significant changes in the positions of distal segments and terminal domains relative to the vertex. Such a propagation would not be possible without a high degree of structural compliance and hence similarity of all segments along the leg. Indeed, morphologically legs are relatively uniform with the exception of terminal domains that have a globular structure, Figures 3.7(d)–(f). The leg extends as a curved rod that is approximately 2 nm thick and 48 nm long. This is suggestive of a repeating character of the underlying structure of the leg. Consistent with this, the proximal, distal and linker segments were all found to be zigzags of α-helices.[95–97] Each zigzag contains 30 amino acid residues conformed into two antiparallel helices.[96,97] Five zigzags (140–150 residues) constitute a repeat pattern no interruptions of which were found throughout the leg.[96] Moreover, the neighbouring zigzags tend to reversibly cluster into four-helix bundles – a mechanism of setting bigger gaps between successive zigzags. This helps the structure to be contorted by slight twists without large energy costs.[96] The accumulation of the gaps gives the leg as smooth a curvature as required for the assembly of a specific coat.

Similarly, the structure of the terminal domain is very well adapted to its function, which is to maintain the extensive surface of protein interactions. The domain is folded as a seven-blade β-propeller.[95–97] Such folds are based on repeats of a four-stranded antiparallel β-sheet motif, which constitutes a blade.[96] These are very rigid structures and are stabilised by largely hydrophobic bonds between the sheets that twist to pack face to face. Propellers are funnel-shaped domains in which blades are arranged around a central pseudotunnel closed by stringent contacts between terminal blades. The folds are employed by many proteins to mediate multiple and complex interactions in constrained environments.[87] Thus, β-propellers are interior elements of protein architectures and their key function is to ensure specific bindings. For example, commonly found in enzymes they are directly involved in recognition and reaction processes, whereas in regulatory multiprotein complexes they serve as assembly platforms.[87,88]

This relatively simple but very robust architecture places the clathrin type of assembly in a prerogative position for nanodesign. In particular, the efficiency with which clathrin performs the assigned tasks opens new and exciting possibilities for engineering drug-delivery systems.[94] Indeed, the assembly of many triskelion copies into lattices is extremely fast. The entire process from the initiation of membrane binding through cargo entrapment and coat formation to transport of the formed vesicles and subsequent removal of clathrin coats runs under a minute.[88,92] For this reason, the clathrin route is the one to watch. Surprisingly, there is scarce information in the literature as regards the designing of clathrin-derived transfection systems. Partly, this can be attributed to the complexity the protein presents for synthesis, partly to its highly conserved primary structure.[89,96] Some researches, however, imitate the mode of clathrin assembly using small-molecule triskelions that are more straightforward to make.[98–100] Others approached the problem from different directions that involve the development of branched molecules, self-assembled dendrimers and polymeric amphiphiles.

3.1.3.2   Storing Exchangeable

Whilst biological macromolecules are preferable candidates for hosts, the representation of guest types proves to be more diverse. Inorganic materials such as metal ions and minerals are amongst the most ubiquitous guests. One typical example comes from ferritin, a protein-iron complex.[101–103]

Ferritin is designed to store and transport iron to and from the cell cytoplasm. The protein part of the complex – apoferritin – is the encapsulant that assembles into a closed 12-nm sphere with a nearly spherical cavity of 8 nm diameter,[104,105] Figure 3.8(a). Its guest – iron – is fixed in the formed cage as a microcrystalline

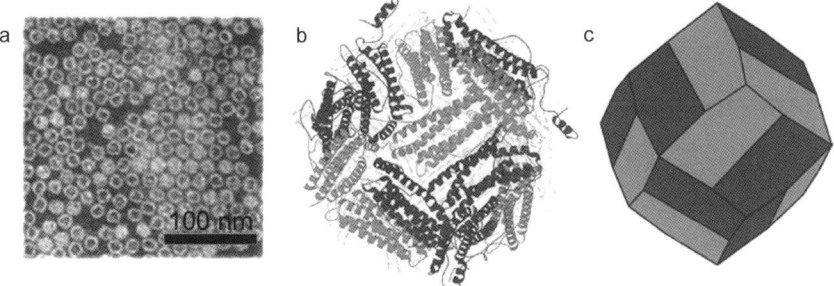

Figure 3.8 Ferritin, (a) Electron micrograph of ferritins seen as white rings 10–12 nm in diameter, (reprinted from Watabe, T. and Hoshino, T. Observation of individual ferritin particles by means of scanning electron microscope. *J. Electron Microsc.* (*Tokyo*), **25**, 31–33 (1976) by permission of Oxford University Press), (b) Ribbon diagram of ferritin showing the H (grey) and L (black) chain four-helix bundles, (c) Schematic orthorhombic representation of ferritin (reprinted from Hamburger, A. E. *et al.* Crystal Structure of a Secreted Insect Ferritin Reveals a Symmetrical Arrangement of Heavy and Light Chains. *J. Mol. Biol.*, **349**, 558–569, Copyright 2005, with permission from Elsevier).

ferric oxyhydroxide phosphate $[FeO(OH)]_8[FeO(H_2PO_4)]$.[103,105] The mineral contains thousands of ferric atoms protected from complexation with water. The principal role of apoferritin is thus to concentrate iron to levels required by living cells. Apoferritin is a globular shell that is assembled from 24 polypeptide subunits each folded as a four-helix bundle.[105,106] The possible rationale for bundles is to arrest or store water whose ordered chains can be observed along the bundles. The protein shell can be built from one or two types of the bundles conventionally classified as heavy (H) and light (L). Structural variations in ferritins from different organisms and tissues as well as differences in rates of microcrystalline formations are based on the relative ratios of these types.[103,107,108] However, notwithstanding certain dissimilarities all ferritins assemble as dodecahedral shells, in which 12 rhombic faces are shaped by 12 dimers of polypeptide subunits,[103] Figure 3.8(b). Because of the anisotropic character of the bundles the dodecahedron structure is somehow stretched and therefore is best thought of as orthorhombic.[103,107] Adopting such a structure bestows all the necessary properties exhibited by ferritins. For instance, interactions between subunits in each dimer form grooves along the axis of the subunits. Grooves cluster into a network lining the inner side of apoferritin, suggesting that these play an important role in the nucleation of the iron core. Most strikingly, the adopted architecture renders the assembled cage holey or porous owing to the formation of channels at the intersections of three and four subunits, Figure 3.8(c).

Ferritin is believed to control in-and-out transport of iron through these channels that lie on three- and four-fold symmetry axes of the dodecahedron,[103,105,108] Figure 3.8(c). Depending on the location of a channel it can be three- or four-fold. The channels are similar in size (0.3–0.4 nm) and differ by polarity thereby two different functions are performed.[109] The polarity is determined by the nature of amino acid residues that are placed along the walls of the channels.[105] Negatively charged carboxylic (aspartates and glutamates) and hydrophobic (leucines) residues were identified to cover three- and four-fold channels, respectively.[103,107] As shown below, this choice proves to be categorical for the normal functioning of ferritin.

The sequestered iron in ferritin is concentrated as a ferrihydrite – a crystalline lattice, with each Fe(III) atom caged by five oxygen ions and a phosphate.[105] Phosphate groups bind residues of the inside of the apoferritin shell and hence are located on the outside of the microcrystalline core.[102] The lattice is a continuous structure and iron atoms cannot be released unless the lattice is broken. To dissolve the lattice ferritin reduces iron from ferric Fe(III) to ferrous Fe(II). The latter that are subsequently entrapped by water molecules become hydrated $Fe(H_2O)_6^{2+}$, *i.e.* soluble and ready for release into the cell cytoplasm.[103,105] Positively charged hydrated irons are favoured for interactions with negatively charged three-fold channels, and so are favoured for the passage through the channels. Hydrophobic four-fold channels do not favour and are "closed" for ferrous ions. The exact role of the hydrophobic channels is not, however, fully understood.[109] It is generally proposed that four-fold channels are those whereby electrons are transferred from outside the protein shell into the core.[101,103] This is meant to reduce Fe(III) or solubilise the crystalline lattice for ferrous ions to be then released by three-fold channels.

Ferritin is a truly outstanding nanoconstruction; the only known natural system that reversibly converts metal ions to minerals on the nanoscale. Given analogous ion–solid transitions in many biomineralisation processes it comes as no surprise that apoferritin is being explored for the conversion of other metals and minerals.

3.2 Reacting Nano

The repertoire of self-assembled encapsulants offered in Nature is indeed very rich and is by no means exhausted in the previous sections. Yet, one notable property that puts all such assemblies into a single category irrespective of origin and chemistry is the highly efficient adaptation of their ultrastructure to intramolecular dynamics. The impact of intramolecular dynamics potentially detrimental to the sustainability of confined spaces is restricted by highly conserved assembly modes. Different encapsulating forms choose different modes and the choice is a direct response to requirements set by encased materials. Subsequently, encapsulants correlate their own assembly with strict geometrical parameters – a simple solution to minimising potential errors. In fact, assembly mutants or misassembled encapsulants lose assigned functions completely. This quality is unique in self-assembly and endemic to encapsulants, which renders their adjustment very conserved and in return grants them the ability to control the size and structure of guest components. Altogether, this has been enthusiastically employed in synthesising discrete nanomaterials – nanometre-sized products of chemical reactions. As expected, natural encapsulants are being seen as nanophase reaction vessels,[110] which has become an accepted practice and has already given rise to a good variety of reactors, capsules and containers.[1–2,4,111]

3.2.1 Clustering Spherical

A general trend in preferring one type of encapsulant for templated synthesis to another is the relative ease of use or reconstitution of a candidate. Since all encapsulants are fairly complex, favouring chemical robustness and versatility is understandable. Though these qualities are not so uncommon amongst natural systems, developments based on best studied protein cages and viruses are also best represented.

3.2.1.1 Contriving Consistent

For example, the above-described ferritins are notably influential in the field.[102,110] The exploration of ferritins for the synthesis of nanomaterials started with the observations that metal ions other than iron can also be bound in and by the apoferritin cage.[110] It was found that amongst many Ni^{2+}, Cd^{2+}, Mn^{2+}, UO^{3+}, Tb^{3+}, VO^{4+} bind specifically at the core sites of the cage.[101,102] Furthermore, Tb^{2+} and Zn^{2+} appear to compete with iron,[112] whereas Fe^{2+} was shown[113,114] to displace Mn^{2+} and VO^{4+}. Interestingly, *in situ* saturation and subsequent oxidation of Fe^{2+} to Fe^{3+} proved to generate new binding sites for Mn^{2+} and VO^{4+}, indicating that the core formation occurs in the interior of the protein.[114,115] In parallel with this, other studies revealed that three-fold channels primarily responsible for the transport of iron can be involved in the initial stages of iron sequestering as the neutralisation of carboxylate groups by esterification prevents oxidation of ferrous iron.[101,116] Taken together this raised a view that different ions at various ratios can be used to equally seed and regulate the formation of the native and foreign cores.

In this vein, elegant examples reported by Mann and coworkers prompted the field.[110] Back in the early 1990s, the researchers were first to describe horse spleen ferritin as a biological container.[117] They demonstrated that *in vitro* reconstituted ferritins can generate various mineral cores of sizes and crystalline patterns identical to those of native iron cores,[118–120] Figure 3.9(a).

Interestingly, it was revealed that both empty and iron-loaded apoferritins can be used.[117,121] For instance, *in situ* reactions of native ferrous cores with sulfides (H_2S/Na_2S) yielded iron sulfide crystalline particles.[121] Similar monodisperse particles were generated by *in vitro* reconstitution of Fe(II) with ferritin in the presence of phosphate, arsenate, molybdate, and vanadate.[122] Furthermore, incubations of apoferritins with manganese chloride at mildly basic pH and room-temperature triggered redox reactions leading to discrete

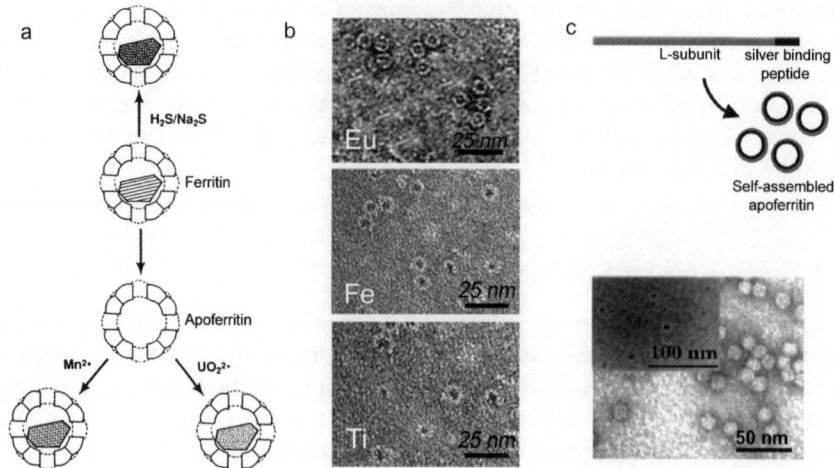

Figure 3.9   Ferritin nanoreactor; (a) Schematic representation of ferritin-supported
nanophase materials; iron sulfide (top), manganese oxide (left) and uranyl
oxyhydroxide (right) (reprinted by permission from Macmillan Publishers
Ltd: Meldrum, F. *et al.* Synthesis of inorganic nanophase materials in
supramolecular protein cages. *Nature*, **349**, 684–687, Copyright (1991)),
(b) Electron micrographs of photomineralised oxyhydroxides; from top to
bottom europium, iron and titanium. Electron-dense regions (mineralised
metals) are surrounded by electron lucent protein shells, (reprinted with
permission from Klem, M. T. *et al.* Photochemical mineralisation of
europium, titanium, and iron oxyhydroxide nanoparticles in the ferritin
protein cage. *Inorg. Chem.*, **47**, 2237–2239. Copyright (2008) American
Chemical Society), (c) Schematic representation of ferritin L subunit fused
with a silver-binding peptide with its subsequent assembly into apoferritin
(top). Electron micrographs of empty apoferritin cages and loaded with
mineralised silver (inset), (reprinted with permission from Kramer, R. M.
et al. Engineered Protein Cages for Nanomaterial Synthesis. *J. Am. Chem.
Soc.*, **126**, 13282–13286. Copyright (2004) American Chemical Society).

precipitations of manganese oxide.[119] Specific ion uptakes and subsequent core
deposition were also achieved for uranyl[121] and cobalt oxyhydroxide,[123,124]
nickel and chromium hydroxides,[125] iron-cobalt alloys[126] and iron oxide
magnetites.[127,128] The latter provided an example of "magnetoferritin"[127] – a
semisynthetic protein that is able to generate and host ferromagnetic
particles.[128]

Furthermore, a team led by Douglas took a somewhat unconventional
step in applying ferritins to nanomaterial synthesis, Figure 3.9(b). They argued
that spatially restricted and isolated iron cores can provide as unique properties
as those displayed by apoferritin cages. The argument is supported by that
ferritins exhibit reductive and oxidative activities under excitation by visible
light, with the ferrihydrite core being a photochemical element.[129] A series of
studies confirmed the photocatalytic function of ferritin cores.[130] For example,

photoreduction of Cu(II) resulted in the formation of copper colloids presenting electron-dense particles with narrowly distributed sizes.[131] By subtle increases in Cu(II):ferritin ratios it was possible to vary the size of particles from 4 to 30 nm.[124] Also, photolysis of Ti^{4+} followed by oxidation was found to lead to the formation of titanium oxyhydroxide cores within the protein cage, Figure 3.9(b). The Ti-loaded ferritins exhibited a catalytic behaviour similar to that of native ferritins.[132] The findings give important evidence to the view that mineralisation of apoferritins and its regulation can be photolysis mediated, which may instigate investigations on the feasibility of ferritin-derived systems for the remediation of toxic metals in the presence and absence of light.

Apoferritin cages were also proposed for catalytic hydrogenation.[133] A Pd cluster synthesised in a protein cage by reduction of Pd(II) ions was demonstrated to catalyse the hydrogenation of olefins.[133] A number of kinetic studies aimed at the elucidation of mechanisms underlying metal deposition in ferritin cores provided further insights into the functional and structural properties of ferritins.[114] To assess Fe(II) uptake competition assays in five different ferritin proteins, varying in channel structure, oxidation and nucleation sites were conducted using a kinetically inert Cr(III) inhibitor.[134] The studies proved that the inhibitor binds at functional sites and plugs apoferritin channels whereby iron entry and following mineralisation are arrested. Partly in contrast to this, further comparative studies of wild-type and mutated ferritins suggested that nucleation sites that were long thought to play a key role in core formation can be excluded without impairing iron oxidation and mineralisation.[135]

Related to this, other researches demonstrated that the interior of apoferritins can be rationally altered to give the protein desired but noninherent capacities.[136] Genetically engineered L subunit of human ferritin with its cavity lined with silver-binding sites was shown to generate silver nanoparticles,[136] Figure 3.9(c). The work is complementary to and derives from studies on short peptide sequences that selectively recognise specific metal ions.[137,138] Decorating the interior of natural protein encapsulants with such peptides provides an effective approach for designing chimeric cages with novel attributes for nanomaterial synthesis.[139]

Likewise, novel characteristics can be rendered through the decoration of outer surfaces of ferritin shells. Ferritins with coats derivatised with biotinylated ligands self-arrange into three-dimensional networks through interactions with streptavidin (a biotin-binding protein).[140] As ordered clusters of many ferritins the assembled arrays exhibit enhanced properties of a single protein.[141] This may be particularly contributive to microelectronics as ferritin arrays may be of use as energy-storage units in nanoscale batteries.[142] This would serve several advantages over common batteries including power density, storage and discharge rates, compactness and energy harvesting. Eventually, such technology would lead to more efficient, cheaper and environmentally cleaner energy sources. Genetically modified ferritins can be used in medicine, for instance, as probes in noninvasive magnetic resonance imaging (MRI).

An interesting proposal in this regard is concerned with the use of ferritins as endogenous reporters in mapping gene expression in tumor cells.[143] In this study ferritins were engineered to be responsive to the antibiotic tetracycline. The expression of modified ferritins was thus regulated by the administration of the antibiotic into cells expressing the protein. Tumour cells transplanted into living mice and surrounded by normal cells were tracked by MRI when an increase in iron uptake and hence in ferritins expression occurred. In contrast, tetracycline dispensed into the cells inhibited the expression. The obtained results hint at a technique for monitoring gene therapy treatments in which targeting and activation of therapeutic genes could be detected by MRI.[143]

3.2.1.2 Scaling Hosting

The overwhelming popularity of ferritins in materials science is a result of the unique fit between the ferritin structure and function. Partly, this is due to ferritins being extremely specialised constructions, partly to their unequivocal assembly mode. Among all protein cages ferritins are constructs to match in the synthesis of nanomaterials and with very few exceptions[144,145] remain the system of choice in nanofabrication.

That said, however impressive the wealth of ferritin applications its dominance can be easily challenged by viruses.[146] The reason for this lies in the main drawback of ferritin cages – their limiting setting in shape and size. In marked contrast, viruses are structurally polymorphic and are exemplified by a vast range of forms and dimensions. Extensive research conducted over the last decade on probing viruses in applications relying on nanoscale precision and reproducibility explicitly supports this.[146]

Viruses are finding use in the generation of semiconductors and liquid crystals, nanowires and tubes, mineral films and fibres.[146] With this, general trends in the synthesis of virus-mediated nanomaterials strictly correlate with the morphological types of viruses; that is, icosahedral viruses are favoured for spherical particulate materials, whereas helical or rod-shaped viruses are preferred for anisotropic and rod-like assemblages. Of sphere-shaped viruses two structurally similar forms have attracted most efforts. These are cowpea chlorotic mottle and mosaic viruses, abbreviated as CCMV and CPMV, respectively,[147,148] Figures 3.10(a) and (b).

These harmless plant viruses are readily extracted from an infected cowpea in gram quantities,[149] which essentially promoted their inclusion into laboratory practice. However, this is not the sole reason for their compatibility with nanomaterial syntheses. The two comprehensively studied at atomic resolution were first icosahedral viruses reconstituted *in vitro*.[147–148,150] Both are remarkably stable, and show an unparalleled robustness in integrity of assembly. For instance, CCMV has a unique ability to "swell", with its size increased by approximately ten per cent, Figure 3.10(c). The swelling is a direct consequence of the formation of sixty separate 2-nm gates on the virus surface.

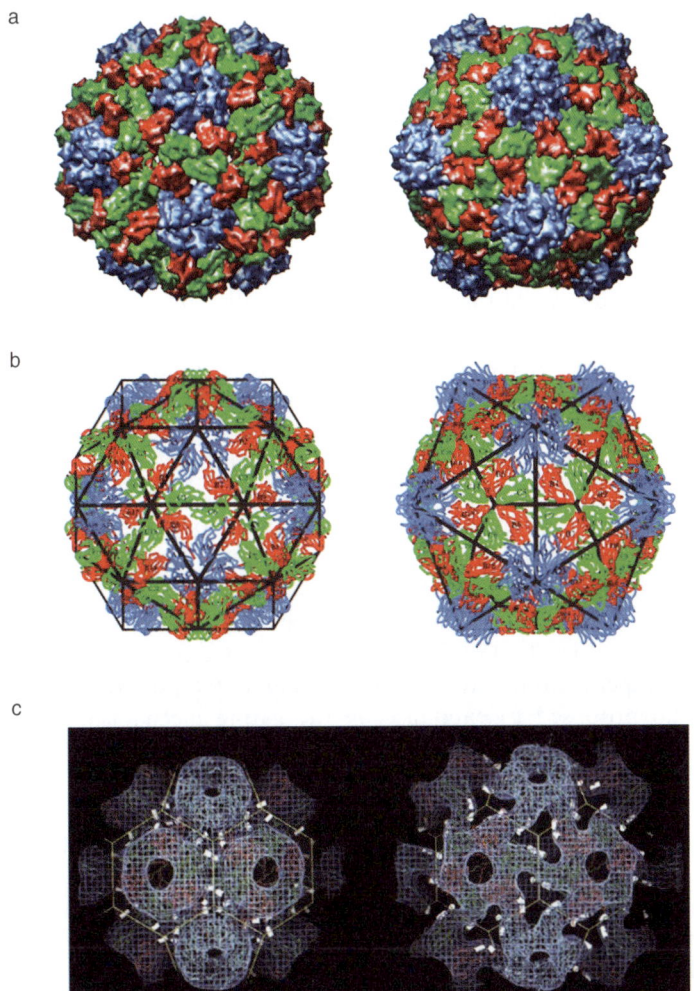

Figure 3.10 Cowpea chlorotic mottle and cowpea mosaic viruses, CCMV and CPMV; (a) Space-filling models of CCMV (left) and CPMV (right) capsids showing subunit organisation. Pentamers and hexamers are formed by three differently coloured domains of one asymmetric icosahedral subunit in each case, (b) $T = 3$ lattices, the black cages denote the edges of truncated icosahedra, (reproduced from VIPER (http://viperdb.scripps.edu)[15]), (c) Cryo-electron density maps of the CCMV capsid at pH 4.5 (left) and pH 7.5 (right, swollen form), (reproduced from Speir, J. A. *et al.* Structures of the native and swollen forms of cowpea chlorotic mottle virus determined by X-ray crystallography and cryo-electron microscopy. *Structure*, **3**, 63–78, Copyright 2005, with permission from Elsevier).

The mechanism allows "breathing", thereby CCMV controls its entrapped content – three single-stranded RNA molecules that are encapsidated separately. The shell of CCMV is thus perforated similarly to that of ferritin, Figures 3.10(a) and (b). However, unlike that of ferritins the gating of CCMV is reversible; that is, the opening and closing of the shell is a repetitive and sequential process. This alludes to the fact that the open–close mechanism should be regulated by external stimuli. This is true in that the process is stimuli-selective: the virus is thermally inert but undergoes a native–swollen transition at different pH, Figure 3.10(c). Consequently, since the process is reversible the assembly of the virus can be controlled by changing pH. Technologically it is attractive, particularly when combined with the ability of a disassembled virus to fully reassemble without its RNA components. The swelling mechanism of easily purified and assembling hollow virions can thus be routinely used. The CCMV virion is composed of prominent pentameric and hexameric capsomers built from 180 identical protein subunits that arrange into a 29-nm wide particle with $T=3$ quasisymmetry,[147] Figure 3.10(b). The interior radius of the particle is averaged to 10 nm, however, positively charged portions of coat proteins project further to radii of 9 nm. This creates a basic cavity of 18 nm in diameter, which is approximately twice the inner diameter of ferritin. Based on these and other morphological and functional resemblances shared with ferritin CCMV was envisaged as a template for mineralisation. Indeed, an experimental confirmation for that was reported by Douglas and Young who proposed to capitalise on the gating mechanism of CCMV in mineral growth. In their exquisitely executed work dated back a decade[151] the researches hypothesised that assembled CCMV can be seen as a dynamically transient host–guest system and it should be possible to couple the properties of such system with pH-dependent inorganic polymerisation to allow for controlled particulate mineralisation.[151] To demonstrate this, a series of reactions was performed. Emptied virions were incubated in the presence of aqueous molecular tungstate (WO_4^{2+}) under conditions where the virus is in its open or swollen form (pH > 6.5), Figure 3.11.

The incubation was then followed by lowering pH (pH < 6.5) to induce the formation of water-insoluble paratungstate polyanions ($H_2W_{12}O_{42}^{10-}$) from tungstate ions that at this stage had been concentrated within virions. Concomitantly, lowered pH caused the pores to close and the basic cavity of virions facilitated the entrapment and subsequent crystallisation of the polyanions. The process proved to be highly selective, with single-crystal materials being formed exclusively within the confines of viral cages.[151] Additionally, the electrostatic aspects of host–guest interactions in the viral core were re-emphasised in this and later studies. Firstly, a similar experiment was carried out with an anionic organic polymer that was loaded into empty virions at high pH to become encapsidated at lowered pH. Secondly, protein segments that shape the interior of the virus were mutated by replacing basic lysine and arginine residues with glutamates. This rendered the core acidic that is remarkably reminiscent of that of native apoferritin. Not only had the mutants assembled into wild-type virions they were also shown to mimic ferritins in

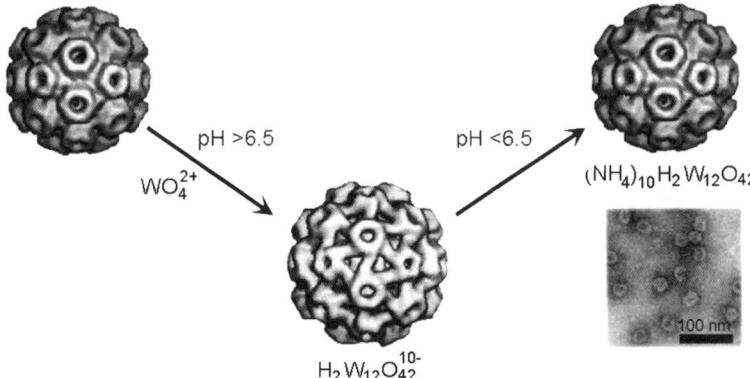

WO_4^{2+} pH >6.5   pH <6.5 $(NH_4)_{10}H_2W_{12}O_{42}$

$H_2W_{12}O_{42}^{10-}$

Figure 3.11 CCMV nanoreactor; Schematic of the mineralisation of the CCMV shell using 3D image reconstructions of the unswollen (left and right) and swollen (centre) forms. Swelling occurs at the three-fold axis and leads to the formation of sixty 2-nm pores. An electron micrograph of para-tungstate particles surrounded by intact virus cages (bottom right), (reprinted by permission from Macmillan Publishers Ltd: Douglas, T. and Young, M. Host-guest encapsulation of materials by assembled virus protein cages. *Nature*, **393**, 152–155, Copyright (1998))

hosting the formation of crystalline ferrihydrate nanocrystals.[152] Thirdly, mutants of exterior portions of the same proteins were demonstrated to promote self-assembled monolayers of CCMV on gold surfaces.[153] Virions assembled from these mutants became decorated with exposed thiols of cysteine residues to covalently bind to amorphous gold. Interestingly, only capsids decorated on one side and chemically passivated on the other were able to self-assemble into extended arrays, whereas virions with symmetrically dispersed thiols tended to aggregate.[154]

The combination of site-specific mutagenesis with chemical derivatisation of viruses has been exploited as an efficient tool for creating addressable nanoscale blocks and platforms. In this regard, the properties of the other cowpea virus, CPMV, proved to be particularly contributive. The virion of CPMV is an icosahedral cage assembled from 60 asymmetric units each containing two subunits, large and small. The assembly adopts icosahedral $T=3$ symmetry with three- and five-fold axes, Figure 3.10(b). A pore centered at each five-fold axis is formed by five units, although the exact role of the pore is unknown. However, its tiny size appears to be a deterrent to diffusion of materials larger than 1 nm.[155] In a set of experiments research teams led by Finn and Johnson[149,156,157] used thiol-selective reagents to probe noncrosslinked cysteine residues in wild-type virions. No reaction was detected with Nanogold, a reagent of 1.4 nm in size.[158] Given the total lack of thiols on capsid surfaces[156] this is as expected. However, adducts of the native viral particles with ethyl mercury phosphate having dimensions of a fraction of a nanometre were readily formed. The X-ray diffraction analysis of the loaded virions revealed that the reagent coupled to a single cysteine residue in the interior surface of the large subunit. Furthermore, an organic dye fluorescein condensed with the same residue in a concentration-dependent manner. 60 attached dye molecules were detected. This strongly suggests that every single interior cysteine residue is equally reactive. The researchers capitalised on this finding to demonstrate that CPMV can be of use as a synthetic platform.[156,158] Exterior mutants of a single cysteine insertion were produced and treated with increasing concentrations of rhodamine, a red-fluorescent dye. The interior cysteine remained active in mutants that prompted a sequential derivatisation of inner and outer surfaces with different molecules under controlled conditions. Gratifyingly, mutant particles decorated with both dyes were produced.[159] Additionally, the treatment of the mutant with Nanogold unreactive towards the native virus yielded gold particles located at the positions of the inserts on the exterior,[160] Figures 3.12(a) and (b). Intriguingly, CPMV itself was shown to be prone to self-organisation into hexagonal crystals, with a typical crystal containing 10^{13} particles ordered as open lattices,[149] Figure 3.12(c). On the other hand, derivatised with biotin molecules the virus was able to interact with biotin-binding proteins to self-organise into networks analogously to those formed by biotinylated ferritins[157] or multilayered arrays on solid supports.[161] Three-dimensional structures of gold nanoparticles[162] as well as monolayers of mutants on gold surfaces[163] were found to benefit equally from the derivatisation of CPMV. The exterior of CPMV was implemented by other molecular types including peptides,[164] carbohydrates,[165] polymers[166] and oligonucleotides.[167]

A somewhat nonstandard route in attaining novel virus-based bionanomaterials has been recently introduced by Dragnea and coworkers. This team proposed to use inorganic nanoparticles as templates for the assembly of virus capsids – an approach reversed to those discussed thus far.[168] A brome mosaic virus (BMV), reminiscent of the cowpea viruses, was used as a model. BMV is a $T = 3$ virus whose capsid assembled from pentameric and hexameric subunits is a compact icosahedral lattice at pH 5.[169] Reversible expansion of the capsid occurs at neutral pH, while further increases cause its complete dissociation.[169] The assembly can be readily reestablished at native pH and ionic strength. Electrostatic interactions between RNA and protein subunits appear to mediate the assembly of the capsid and can be mimicked. Thus, functionalised gold particles were used to initiate the assembly of the virus by imitating the interactions of the native RNA. Assembled virus-like particles exhibited symmetric icosahedral packing and pH-induced swelling transition both properties being intrinsic to the native virus. Furthermore, varying the gold core diameter it was possible to control the structure of the particle. Particles of varying diameters that could be ascribed to three symmetry classes were generated.[170]

Another laboratory led by Manchester was actively engaged in establishing the *in vivo* behaviour of cowpea mosaic viruses.[171,172] The virus whose replication is normally restricted to plant hosts appeared to be able to infect mammalian organisms inclusive of humans.[171] The binding of the virus to mammalian cells was found to be mediated by a conserved protein identified on

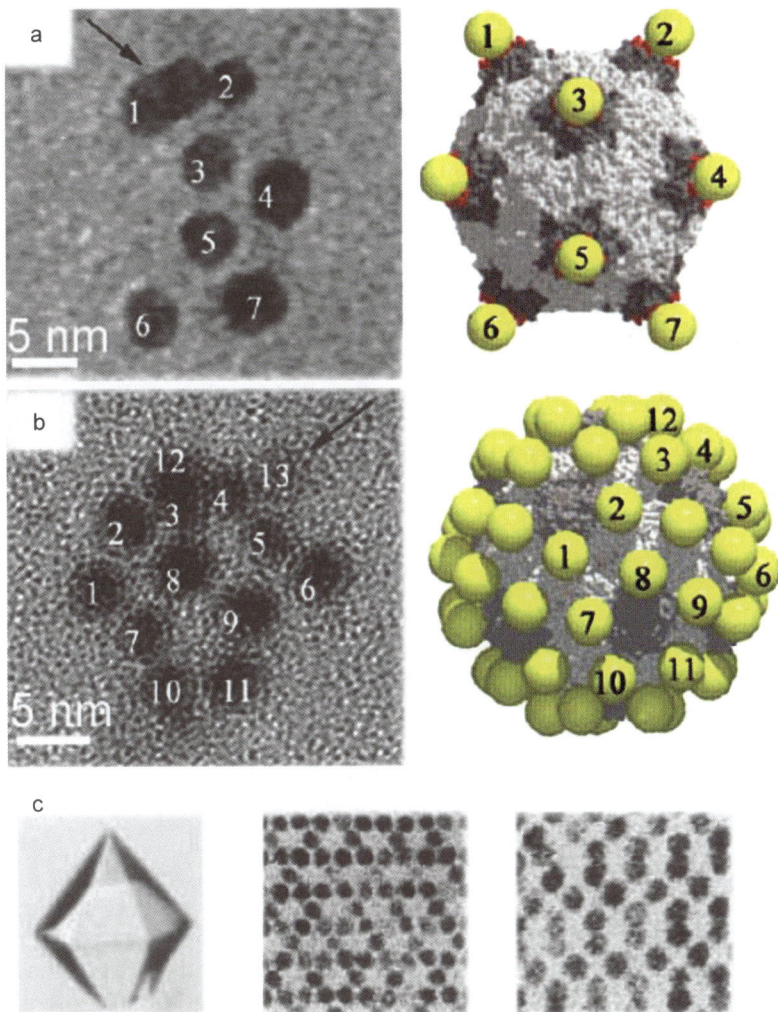

Figure 3.12 CPMV as nanoscale 3D scaffold and building block; (a) and (b) Electron micrographs of mutants decorated with gold nanoparticles (left) and their corresponding models (right), (a) One gold particle per five-fold axis indicated by the arrow, (b) Five gold particles centred about the five-fold axis. The arrow indicates a nanoparticle on the other side of the virus, (reprinted with permission from Blum, A. S. *et al.* Cowpea Mosaic Virus as a Scaffold for 3-D Patterning of Gold Nanoparticles. *Nano Lett.*, **4**, 867–870. Copyright (2004) American Chemical Society), (c) Electron micrographs of an hexagonal CPMV crystal (left) sectioned perpendicular to the *c*-axis (middle) and the *a*-axis (right) containing 10^{13} particles (Wang, Q. *et al.* Icosahedral virus particles as addressable nanoscale building blocks. *Angew. Chem. Int. Ed.*, **41**, 459–462 (2002). Copyright Wiley-VCH Verlag GmbH and Co. KGaA. Reproduced with permission).

plasma membranes. Although the main emphasis here is on fundamental aspects of virus traffic pathways through mammalian cells a number of bio-medical applications such as targeted drug delivery, diagnostic noninvasive imaging and vaccine development was envisaged and also pursued.[173] For example, the same group described a new strategy for continuous imaging of tissues deep inside living organisms.[174,175] Fluorescently labeled CPMV were used as imaging probes for the intravital visualisation and monitoring of human tumor neovascularisation.[174]

The impact cowpea mosaic viruses have had and are most likely to continue having on advancing nanotechnology is significant. Nevertheless, the evident success in employing CCMV and CPMV in materials science has inspired pursuits of broadening the range of viral icosahedra amenable to derivatisation. Other spherical viruses such as bacteriophage MS2,[176] turnip yellow mosaic virus (TYMV),[177] retroviruses[178] have been investigated as addressable plat-forms and bionanoblocks for bio- and nanotechnology.

3.2.2   Following Linear

Applying virus-based designs for nanomaterials appears to be surface-oriented; that is, inner and outer surfaces of viral capsids are predominantly targeted to render novel physical and chemical properties and functions. This remains the main tendency in engineering templated nanomaterials. The same applies to filamentous viruses. Although tubular forms are not as abundant as spheres in the viral kingdom the application of filamentous viruses in nanomaterial syntheses is fully complementary to that of viral icosahedra.

3.2.1.1   Channelling Inner

Many reports on templating the assembly of various nanostructured materials using TMV and M13 bacteriophage have clearly shaped this area of the field. Following original studies by Mann and coworkers on adapting TMV structure for biomineralisation[179] the templating capacity of filamentous viruses to direct crystallisation with the nanoscale precision was comprehensively confirmed by a number of experiments.

As for any other virus TMV is characterised by repeated patterns on its surfaces that are formed by charged residues. Understandably, this can be and has been seen as advantageous in ascribing nucleation sites for seeding crystal growth. Such sites can also be genetically[180] or chemically[181] "redesigned" to accommodate materials with specific requirements. In this regard, both native and recombinant TMV were successfully employed for the deposition of metal nanoparticles (platinum, gold, silver),[182] clusters[183] and ultrathin films,[184] in light harvesting[185] and surface immobilisation.[180,186] Furthermore, sites on either surface of the virus, whether exterior or interior, can be selected for targeted installation of material components. This can be accomplished via chemical modifications[181] or by varying mineralisation conditions.[183]

For example, platinum complexes reduced at acidic pH specifically bound to and precipitated on external surfaces of TMV capsids as opposed to photo-chemically reduced silver (I) salts that yielded aligned nanoparticles loaded in the internal channel of the virus.[182] As a consequence, unequivalent size distribution of formed nanoparticles can be observed. External particulate platinum was composed of 10-nm particles and encased silver particles were restricted to the diameter of the channel, *i.e.* 4 nm. Nanoparticles aligned inside the virus cavity are pictured as precursors for metal nanowires – components much awaited in electronics.[182,187] Still, the contingency of this concept would depend on the availability of finer-defined and continuous alignments of particles.

Encouraging in this respect is a series of concepts that takes the application of atomic layer deposition to TMV for assembling cobalt and nickel wires.[187,188] The wires being just a few atoms in diameter can be made extended to several micrometres, Figure 3.13(a). Perfectly compatible with an assortment of chemical processes[179] TMV can also be externally decorated templating nanotubular structures of different compositions that appear to be particularly promising as semiconducting materials.[189]

Otherwise successful, the reported approaches are not totally exclusive of drawbacks.

To name one, the dimensions of all of the outlined materials are limited by those of TMV. This stimulates searches for ways of producing longer but independent structures. One direction to approach this may be a general concept of virus supra-assemblies thus far applied to icosahedral viruses and ferritins. But unlike spherical assemblies TMV exhibit high aspect ratio and hence are inherently able to form 1D objects. Longitudinal fibrillar assemblies of abutting TMV rods extending several micrometres, TMV "polymers", can indeed be generated, for example, by *in situ* polymerisation of polyaniline on the TMV surfaces.[190] Given that the tendency of TMV to aggregate in the presence of specific metals is believed to be a function of virus concentrations and pH[191,192] it should also be possible to tailor the composition and dimensional characteristics of the materials.

Concurrent with TMV-based developments, M13 phage has been harnessed as a supporting scaffold for a similar collection of nanomaterials. The main accent, however, has been made on external surfaces of the virus.

3.2.1.2 Converting Outer

The Belcher laboratory has led this theme for several years presenting a comprehensive exploration of the virus. Genetically engineered M13 constitutes the streamline approach of the group, which has developed from an original study by the laboratory on using combinatorial phage-display libraries for identifying surface-active peptides.[138]

Phage displays present a popular selection technique in which a random library of about 10^9 peptides or proteins is expressed on the coat of a phage virion, with each phage displaying a different sequence. Because genes that encode the library are within the virion, each sequence is linked to its gene.[193–195]

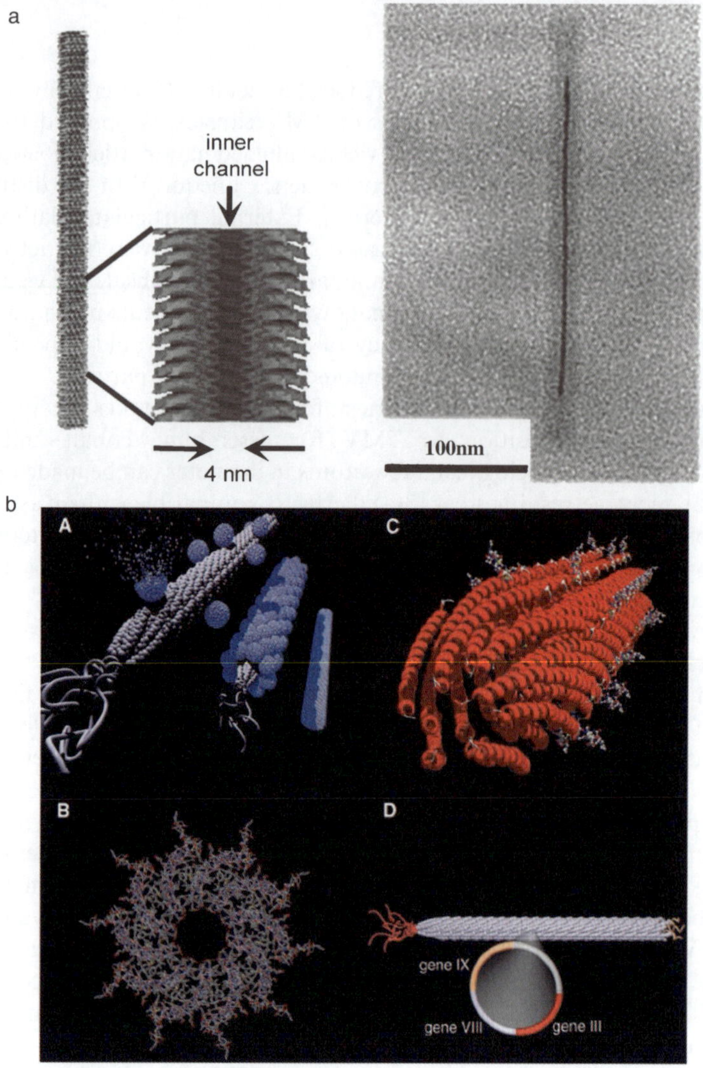

Figure 3.13 TMV- and M13-templated nanowires; (a) 3D reconstruction of TMV (left) and an electron micrograph of TMV filled with a cobalt wire 200 nm in length and 3 nm in diameter (right), (reprinted from Zhu, Y. *et al*. Automated Identification of Filaments in Cryoelectron Microscopy Images. *J. Struct. Biol.*, **135**, 302–312, Copyright 2001, with permission from Elsevier; Uchida, M. *et al*. Biological containers: protein cages as multifunctional nanoplatforms. *Adv. Mater.*, **19**, 1025–1042 (2007). Copyright Wiley-VCH Verlag GmbH and Co. KGaA. Reproduced with permission; reprinted with permission from Knez, M. *et al*. Biotemplate Synthesis of 3 nm Nickel and Cobalt Nanowires. *Nano Lett.*, **3**, 1079–1082. Copyright (2003) American Chemical Society), (b) Scheme of a nanowire synthesis templated by M13 including nucleation, ordering and annealing of virus-particles assembly (A), with virus symmetry allowing the aggregation of nucleated particles along all, x, y and z directions (B) and (C) Schematic showing the genetically modified capsid at ends encoded by the phagemid DNA (D), (From Mao, C. et al. Virus-Based Toolkit for the Directed Synthesis of Magnetic and Semi-conducting Nanowires. *Science*, **303**, 213–217 (2004). Reprinted with permission from AAAS).

This, in conjunction with different binding affinities of expressed peptides to a given target, allows for a selective partitioning of bound phages from unbound constructs, and eventually for the identification of active peptide sequences.

The researchers hypothesised that such libraries can be used for identifying peptides that would be capable of specific binding to structurally defined inorganic surfaces.[138,196–198] To confirm this path experimentally, MS13 phage, a standard tool for phage displays, was chosen, with MS13-displayed peptides found to have high selectivity for binding to metallic and polymeric surfaces. Furthermore, it was established that peptide sequences can discriminate between the crystallographic orientation and composition of chemically related surfaces.[138,199] Accordingly, a next step taken was to use MS13 mutants based on the expression of pIII minor protein to nucleate and control the growth of inorganic nanomaterials.

Peptides with selected affinities to zinc and cadmium sulfides were used for the generation of III-V and II-VI semiconductor nanocrystals, *i.e.* quantum dots. Similarly, magnetic nanocrystals grown on M13 from a ferromagnetic phase of FePt or cobalt ions controlled the crystallisation of FePt and Co-Pt alloys with an appreciable degree of ordering.[196,200,201]

Related to this is an intriguing observation made on different MS13 constructs able to specify nucleation of either the sphalerite or wurtzite structure of semiconductor nanocrystals.[138] This feature can be coupled with the aniso-tropic nature of MS13 particles to fabricate MS13/ZnS liquid-crystalline hybrids. The hybrids resulting from nanocrystal-mediated assembly of phage mutants neatly align over centimetre length scales into multilayered films.[202,203] Owing to the integrated alignment of nanocrystals within the films such hybrids were envisaged as miniaturised electronic devices, which was supported somewhat by other designs.

In one study[204] gold-binding peptides were genetically incorporated into the MS13 coat to allow for hybrid gold-cobalt oxide nanowires that are known to possess improved battery capacity. Thus, the mutant assembled on 2D poly-electrolyte multilayers facilitated the nanofabrication of microscopic lithium ion batteries. In another study, peptides nucleating zinc and cadmium nano-crystals were expressed on the major coat protein (pVIII) to saturate the exterior of virions.[196] The peptides nucleated and organised II-VI nanocrystals clustered along the MS13 cylinder into quantum dot nanowires.[196,205] Both homogeneous and heterogeneous nanowires consisting of one and two metal types, respectively, were synthesised, Figure 3.13(b).

A recent tendency in the synthesis of MS13-enabled materials, which appears to be common for all virus templates, is to take advantage of supraviral assemblies. One route towards this is the modification of the outer surface of phages with peptides or polymers.

M13 can be bacterially expressed with streptags (streptavidin-binding sequences)[206] and mixed with functionalised streptavidin to give a range of nanostructures spanning from liquid crystals[207] through tethered archi-tectures[208] and nanowires[209] to precursors of magnetic nanorings.[210]

As proposed by Yang *et al.*,[211] highly charged M13 can be used as poly-electrolyte components in complexation with cationic lipid membranes. M13 phages were found to form alternately layered structures analogous to those produced by simple DNA-membrane lamellar complexes.[212] However, in such stacked lattices virus layers were attributable to pores with sizes ten times larger in cross-sectional area than those formed by DNA. This may be of use in packaging macromolecules or even nanosized species. Indeed, where DNA co-organises with "point-like" divalent ions (Mg^{2+}, Ca^{2+}),[212] MS13 is able to template nanoscopic arrays of 1-2 nm cations. Additionally, Yoo *et al.*[213] evinced that such monolayers can serve as tunable scaffolds in the nuclea-tion, growth and orientation of nanoparticles and nanowires over multiple length scales.

Thus far, materials templated only by a few viruses have been considered. Admittedly, this is the current state-of-the-art. Examples based on other or related viruses have just started to emerge in the literature, however, more in the context of confirming similarities than in that of qualitative originality. Nevertheless, the number of such develops is steadily growing to make the future of virus-directed synthesis confidently promising. This is also justified by the quick start of the field that has achieved the status of a truly separate discipline in record times taking a niche between physicochemical sciences and biotechnology.

The potential of the templated synthesis of nanomaterials is enormous and is being expanded almost on a daily basis. New materials arising from various chemical reactants involve more active research and attract stronger interest from industrial sectors. Altogether, this raises captivating expectations for the quick transfer of virus-templated materials to applications that may include but are not limited to conductive thin-film electrodes, miniaturised batteries, solar cells, light-emitting diodes and the like.

3.3   Repairing from Inside

In 2006 Foresight pronounced six challenges faced by humanity that can be addressed through the progress of nanotechnology.[214] One of these is "increasing the health and longevity of human life". An application anticipated to provide a solution to this challenge was defined as gene therapy. Loosely speaking, gene therapy (GT) is the approach of using genes as drugs.[215–217] Because many diseases result from genetic causes GT gives an attractive strategy for treatment. In concept, single-gene disorders such as cystic fibrosis or haemophilia can be cured by correcting a defected gene.[216,218] It is both appealing and seemingly feasible, as can be judged by ongoing clinical trials. In practice, GT has a number of unresolved issues regarding administration and potential side effects.[216–218] In reality, efficient and safe gene delivery remains a stumbling block for more than 1000 clinical trials starting from the first human gene-therapy test reported back in 1989.[219,220]

Viruses initially proposed as delivery vectors retain their predominant lead.[218,219] This choice is of no surprise given that viruses are Nature's most efficient transfectors.[221,222] Using the same reasoning, however, it can be argued that viral hosts represented by nearly all types of tissues have evolved effective mechanisms to combat viral invasions.[223] Therefore, the success of the therapeutic application of viral vectors is two-fold; efficient gene delivery and the circumvention of host defence. The latter is a lesser issue for an alternative class of gene delivery systems that are based on artificial or nonviral vectors.[222–224] However, as may be expected, "artificial" lacks the efficiency of Nature's designs. Nevertheless, the fast delivery of the concept into laboratory practice prompted a more ambitious term reflecting the purpose – an artificial virus.[34,222,223]

3.3.1 Uninviting Levy

Establishing leads to an artificial virus has become a central problem in gene therapy.[34,225,226] Having recently emerged, the field is vast and accounts the most impressive number of reports of new designs and solutions.[34,223,226] However, an obvious question to be asked remains – why. Why virus?

The concept of an artificial virus is built in the fundamentals of GT. GT targets an infected or incorrectly working cell. Cell targeting is performed at a subcellular or nanolevel and can therefore be expected to be best achieved by nanosized systems.[34] However, it is also true that nucleic acids are nanoscopic objects and are able to transverse the plasma membrane on their own.[227]

Indeed, no specialist delivery is required to achieve expression of DNA or RNA vectors directly injected into skin or muscle tissues.[228] Moreover, the approach can serve as a stepping stone for therapeutic methods that may be gathered under the heading of DNA vaccines.[227] Vaccines against viral infections and some tumours can be generated by site-specific injections of engineered DNA plasmids encoding viral antigens and inhibitory genes, respectively.[227] This is very attractive and advantageous over viral vectors as anti-DNA antibodies are not found in experiments involving direct DNA injections.[229] The readministration of naked vectors for repeated therapy or where escalation is necessary would thus become possible. This otherwise strengthening of the methodology is not, however, persistently sufficient for DNA vaccines to transfer into clinical practice. The reason is still the insufficient expression of naked genes.[230,231]

A number of physical techniques were proposed to improve this. For instance, in the so-called "gene gun" method gold nanoparticles coated with reporter genes are used to bombard mammalian cells.[229,231] Nanoparticles act as high-density carriers that localise genes in a target cell. Though gene transfer was observed in various cell lines ranging from epithelial to lymphocytes, only transient expression could be detected.[231] In other methods such as electroporation[230] where expression is stimulated by high-frequency and low-voltage electric pulses, also known as electrogene therapy, the level of expression was

100-fold higher.[232] However, it was also noted that the observed low levels of circulation of a therapeutic protein may hinder the adaptation of the technique for the treatment of other diseases such as haemophilia that requires higher amounts of injected DNA.[227,232] The restricting point here is the configuration of electrodes that ultimately dictates the per-site efficiency of DNA electroporation. A compromise can be multiple stimulations leading to increased volumes of injection. Yet, even this may not be sufficient to reach transducing efficiencies typical for viral vectors because naked nucleic acids being exposed to serum nucleases are degraded in biological fluids.

In contrast, packaging genes ensures a minimalistic approach to furnish two primary functions necessary for gene expression in recipient cells; that is, protection and delivery of a genetic material. Following this route, most viral vectors are little more than viruses with modified genomes and hence constructing a viral vector comes down to a partial recoding of its genome, whereas the native virion structure and composition that provide means for cell targeting and transfection are preserved.

Nonetheless, otherwise best transfection systems viruses have a number of properties undesirable for GT vectors.[233] These include insertional mutagenesis, which can trigger the development of cancers, loading capacity restricted by the size of native NAs, immune response (*e.g.* inflammatory) and the lack of cell-specificity. All in all, this alludes to the development of an alternative type of transfection systems.

Naturally, a challenger to virus-derived systems has to be safer, neither toxic nor immunogenic, straightforward to produce, and most of all should be able to transduce cells specifically and at least as efficiently as viral vectors.[234] In this context, nanometric self-assemblies capable of forming protecting coats or shells around nucleic acids such as cationic polymers, lipo- and polysomes, amphiphiles, micelles and dendrimers have quickly become chief candidates for an artificial virus – a magic bullet in molecular medicine, Figure 3.14.

3.3.2   Necessitating Exterior

3.3.2.1   Antagonising Dressing

In its simplest form a synthetic virus is a nucleic acid complexed with an oppositely charged polyelectrolyte. As a polyacid NA carries a large negative charge that becomes neutralised upon condensation with a polycation. This type makes use of the bulk of electrostatic interactions between the opposite ionic forms. A number of polycations have been used to envelope nucleic acids. Polyethylenimine (PEI) and poly L-lysine (PLL) are typical and most popular examples,[34,223,235] Figure 3.15(a).

PEI and PLL readily bind NAs with forming the so-called polyplexes. Polyplexes are sized in the range of 50–300 nm depending on the molecular weight of the polycation used and its charge ratio with the complexed NA,[236,237] Figure 3.15(b). Smaller < 30 nm polyplexes can also be produced by

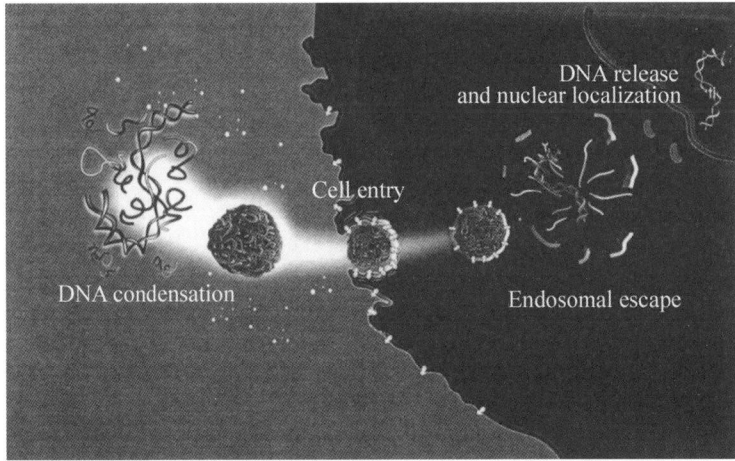

Figure 3.14 Magic bullet of artificial virus, (reprinted from Zuber, G. *et al.* Towards synthetic viruses. *Adv. Drug Deliv. Rev.*, **52**, 245–253, Copyright 2001, with permission from Elsevier).

choosing appropriate complexation regimes, but these are rare and require specific conditions.[225]

The transfection of polyplexes is passive and largely relies on their buffering properties,[236] Figure 3.15(c). Both PEI and PLL are high cationic density polymers that are built of residues able to be protonated at virtually any pH including physiological. This is believed to facilitate efficient transfection through the plasma and endosome membranes that in turn may permit the delivery of NAs into the cytoplasm following the endosomal escape of the material and hence the avoiding of its degradation in lysosomes. The enticing simplicity this type represents has been a major reason for the popularity of polycation-based NA-carriers since the very inception of the concept.[225] However, traditionally low transfection activity, short circulation in blood-stream and the polydispersed character of the final forms are among a few to render this system inefficient. Nevertheless, the polyelectrolyte concept has not been entirely abandoned.

Over the last fifteen years PEI and PLL earned a status of gold standard synthetic vectors employed as core platforms for further improvements. A simple strategy reported by Seymour and coworkers reflects one of general trends applied to optimising the polyelectrolyte concept.[238–240]

The researchers set out to reduce the toxicity of polycations, however, without compromising the stability of formed polyplexes under physiological conditions. To achieve this, a linear reducible polycation (RPC) was designed as a short PLL block modified at the termini with cysteine residues. Following mixing with a nucleic acid a high molecular weight vector is formed by the oxidative polycondensation of RPC.[240] This allowed for the cleavage of the vector in the reducing environment of the cell that was accompanied by

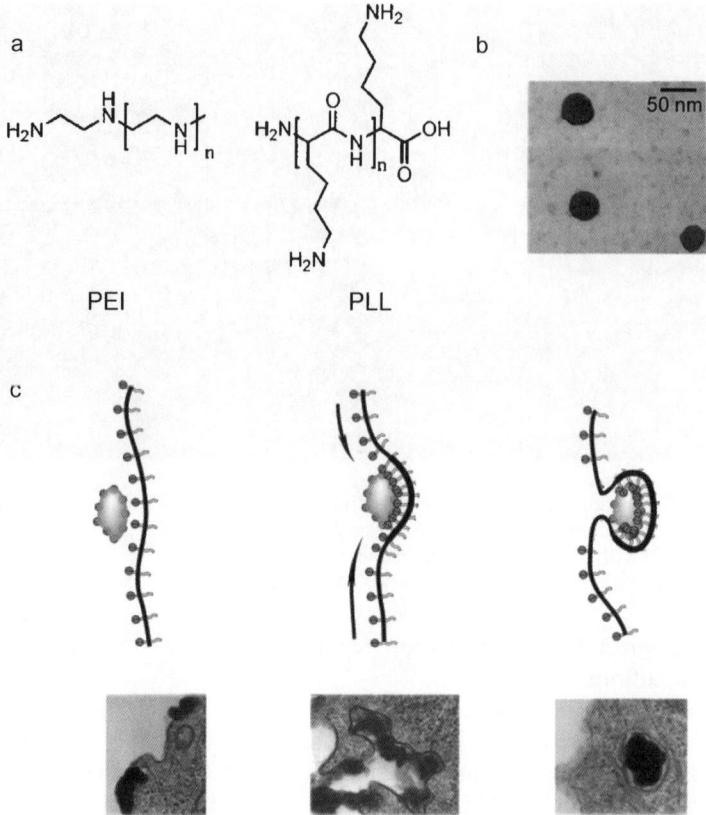

Figure 3.15 Polyethylenimine (PEI) and poly L-lysine (PLL); (a) Chemical structures
of the polycations, (b) Electron micrograph of PLL/DNA polyplexes,
(Ward, C. M. *et al*. Modification of pLL/DNA complexes with a multivalent hydrophilic polymer permits folate-mediated targeting *in vitro*
and prolonged plasma circulation *in vivo. J. Gene Med.*, **4**, 536–547
(2002). Copyright Wiley-VCH Verlag GmbH and Co. KGaA. Reproduced with permission), (c) Schematics and corresponding electron
micrographs of the receptor-mediated cellular uptake of complexed
DNA showing gradual membrane zippering around cationic particles
followed by the formation of size-adjusted intracellular vacuoles
(Kopatz, I. *et al*. A model for nonviral gene delivery: through syndecan
adhesion molecules and powered by actin. *J. Gene Med.*, **6**, 769–776
(2004). Copyright Wiley-VCH Verlag GmbH and Co. KGaA. Reproduced with permission).

significant decreases in the toxicity of PLL. Such a methodology renders
polycation vectors responsive to the intracellular environment, which facilitates
the release of NA into the cytoplasm.

Similarly, McKenzie *et al*.[241] proposed reductively activated peptides that are
also based on short PLL. In this design cysteine groups are inserted intermittently to form randomised interpeptide disulfide bonds upon oxidation. In

both cases hundred-fold increases in gene expression were observed compared with standard PLL vectors.[241] However, the lack of the endocytic escape of the polyplexes was also reported. This may be a result of decreasing the buffering capacity of polycations caused by the inclusion of inert cysteine residues, as was confirmed by subsequent studies. RPC with insertions of histidine residues known to defy rapid pH fluctuation showed superior cytoplasmic delivery of various nucleic acids over that of the cysteine-containing RPC.[226,238] Histidines are also advantageous in that they can be readily and reversibly modified that makes vectors on their basis able to be further improved by the incorporation of targeting ligands.[238] Ligands selective for certain cellular sites enable targeted gene delivery and the development of polycations that can be specifically furnished with cell-homing domains offers a step forward towards multi-functional systems.

Yet again, main objectives in the development of safer and more efficient polycation-based vectors continue being focused on transfection efficiency and cytotoxicity. Both are correlated with the sizes and topologies (linear, branched or grafted) of polycations. Larger polymers provide better protection for NAs against nucleases, which, however, comes at the expense of toxicity to be higher for larger net charges. These properties are of particular importance for *in vivo* experiments that ultimately determine the feasibility of a given gene-transporting system.[222] Although polyplexes were demonstrated as relatively safe and effective for gene transfer through a range of administration routes including direct nose and lung instillation,[242] kidney perfusion,[243] intracerebral and intravenous injections,[244] the question mark remains over the toxicity and stability of synthetic polycations *in vivo*.

This stimulates and sustains steady interest in the exploration of naturally occurring biopolymers. For example, gelatine and chitosan that are ubiquitous to various organisms are enzymatically more stable and appear to be less toxic than PEI and PLL. Nanometric particles formed by these biopolymers in complexation with DNA can be delivered via a number of routes including intragastric administration, not typical of synthetic polycations, which holds promise for oral vaccination.[245] However, little is known about the biocompatibility of such biopolymers with different tissue types.

An alternative to this is the engineering of graft hybrids of natural and synthetic polycations. These exhibit better physicochemical properties compared to homopolymers and are more stable and less toxic than the synthetic ones. They also have higher buffering capacities than purely natural components. Such a combination increases the chances for the polycations to bypass processes that affect the disposition of vectors and again particularly after *in vivo* administration.[246] A further improvement can be done through PEGylation – a process of derivatisation of macromolecular systems with the strands of polyethylene glycol (PE(G) Hydrophilic PEG can alter some and intensify other properties of vectors and is used to reduce cytotoxicity or to impart greater biocompatibility to poorly biodegradable polycations.[247]

Taken together, the outlined properties make polycations comfortably amenable to empirical optimisation. However, polydispersity – innate property of polycations – presents a major hurdle for a polyelectrolyte version of an artificial virus and is marked by a magnitude of different shapes and sizes of formed polyplexes.[221,226] Partly, this is attributed to the fact that polycations used for the complexation of nucleic acids are heterogeneous polymers, partly to the nature of resulting morphologies that is strongly affected by the way the complexes are prepared. For instance, kinetically driven rapid mixing may give complexes that are smaller and more defined than those formed through slow and thermodynamically sustained titration. Strictly speaking, this leads to the lack of control over the size and loading capacities of polycation-based vectors. In other words, where the performance of polyplexes is gratifying in protecting naked nucleic acids their delivery capabilities can be dramatically affected. In this regard, a related approach – the use of cationic peptides – could offer better complementary solutions.

3.3.2.1.2 Renting Occasional. As discussed in the previous chapters, DNA in the cell or in viral capsids is associated with proteins and is condensed through the cationic residues presented on the protein surfaces. More detailed studies revealed that the residues compile conserved motifs of short peptide sequences (<20 amino acids) primarily responsible for binding DNA. Given that the highly controlled synthesis of peptides of this size range is well established and relatively straightforward a number of natural and synthetic peptide sequences have been proposed as gene carriers.

For instance, major components of nucleosome histones and histone-resembling proteins[248] contain multiple repeats of AAKP[248] and SPKK[249] sequences. Synthetic peptides composed of these repeats were shown to wrap[248] and condense[250] DNA to mediate its delivery into cells, Figure 3.16. A unique mechanism of cell entry mediated by these peptides as well as the whole proteins has been recently proposed as "histone-mediated transduction"[251] providing evidence for its energy- and receptor-independent manner.

Miller and coworkers thought along the same lines and looked into cationic sequences derived from adenovirus.[252] Arginine-rich sequences of virus core complex peptide μ (mu) and protein V were probed for precondensation of transgenic DNA – a kinetically favourable route to controlling DNA complexation and transfer. At certain ratios the peptides condensed with DNA and generated negatively charged complexes that upon further complexation with cationic lipids produced virus-alike nanoparticles.[253] The particles were uniformly sized (120 ±30 nm) and stable to aggregation. In subsequent studies, the precondensation effect was related to the stabilisation of the vector in biological fluids as a function of chemical potential and self-assembly.[254] The phenomenon is reminiscent of that observed in protein folding and its further elucidation may open up ways to more efficient vector formulations and transfer.

Other emerging technologies using peptides concern the development of enhancers of endosomal release following endocytosis. This is considered

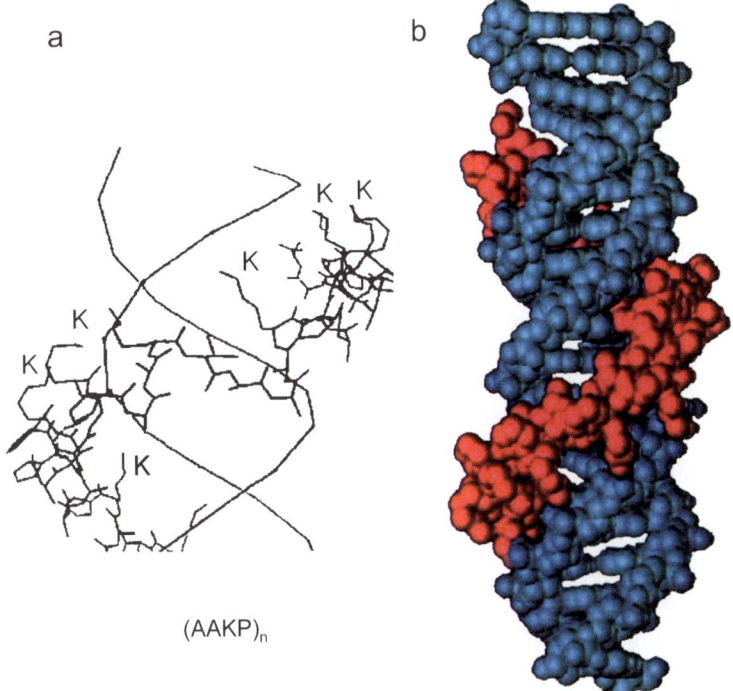

(AAKP)ₙ

Figure 3.16 Peptide-wrapped DNA; (a) The phosphorous traces of DNA (thin lines) shown as relative to a histone-derived peptide, (AAKP)n, running from lower left to upper right. K indicates some lysine residues, (b) A space-filing diagram of DNA (blue) wrapped with the peptide (red), (reprinted from Medvedkin, V. N. *et al.* Interactions of (Ala*Ala*Lys*Pro)n and (Lys*Lys*Ser*Pro)n with DNA. Proposed coiled-coil structure of AlgR3 and AlgP from Pseudomonas aeruginosa. *Protein Eng.*, **8**, 63–70 (1995), by permission of Oxford University Press).

by many as a limiting factor in the entry of exogenous NAs into the cytoplasm and one of the major barriers in gene transfer.[221,226] Without a means of escaping endosomes that develop into NA-digesting lysosomes an exogenously delivered gene cannot be expressed as this invariably requires its release into the cytoplasm and transfection into the nucleus. The importance of endosomal escape has been demonstrated by a number of studies using lysosomotropic agents.

For example, an antimalarial drug chloroquine can considerably enhance transfection in different nonviral vectors.[255] Although the mechanism of its action is not entirely understood chloroquine is widely used as a standard endosome-release reagent in gene-transfer experiments.

Similar effects have been achieved with the use of cationic membrane-active peptides abundantly found in a wide range of organisms as a part of the innate immunity against microbial invasions. Given the similar organisation of endosomal and bacterial membranes both natural[256] and synthetic[257] peptides have been proposed as endosome-disruptive agents.

A relatively recently established class of cell-penetrating peptides also attracts widespread attention.[258,259] These short amphipathic peptides effectively traverse the plasma membrane bypassing its permeability barrier and to deliver a repertoire of cargoes ranging from small molecules to nanoparticles. Cell-penetrating peptides are prototypical arginine-rich sequences ($<$ 10 amino acids) of the so-called protein transduction domains including known antennapedia homeodomain of Drosophila,[260] the transactivator of transcription (TAT) protein identified in HIV,[261] DNA-binding protein VP22 from herpes simplex virus,[262] and neuropeptides galaninand mastoparan.[263] Interestingly, structurally the peptides are very similar to host defense peptides, viral core sequences and pro-apoptotic motifs of dependence receptors suggesting a uniform membrane-associated mechanism of action.

Concomitantly, Wagner and coworkers set out to mimic processes whereby viruses destabilise endosomes.[264] Viruses use natural endosomal acidification to trigger conformational changes of their proteins required for membrane fusion and lysis of endosomes. Relying upon this phenomenon the researches constructed PLL vectors conjugated with the amino terminal sequence of influenza virus hemagglutinin HA-2 that was found to intensify gene transfer in a series of cell lines.[265] Similarly, Gottschalk *et al.* aimed at designing peptides that would emulate viral DNA-binding proteins such that virus-like vectors can be constructed.[266] In their original design a cationic lysine-rich sequence was used to condense DNA, whereas an amphipathic and mildly anionic peptide was made to serve as a pH-dependent endosome-release agent. A vector was constructed in two steps: DNA was first mixed with the condensing peptide to give a positively charged polyplex followed by an "activation" of the vector via adding the endosomal releasing agent. Gratifyingly, high levels of expression were reported in various cell lines transfected by the vector.[266]

One of the strong sides of using peptides in gene delivery is local targeting. Unless specifically targeted, cellular uptake of both bare and cationic or complexed NAs is a random and statistic process. Therefore, incorporation of homing motifs that can be recognised by cellular receptors into a vector looks appealing and promising. Indeed, vectors derivatised with cell-adhesion motifs such as RGD[267] from fibronectin or YIGSR[268] of laminin that are ubiquitous for many cell surfaces substantially augment receptor-mediated gene transfer.[269] Also, nuclear localisation signal sequences that "tag" proteins into the nucleus have been proposed as nucleus-targeting components. So also have core proteins of some viruses been shown to facilitate in targeting nucleolus, which leads to increased expressions of therapeutic proteins.[270] A number of targeting strategies using peptides directed at various cellular targets including tumour and vascular have been applied to different nonviral vectors. This has been found equally advantageous for another class of NA-complexants – liposomes.

3.3.2.2 Phasing Wet

Chemically liposomes are related to polycations. First described by Bangham[271] in the early 1960s these have been proposed and extensively studied as nonviral vectors since the introduction of the lipofection DNA–transfer procedure by Felgner *et al.*[272]

Liposomes are best thought of as closed spherical structures of lipid bilayers that contain an aqueous core or cavity, Figure 3.17. Correspondingly, those liposomes that are loaded or complexed with NAs are referred to as lipoplexes.[34,221] Two types of liposomes are distinguished – cationic and anionic. Anionic formulations initially proposed for drug and gene delivery were abandoned soon after due to their poor and extremely inefficient encapsulation of NAs.[226] In marked contrast, cationic lipids were found to readily interact along negatively charged NA helices on forming vesicular aggregates that subsequently convert into bilayered lipoplexes when the critical density of lipids is reached.[273] Similarly to polycations, cationic lipoplexes are easily fused with the plasma membrane and can bypass the lysosomal degradation route. Although liposomes do not encapsulate NAs in the way viruses do, nanometric particulate lipoplexes are assembled from lipids binding along NAs,[273] Figures 3.17(b) and (c). The resulting tight packing follows the shape of and is maintained by the size of a to-be-wrapped NA. However, the size distribution of lipoplexes is very broad and ranges from 20 nm to several micrometres in diameter, Figure 3.17(b). The reason for that should be sought in the nature of liposome constituents.

3.3.2.2.1 Facing Concentric. A cationic lipid is an amphiphilic molecule consisting of a hydrophobic hydrocarbon tail and a polar (cationic) head group, Figure 3.17(a). Self-assembly of such amphiphiles is driven by the dual preference for solvent, and in water occurs as a result of reaching a critical concentration of hydrophobic interactions. Therefore, molecular parameters of amphiphiles such as hydrophobic volume, hydrocarbon tail length and head group area determine the optimal type of self-assembly in any given case. The mutual relationship of these parameters and their impact on self-assembly is given in the surfactant concept.[274] The concept operates with a surfactant parameter, S, which is defined as:

$$S = v/l^* a_0$$

where v is the hydrophobic volume specified by the hydrocarbon tail, l is the length of the tail, and a_0 is the effective area per head group.

S sets geometrical constraints on the packing of amphiphiles and hence restricts the number of potential self-assemblies. In short, S relates the properties of amphiphiles to average curvatures of their preferred assemblies. Curvatures can be positive and negative. Positive curvatures associate with assemblies that are curved around the hydrophobic part (normal phases), whereas negative denotes curvatures around polar groups (inverted phases).

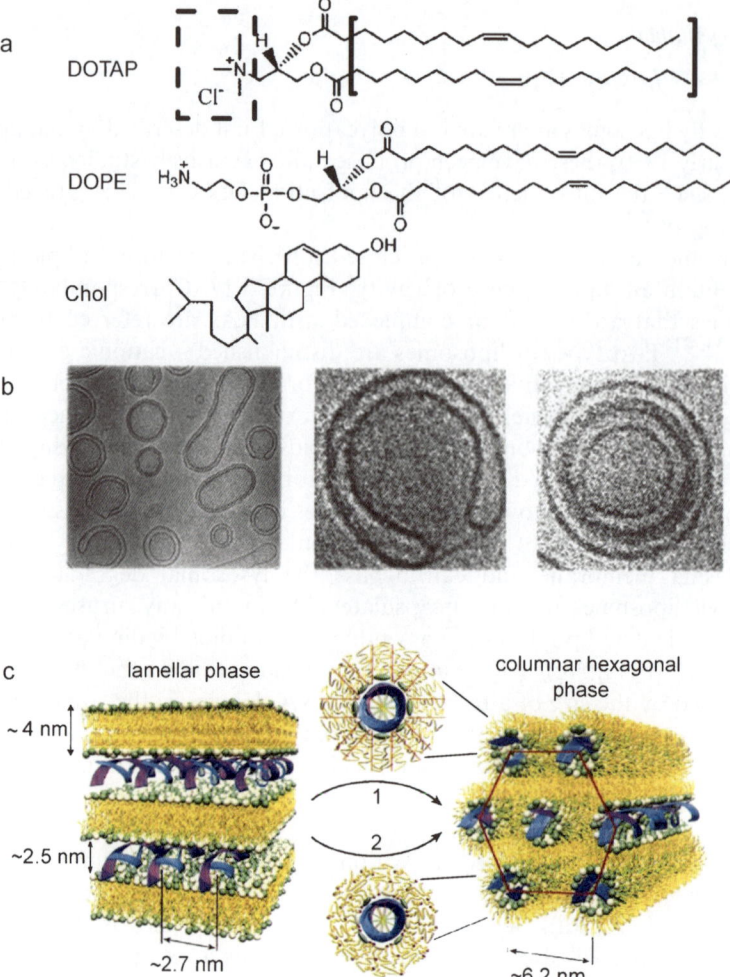

Figure 3.17 Liposomes; (a) Chemical structures of common liposome constituents. Cationic lipid 1,2-dioleoyl-3-trimethylammonium propane – DOTAP, neutral lipid dioleoyl phosphatidylethanolamine – DOPE, and cholesterol – Chol. The polar head and hydrophobic tail of DOTAP are indicated by a dashed rectangle and square brackets, respectively, (b) Electron micrographs of DOTAP:Chol liposomes without DNA (left and middle) and with DNA packed between two bilayers of DOTAP:Chol (right), (reprinted by permission from Macmillan Publishers Ltd: Templeton, N. S. *et al.* Improved DNA: liposome complexes for increased systemic delivery and gene expression. *Nature Biotech.*, **15**, 647–652, Copyright (1997)), (c) Schematic representation of lamellar-columnar phase inversion. The lamellar phase of DNA (purple-blue) sandwiched between cationic DOTAP bilayers resulting in a multilayered lipoplex assembly (left), and a columnar inverted hexagonal phase (right) converted from the lamellar phase via two distinct pathways: (1) generation of negative curvature induced by DOPE (cone shaped as opposed to cylindrical DOTAP, centre top), and (2) loss in membrane rigidity caused by DOPC (*N*-methylated DOPE) and the cosurfactant hexanol, (from Koltover, I., Salditt, T., Radler, J. O. and Safinya, C. R. An Inverted Hexagonal Phase of Cationic Liposome-DNA Complexes Related to DNA Release and Delivery. *Science*, **281**, 78–81 (1998). Reprinted with permission from AAAS).

Thus, optimal stabilities of different assemblies in water can be deduced using a simple rule: the smaller the S value the more curved and contained are the assemblies formed. That is, $S < 1/3$ gives spherical, $S = 1/3–1/2$ cylindrical and $S = 1/2–1$ lamellar assemblies. For instance, amphiphiles with two long alkyl tails would have larger S and yield lamellar structures, Figure 3.17(c). Spherical structures assembled from more polar amphiphiles would shift to less curved shapes with increased amphiphile concentrations as this would lead to decreased effective areas per head group. Although the organisation of liposomes can be explained by the surfactant concept, it is very approximate.

Liposomes are composed of concentric membranes each 4 nm thick, Figure 3.17(b). The number of such membranes varies even within one liposome type. This restricts the prediction of size and shape of the formed liposomes and cannot be addressed by the surfactant theory. In general, liposome-structure prediction remains challenging and constitutes the major drawback of the type as a basis for an artificial virus. On the other hand, finding a solution to the problem is instigated by the popularity of cationic lipoplexes that have given promising results in the treatment of cancer and cystic fibrosis, and by the continuingly growing commercial interest in such formulations.[225] Amongst others available on the market, the best known are quaternary ammonium amphiphiles such as Lipofectin, LipofectAmine, Cellfectin, Transfectam, LipoTaxi, and TransFast.[275] Lipoplexes based on these amphiphiles have been proven safe and efficient for local delivery. All are in systematic use in gene-transfer studies and are key tools for ongoing human clinical protocols. However, the incomplete understanding of factors that determine the transfection of lipoplexes *in vivo* complicates the structure prediction even further.

3.3.2.2.2 Encircling Between. The structure of lipoplexes is dictated by the composition of NAs and lipid amphiphiles and their stoichiometry, with both being responsive to external stimuli.[276] Natural polycations such as polyamines spermine and spermidine can condense DNA into toroids and rods of 70–100 nm irrespective of its size.[277] These interactions are, however, unstable and reversed under physiological conditions. In relation to the natural amines the discussed amphiphiles are lipoamines, *i.e.* amines with hydrophobic tails with a strong tendency to self-aggregation. The aggregated tails, which are physically linked to DNA-binding cationic head groups, mediate the stable packing of nucleolipidic nanoparticles. These nanoparticles are not always discrete and different sizes can be observed depending on amphiphile-DNA ratios.[276] Although more stable than polyamines these undergo relatively fast decomposition in the bloodstream. Structurally liposomes are different from lipoplexes mainly because NAs are not encapsulated but instead glued within a lipid bilayer of amphiphiles.[221,225] Furthermore, the number of complexed NA molecules within a lipoplex can vary as opposed to viruses that host one nucleic acid per capsid.

Altogether this provided a basis for the need to develop a general structural model of lipoplexes.

In an original "beads-on-string" concept proposed by Felgner *et al.*[272] one DNA strand is decorated with several liposomes. The model was executed from electron microscopy observations, with a variety of structures being found, including DNA-sized strings and liposome fusions, oligolamellar and tubular structures. Later, a combined *in situ* microscopy and X-ray diffraction analysis performed by Safinya and coworkers initiated the re-examination of the model.[278]

Notably, it was found that the addition of linear or plasmid DNA induced a topological transition of < 200 nm liposomes into collapsed condensates of one-micrometre globules. The globules (lipoplexes) were established to occur as a result of DNA screening the electrostatic interactions between lipid bilayers that surprisingly led to condensed and highly ordered multilayers.[279] Upon complexation with liposomes DNA thus became sandwiched in between the planar arrays of lipid bilayers, Figure 3.17(c). The arrays were found to be 6.5 nm thick and highly periodic.[279] Another but less pronounced periodicity of 4 nm was attributed to the thickness of a lipid bilayer separating DNA helices within the plane of a sandwich.[278] Thus, 2.5 nm is left to the aqueous cavity that hosts DNA, which is fully consistent with and just sufficient for a single layer of DNA (2 nm) plus its hydration shell, Figure 3.17(c).

While the mechanism of the formation of lipoplexes can be explained by the model an explicit answer to how structure of lipoplexes can be predicted and rationally controlled is yet to be given. As a specific example, the size dependence of lipoplexes revealed as a function of the total lipid to DNA weight ratio yields globules coexisting with excess liposomes that can become chained and grow larger with the addition of more DNA.[278,279] This confirms that the complexation of NAs by cationic lipids is neither quantitative nor discrete, which makes the drawbacks that are characteristic for polyplexes equally topical for lipoplexes.

3.3.2.2.3 Singling Out Unique. In resonance with this, a procedure proposed by Behr and coworkers[222] may reinforce the search for an optimised model. The researchers focused on monomolecular condensation of DNA.[280] To realise this, a series of cationic thiol-containing detergents was designed.[281,282] Detergents are used to reversibly condense DNA at concentrations well below their critical micelle-forming concentrations (cmc) with forming mononuclear lipoplexes that then become stabilised by oxidation of thiols. The rationale of the concept thus consisted in the detergent having a high cmc at the condensing step and a very low cmc at the final step that follows oxidation. This is believed to prevent cationic micelle-mediated aggregation of the anionic DNA particles and at producing lipoplexes sufficiently stable for delivery to the cell nucleus. Discrete (30 nm) DNA monomolecular complexes were indeed formed and the design proved to be fairly successful *in vitro*,[282,283] Figure 3.18(a).

Although the complexes were more stable and exhibited better diffusion properties than those characteristic of other polycations and cationic lipids, decomplexation of the formed lipoplexes in the cell and their poor stability *in vivo* exposed common drawbacks of lipid amphiphile-based systems.[281]

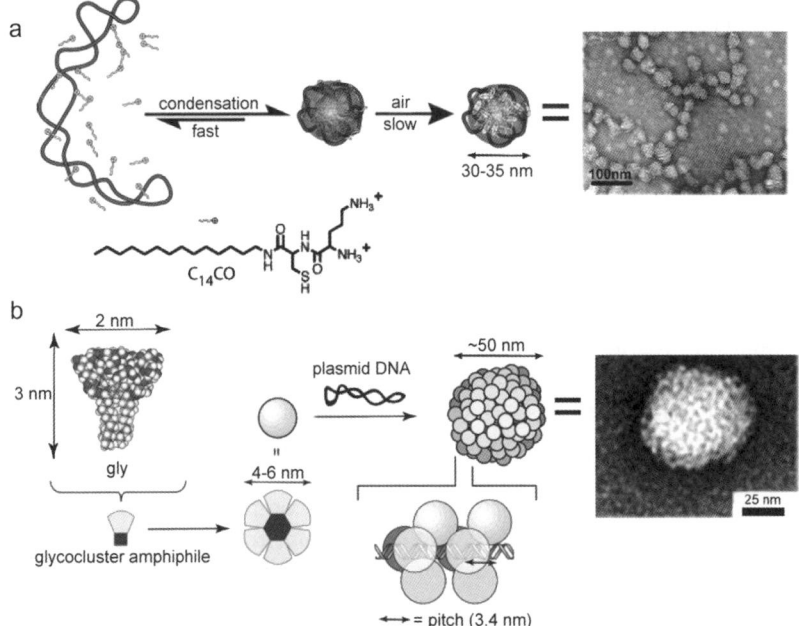

Figure 3.18 Self-assembled amphiphiles; schematics of DNA-complexation by cationic thiol-containing (a) and neutral glycocluster (b) amphiphiles leading to the formation of monomolecular DNA nanoparticles (a) and "glycoviruses" (b) Electron micrographs are shown correspondingly, (reprinted with permission from Chittimalla, C. *et al.* Monomolecular DNA Nanoparticles for Intravenous Delivery of Genes. *J. Am. Chem. Soc.*, **127**, 11436–11441. Copyright (2005) American Chemical Society, and Nakai, T.*et al.* Remarkably Size-Regulated Cell Invasion by Artificial Viruses. Saccharide-Dependent Self-Aggregation of Glycoviruses and Its Consequences in Glycoviral Gene Delivery. *J. Am. Chem. Soc.*, **125**, 8465–8475. Copyright (2003) American Chemical Society).

In this respect, an alternative view on designing DNA-encapsulating amphiphiles was taken by Aoyama et al.[284–286] Instead of traditional polymer and lipid cations a novel class of glycoconjugate amphiphiles, dubbed glycoclusters was proposed.[285] Glycoclusters assembled into micelle-like pentameric nanoparticles having striking resemblances with viral capsid proteins. This was conceived to guarantee mononuclear and stoichiometric particles, that were referred to as glycoviruses,[285] Figure 3.18(b).

Reflecting the main idea of the design the offered argumentation takes a clearly distinctive route. As viewed by the researchers, most NA-packing systems based on cationic lipids or polymers aid or cause NAs to collapse into viral-sized complexes. In contrast, spherical and neutral glycolnanoparticles, that are strongly adhesive to NAs but not to each other, coat genes in a way similar to that of viral capsomers. The complexation of DNA was assumed to result from a tight aligned binding of the nanoparticles along the major groove of DNA. The groove is 1.3 nm wide and has a 3.4-nm pitch, which suggests that four 4-nm glycoparticles can be shared by two pitches without imposing significant steric penalties, Figure 3.18(b).

This gives an optimal packing that creates maximised contacts between the "glycocapsomers" and DNA. The final appearance of glycoviruses was ~50 nm wide spheroids that showed transfection efficiencies an order of magnitude higher than those of Lipofectin.[286] This is both very important and interesting as such neutral glycoviruses have no means for endosomal buffering displayed by cationic systems. In this light, mechanisms that accept cation-mediated endosomal release as a key criterion for successful gene transfer are not applicable to the glycoviruses and may not be prerequisite for artificial viruses in general. Another aspect of gene delivery addressed by the glycoviral concept and highlighted as size-restricted endocytosis effective only for < 100 nm formulations calls for more detailed information on the role of different endocytosis types in the cellular uptake of synthetic vectors.[284] This, along with other fundamental problems, including poor complexation of NAs, short duration of gene expression, inefficient intracellular transfection, lack of target specificity and differences between *in vitro* and *in vivo* outcomes appear to continue being the main hurdles towards a synthetic system that would form a basis for an artificial virus.

3.3.3   Sharing the Balance

That said, the separation of approaches aiming at artificial viruses into two main routes may be formulated.

The first concentrates on finding better complexation formulations for NA-condensing systems based on the already established types. The main emphasis is on *in vivo* applications as this provides a viability test for a balanced set of properties indispensable for a successful vector.[223,246] Because mechanisms of *in vivo* transfer greatly depend on the route of administration and tissue/cell types, their direct impact on the physicochemical properties of a desired formulation is the main focus of the approach.[34,221–223,225–226,234]

The second route is structural and concerns designing virus- or cage-alike systems. The main feature of this emerging strategy, which is different from all of the aforementioned concepts, is that it is strictly directional towards finding an architectural rationale in designing artificial virions. In other words, the approach aims at extracting general rules from the principles of biological self-assembly and applying them into the engineering of the components of synthetic cages *de novo*.

Per se, the task is extremely challenging as it lacks a clear empirical foundation and is tailored on assumptions compromising the vision of a future design. On the other hand, elements of imagination and creativity are more distinctively expressed to impart a rare flexibility of merging views influenced by different disciplines to the approach. As an accepted rule for this type of approach, geometric symmetry is considered as the most critical part in initiating a new design.

3.3.3.1 Driving Symmetrical

One of the best examples of symmetry-driven designs is a nanohedra concept introduced by Padilla *et al.*[287] The concept stands upon a combination of symmetry and polyvalence giving a general strategy for designing self-assembled cages and arrays, Figure 3.19(a). The strategy is to fuse protein domains able to oligomerise with different numbers of partners. A dimeric domain and a trimeric domain, termed as those associating with two and three partners at a time, respectively, were linked together into a larger fusion protein. By having two functionally separate parts that bind their other copies the protein assembles into symmetric objects – nanohedra. The relationship between two virtual symmetry axes, one from each oligomerisation domain, selects a particular arrangement of two domains and defines a resulting architecture. The correct disposition of the domains is thus a prerequisite and can be enforced by a rigid linker, which has to be a part of one persistent fold. This can be achieved if interconnection sites between each domain and a linker adopt the same secondary structure. Using these criteria two naturally dimeric and trimeric proteins with terminal α-helical domains were identified. Bromoperoxidase and the M1 matrix protein of influenza virus were chosen as trimeric and dimeric blocks, respectively. The blocks were connected into a single protein chain via a short helical linker derived from a ribosomal protein. 12 copies of the chimeric ∼50-kDa protein were expected to follow a cubic point symmetry resulting in a tetrahedral cage of 18 nm diameter with an edge length of 15 nm, Figure 3.19(a).

The formation of roughly spherical and discrete particles with diameters between 12 and 15 nm was experimentally observed. Although minor smaller and larger components were also detected suggesting some level of equilibrium-defined polymorphism in the assembly, the observed values were in good agreement with the design.

However, it is unclear what symmetry class the formed nanoparticles belong to and what are their abilities to encapsulate molecules and materials. Surprisingly, the design did not undergo further development. Nonetheless, the concept of nanohedra marks the first example of symmetry-driven design of protein cages and offers an avenue to future designs.

An instantaneous affirmation of this is provided by an analogous approach proposed by Burkhard *et al.*[288] This research group applied basic concepts of icosahedral symmetry to designing regular polyhedra using synthetic self-oligomerising peptides. A key assumption of this design was based on the fact that any regular polyhedron is built of multiple copies of one asymmetric unit and the number of units and the type of a polyhedron should be defined by the least common multiple of different oligomerisation domains comprising the unit, Figure 3.19(b). Because asymmetric units are of tripyramidal arrangements, with each of their edges laying on one of the rotational symmetry axes,

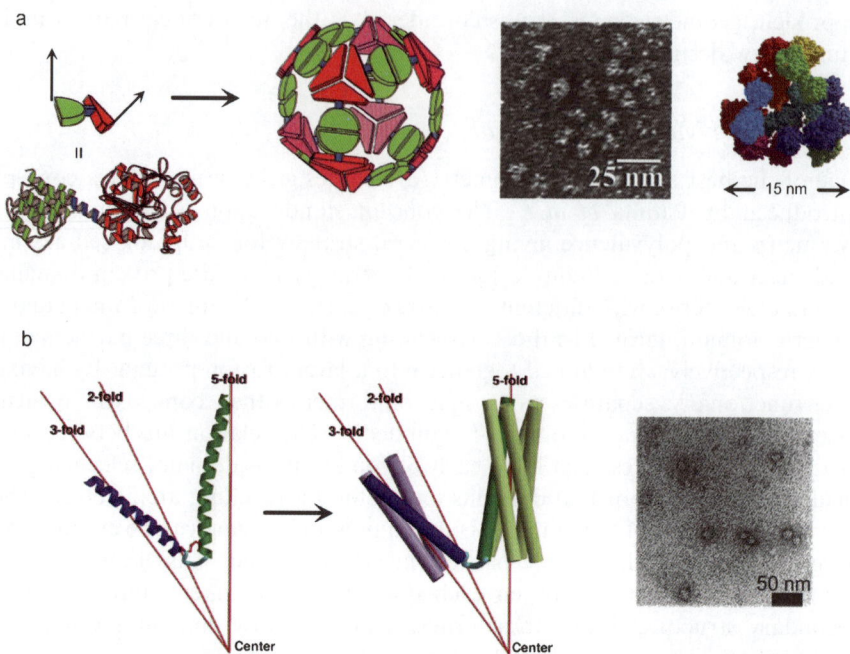

Figure 3.19 Self-assembled nanohedra; (a) Schematic of a fusion protein design consisting of one dimeric (green semicircle) and one trimeric (red triangle) domain linked by a short linker (blue stretch). The domains are specifically oriented to assemble into a cubic cage – discrete nanoparticles and a tetrahedron as appeared in the electron micrograph and stereo reconstruction, respectively, (reproduced with permission from Padilla, J. E. *et al.* Nanohedra: Using symmetry to design self assembling protein cages, layers, crystals, and filaments. *Proc. Natl. Acad. Sci. USA*, **98**, 2217–2221. Copyright (2001) National Academy of Sciences, USA), (b) (Left), internal symmetry elements of a dodecahedron expressed in a 3D building block consisting of two coiled-coil domains with three- (blue) and five- (green) fold symmetries separated by a short linker (cyan). The intrahelical disulfide bridge bending the block is shown in red. Thin red lines indicate rotational symmetry axes. (Right), an electron micrograph of the assembled dodecahedra, (reprinted from Raman, S. *et al.* Structure-based design of peptides that self-assemble into regular polyhedral nanoparticles. *Nanomedicine*, **2**, 95–102, Copyright 2006, with permission from Elsevier).

oligomerisation domains can be superimposed onto the edges and if linked together would yield a 3D self-assembling block, Figure 3.19(b). Accordingly, a block was made as a single peptide chain of two domains linked via a short and reversibly rigidified linker. Pentameric and trimeric α-helical coiled-coil domains were chosen to correspond to five- and three-fold symmetry axes of the polyhedron. The assembly of the peptide into a polyhedron was described as a two-step process. At first, fifteen copies of the block arrange into an "even unit", which then associates with other even units into larger and closed structures. In this notation, the even unit resembles a subunit in virions and determines the morphology and size of resulting nanoparticles. This, in turn, depends on how regular the geometry of the units is and potentially imposes the drawback of structural polymorphism onto the design. Indeed, this was observed. Analytical ultracentrifugation, which is used to probe the distribution of molecular masses, revealed a broad range of masses depending on conditions of assembly. Consistent with this, electron microscopy showed spherical nanoparticles with diameters variable from 15 to 45 nm, which also suggests certain levels of irregularity in the packing of the even units. Interestingly, a carefully adjusted refolding of the peptide was found to aid in tuning more uniform nanoparticles.

Avowedly, the self-association of the even units into particles with a desired polyhedral symmetry is subject to a number of factors that can be adjusted by design. This is true for any designed systems and is in intimate relation with protein folding, which is anticipated to be a major drive for the emergence of similar concepts.

Furthermore, no matter how complex the amino-acid sequence of a given protein is its folding can be described by the known number of conserved secondary structure elements that can be of direct use as autonomous units for constructing nanostructures.[289]

3.3.3.2 Sealing Annular

One example to support this may come from recently reported self-assembled polynanocages.[98]

Unlike the previous two designs, the encapsulating properties of which are to be addressed, the polynanocages were designed as multiply encapsulating materials; that is, copious cavities compose the interior of one assembly making it highly porous, Figure 3.20(a).

The system is unique in that each cavity it hosts is structurally self-maintained and functionally independent. Designing a system of such complexity requires finding a suitable molecular topology that can provide a scaffolding framework and a self-assembling motif flexible enough to encode and sustain the chosen topology.

Based on this recognition, the design employed a dendrimer architecture expressed by the dimeric leucine-zipper motif. The polynanocages were designed in two steps.

First, a noncovalent dendrimer framework was assembled from a single peptide. The peptide to appoint a subunit of assembly was designed to possess two oppositely directing polar faces. This was meant to lead to the formation of networks of electrostatic interactions that favour associations between dimers rather than within a single dimer. Second, distinguishable nanometric cavities were introduced into the framework by incorporating another subunit complementary to the first. Thus, a resulting structure

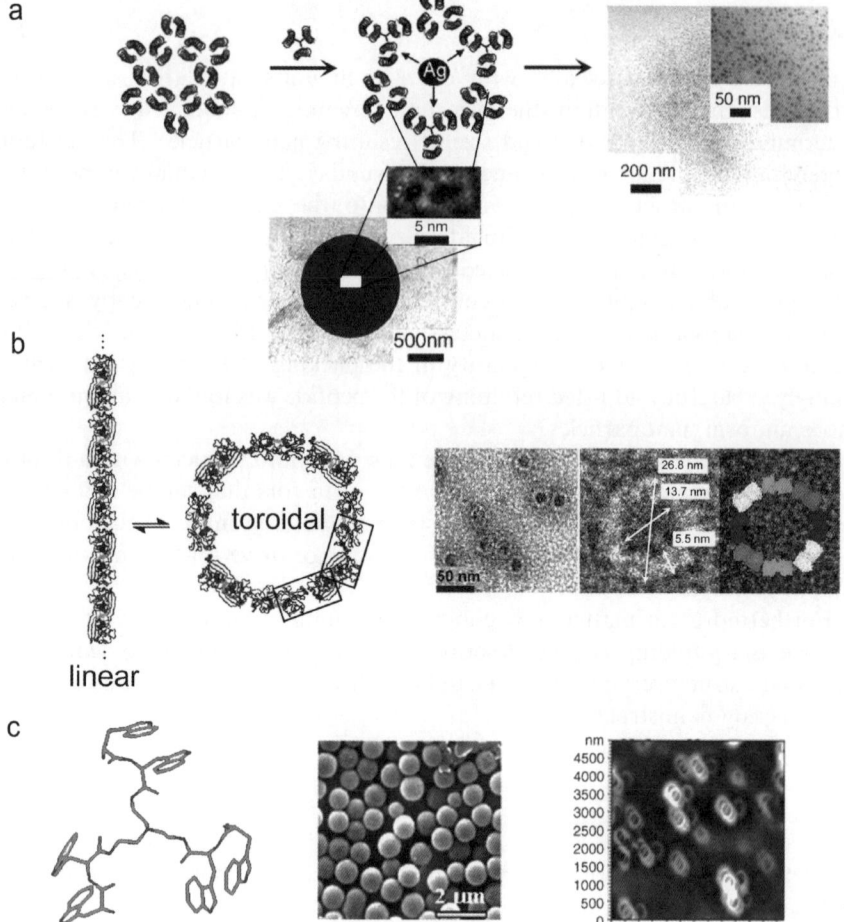

Figure 3.20 Self-assembled nanocages; (a) Schematic of a polynanoreactor design showing a single cage (left) and when extended to host a silver nanoparticle (right). Electron micrographs of an individual polynanoreactor with the inset showing two overlapping rings (bottom), and a spherical spread of silver nanoparticles synthesised within the reactor that was subsequently degraded enzymatically. Mean diameters of both nanorings and nanoparticles are consistently ~5 nm, (Ryadnov, M. G. A Self-Assembling Peptide Polynanoreactor. *Angew. Chem. Int. Ed.*, **46**, 969–972 (2007). Copyright Wiley-VCH Verlag GmbH and Co. KGaA. Reproduced with permission), (b) Schematic of a protein nanoring assembly. Individual proteins are highlighted in boxes (left). Electron micrographs of assembled nanorings, an individual assembly with indicated dimensions and with its space-filling match, (reprinted with permission from Carlson, J. C. T. *et al.* Chemically Controlled Self-Assembly of Protein Nanorings. *J. Am. Chem. Soc.*, **128**, 7630–7638. Copyright (2006) American Chemical Society), (c) The synthetic ditryptophan triskelion (left). Scanning electron and atomic force microscopy images of the triskelion incubated for a week in 90% aqueous methanol showing spherical particles (middle) and multilamellar structures (right), (Ghosh, S. *et al.* Bioinspired Design of Nanocages by Self-Assembling Triskelion Peptide Elements. *Angew. Chem. Int. Ed.*, **46**, 2002–2004 (2007). Copyright Wiley-VCH Verlag GmbH and Co. KGaA. Reproduced with permission).

can be referred to as a supramolecular dendrimer or supradendrimer, Figure 3.20(a).

Dendrimers are symmetrically repetitive molecules with multiple branches radiating from a central core. This gives dendrimers unique characteristics, one of which is especially important in the context of the design. Multiple branching implies multiple branching cells or cores. Structurally static and functionally reserved in most dendrimer designs branching cells can be made dynamic. In other words, these can be converted to multiple cavities; that is, hollow cells, providing branching is sustained. This is the underlying concept of the design.[98] In support of this, dendrimers were shown to be perfectly compatible with self-assembly. Pioneering studies by Zimmerman *et al.*,[290] Percec *et al.*,[291] Tomalia *et al.*,[292] Frechet *et al.*[293] and Meijer *et al.*[294] on supramolecular dendrimer systems give a strong appeal in applying properties of the dendritic state to engineering nanoscale cages. Yet surprisingly, information on dendrimer-based nanoscale reactors in the literature is scarce.

A notable advantage of the reported polynanocages is their polyvalent character, as opposed to other known self-assembled cages. Most, if not all, supramolecular encapsulants irrespective of their origin and nature are monovalent; that is, one cage hosts one cavity. In relation to nanoreactors polyvalence can lead to the amplification of guests or synthesised materials. This was demonstrated using the polynanocages. A conventional citrate-reduction method of ionic to colloidal metals was applied. Briefly, a solution of silver nitrate was added to polynanoreactor preparations followed by the immediate addition of sodium citrate as reducing reagent. Following incubation of the preparation with chymotrypsin, electron microscopy revealed sub-micrometre-to-micrometre spreads of individual and uniformly sized silver nanoparticles, Figure 3.20(a). The shapes of spreads corresponded to that of reactors, whilst the average diameter of nanoparticles compared to that of the cavities. Ultrastructural analysis of individual forms by TEM revealed fusions of cavities as networks of 5-nm rings constituting the porosity of polynanoreactors.

Discrete monomeric protein-based nanorings can also be designed. In one example, Carlson *et al.*[295] demonstrated that dihydrofolate reductase tethered together by a flexible peptide linker can be assembled into stable ring-like structures with diameters ranging from 8 to 20 nm. The sizes of nanorings depend on the length and composition of the peptide linker as well as the affinity of a bivalent ligand that binds the protein and mediates its dimerisation. As judged by electron microscopy, nanorings assembled as discrete arrangements of 4 to 8 dimerised proteins, Figure 3.20(b).

The assembly of proteins into nanorings of similar sizes and in similar fashions is a common phenomenon in Nature, with different systems performing different functions. These may include proteins contributive to DNA homologous recombination such as RAD52 and RuvB, tryptophan RNA attenuation protein that regulates tryptophan synthesis, phage recombinases, DNA helicases and many others. However, selective to very specific functions such proteins are structurally much conserved and are not easily amenable to design.

Ghosh *et al.*[99] tackled the problem of designing nanocages from a slightly different direction and took the triskelion structure of the natural cage clathrin as an inspiration. A tripodal tris(2-aminoethyl)amine, commonly used as a setup scaffold for dendrimers, was derivatised with three copies of a tryptophan dipeptide, Figure 3.20(c).

Tryptophan is a hydrophobic aromatic amino acid with an indole ring in its side chain. Accordingly, it was proposed that two adjacent indole rings forming an assembly motif would enhance the overlap of one branch via π stacking and hydrophobic interactions and that the rigid triskelion arrangement of three motifs would direct the assembly in three dimensions. Consistent with the expectation, the conjugate assembled into hollow nanospheres in aqueous methanol in the course of one week at 37 °C, Figure 3.20(c).

In contrast, triskelions of single tryptophans or smaller and less hydrophobic diphenylalanines gave sporadic structures of ill-defined morphologies. Interestingly, electron and atomic force microscopies revealed that the organisation of the ditryptophan spheres is regular and multilamellar. Several layers were observed within one sphere and the structure of each layer was indicative of a concentric ring imputing the assembly to the vertical coalescence of several rings. The encapsulating properties of the rings were also probed. Tryptophan triskelions were incubated with a dye (rhodamine) to permit fluorescence detection of dye-loaded cages. Fluorescent circular structures detected by fluorescence microscopy affirmed the guest entrapment properties of the nanocages supporting their use as delivery vehicles. However, the assembly of such cages in water and/or physiological fluids appears to be less achievable due to the highly hydrophobic character of the structures.

Other examples that are based on different chemistries and structural principles follow similar paths in designing self-assembling encapsulants. The regularity with which they appear in the literature advocates for the constant pursuit of virus-like particles as the most promising type of synthetic encapsulants.[296]

3.4   Outlook

Principally, the concept of an artificial virus addresses the challenge of the architectural design of viruses. Given the ubiquity of geometric conventions for viral assembly in all nanoscale caged systems the challenge constitutes one major goal in nanodesign identified as finding a synthetic answer to dynamic encapsulating systems.

Admittedly, the concept might undergo design adjustments to be consistent across all nanocage types but clearly it has a bright future. In technological terms, with the coming of an artificial virus the number of nonviral vectors that currently account for about twelve per cent of all clinical trials worldwide[220] is very likely to increase. It is also clear that the route to the rationale of synchronised nanocage assemblies is not as short and straightforward as initially anticipated, and it may take a considerable amount of time and research investment before a more expressive position on the application aspects of the subject is established.

That said, constructing a synthetic approximation of naturally occurring encapsulants will not cease its persistence of innovation to give more exciting and creatively captivating designs that will eventually produce the challenger.

References

1. F. Hof, S. L. Craig, C. Nuckolls and J. Rebek Jr, Molecular Encapsulation, *Angew. Chem. Int. Ed.*, 2002, **41**, 1488–1508.
2. D. M. Vriezema, M. Comellas Aragones, J. A. A. W. Elemans, J. J. L. M. Cornelissen, A. E. Rowan and R. J. M. Nolte, Self-Assembled Nanoreactors, *Chem. Rev.*, 2005, **105**, 1445–1490.
3. T. Schrader and A. D. Hamilton (eds.), *Functional Synthetic Receptros*, Wiley-VCH, Weinheim, 2005.
4. M. M. Conn and J. Rebek, Self-Assembling Capsules, *Chem. Rev.*, 1997, **97**, 1647–1668.
5. V. M. Dong, D. Fiedler, B. Carl, R. G. Bergman and K. N. Raymond, Molecular Recognition and Stabilisation of Iminium Ions in Water, *J. Am. Chem. Soc.*, 2006, **128**, 14464–14465.
6. W. M. Stanley and E. G. Valens, *Viruses and Nature of Life*, Dutton, New York, 1971.
7. W. Chiu, R. M. Burnett, R. L. Garcea (eds.), *Structural Biology of Viruses*, Oxford University Press, New York, 1997.
8. M. Fischlechner and E. Donath, Viruses as Building Blocks for Materials and Devices. *Angew. Chem. Int. Ed.*, 2007, **46**, 3184–3193.
9. C. M. Fauquet, M. A. Mayo, J. Maniloff, U. Desselberger, L. A. Ball (Eds.), *Virus Taxonomy: VIIIth Report of the International Committee on the Taxonomy of Viruses*, Elsevier Academic Press, 2005.
10. F. H. C. Crick and J. D. Watson, *Structure of Small Viruses*, 1956, **177**, 473–475.
11. T. S. Baker, N. H. Olson and S. D. Fuller, Adding the Third Dimension to Virus Life Cycles: Three-Dimensional Reconstruction of Icosahedral Viruses from Cryo-Electron Micrographs, *Microbiol. Mol. Biol. Rev.*, 1999, **63**, 862–922.
12. P. Natarajan, G. C. Lander, C. M. Shepherd, V. S. Reddy, C. L. Brooks and J. E. Johnson, Exploring icosahedral virus structures with VIPER, *Nat. Rev. Microbiol.*, 2005, **3**, 809–817.
13. R. W. Horne and P. Wildy, Symmetry in virus architecture, *Virology*, 1961, **15**, 348–373.
14. J. J. Rux and R. M. Burnett, Adenovirus structure, *Hum. Gene Ther.*, 2004, **15**, 1167–1176.
15. C. M. Shepherd, I. A. Borelli, G. Lander, P. Natarajan, V. Siddavanahalli, C. Bajaj, J. E. Johnson, C. L. III Brooks and V. S. Reddy, VIPERdb: a

relational database for structural virology, *Nucl. Acids Res.*, 2006, **34**, D386–389.

16. R. W. Horne and P. Wildy, Virus structure revealed by negative staining, *Adv. Virus Res.*, 1963, **10**, 101–170.

17. J. E. Johnson and J. A. Speir, Quasi-equivalent viruses: a paradigm for protein assemblies, *J. Mol. Biol.*, 1997, **269**, 665–675.

18. A. Lwoff, T. F. Anderson and F. Jacob, Remarks on the characteristics of the infectious viral particle, *Ann. Inst. Pasteur*, 1959, **97**, 281–289.

19. D. L. Caspar and A. Klug, Physical principles in the construction of regular viruses. Cold Spring Harbour Symp, *Quant. Biol.*, 1962, **27**, 1–24.

20. D. L. Caspar and E. Fontano, Five-fold symmetry in crystalline quasi-crystal lattices, *Proc. Natl. Acad. Sci. USA*, 1996, **93**, 14271–14278.

21. J. T. Finch and A. Klug, Structure of Poliomyelitis Virus, *Nature*, 1959, **183**, 1709–1714.

22. M. G. Rossmann, E. Arnold, J. W. Erickson, E. A. Frankenberger, J. P. Griffith, H.-J. Hecht, J. E. Johnson, G. Kamer, M. Luo, A. G. Mosser, R. R. Rueckert, B. Sherry and G. Vriend, Structure of a human common cold virus and functional relationship to other picornaviruses, *Nature*, 1985, **317**, 145–153.

23. A. Klug, Molecular structure: Architectural design of spherical viruses, *Nature*, 1983, **303**, 378–379.

24. F. A. Eiserling and R. C. Dickson, Assembly of Viruses, *Ann. Rev. Biochem.*, 1972, **41**, 467–502.

25. R. Twarock, Mathematical models for tubular structures in the family of Papovaviridae, *Bull. Math. Biol.*, 2005, **67**, 973–987.

26. R. F. Bruinsma, W. M. Gelbart, D. Reguera, J. Rudnick and R. Zandi, Viral Self-Assembly as a Thermodynamic Process, *Phys. Rev. Lett.*, 2003, **90**, 248101.

27. D. J. Wales, Energy landscapes, self-assembly and viruses, *Comp. Math. Meth. Med.*, 2005, **6**, 107–110.

28. A. Zlotnick, Distinguishing Reversible from Irreversible Virus Capsid Assembly, *J. Mol. Biol.*, 2007, **366**, 14–18.

29. R. Kerner, A stochastic model of icosahedral capsid growth, *Comp. Math. Meth. Med.*, 2005, **6**, 95–97.

30. R. Zandi, P. van der Schoot, D. Reguera, W. Kegel and H. Reiss, Classical Nucleation Theory of Virus Capsids, *Biophys. J.*, 2006, **90**, 1939–1948.

31. R. Twarock, Mathematical virology: a novel approach to the structure and assembly of viruses, *Philos. Transact. Royal Soc. A: Math. Phys. Eng. Sci.*, 2006, **364**, 3357–3373.

32. R. Twarock, A tiling approach to virus capsid assembly explaining a structural puzzle in virology, *J. Theor. Biol.*, 2004, **226**, 477–482.

33. R. Zandi, D. Reguera, R. F. Bruinsma, W. M. Gelbart and J. Rudnick, Origin of icosahedral symmetry in viruses, *Proc. Natl. Acad. Sci. USA*, 2004, **101**, 15556–15560.

34. E. Mastrobattista, M. A. E. M. van der Aa, W. E. Hennink and D. J. A. Crommelin, Artificial viruses: a nanotechnological approach to gene delivery, *Nat. Rev. Drug Discov.*, 2006, **5**, 115–121.

35. G. Stubbs, Fibre diffraction studies of filamentous viruses, *Rep. Prog. Phys.*, 2001, **64**, 1389–1425.

36. H. Feldmann and M. P. Kiley, Classification, structure, and replication of filoviruses, *Curr. Top. Microbiol. Immunol.*, 1999, **235**, 1–21.

37. J. N. Culver, Tobacco Mosaic Virus Assembly And Disassembly: Determinants in Pathogenicity and Resistance, *Ann. Rev. Phytopathol.*, 2002, **40**, 287–308.

38. H. W. Ackermann, 5500 Phages examined in the electron microscope, *Archiv. Virol.*, 2007, **152**, 227–243.

39. D. L. D. Caspar, Assembly and stability of the tobacco mosaic virus particle, *Adv. Protein Chem.*, 1963, **18**, 37–121.

40. H. Fraenkel-Conrat and R. C. Williams, Reconstitution of activetobacco mosaic virus from its inactive protein and nucleic acid components, *Proc. Natl. Acad. Sci. USA*, 1955, **41**, 690–698.

41. K. Namba and G. Stubbs, Structure of tobacco mosaic virus at 3.6 A resolution: implications for assembly, *Science*, 1986, **231**, 1401–1406.

42. R. Díaz-Avalos and D. L. D. Caspar, Structure of the Stacked Disk Aggregate of Tobacco Mosaic Virus Protein, *Biophys. J.*, 1998, **74**, 595–603.

43. A. Klug, The tobacco mosaic virus particle: structure and assembly, *Philos. Trans. R. Soc. Lond. B*, 1999, **354**, 531–535.

44. A. C. Bloomer, J. N. Champness, G. Bricogne, R. Staden and A. Klug, Protein disk of tobacco mosaic virus at 2.8 Å resolution showing the interactions within and between subunits, *Nature*, 1978, **276**, 362–368.

45. P. J. G. Butler, Self-assembly of tobacco mosaic virus: the role of an intermediate aggregate in generating both specificity and speed, *Philos Trans R Soc. Lond B Biol. Sci.*, 1999, **354**, 537–550.

46. C. Sachse, J. Z. Chen, P.-D. Coureux, M. E. Stroupe, M. Fändrich and N. Grigorieff, High-resolution Electron Microscopy of Helical Specimens: A Fresh Look at Tobacco Mosaic Virus, *J. Mol. Biol.*, 2007, **371**, 812–835.

47. P. J. G. Butler, The Current Picture of the Structure and Assembly of Tobacco Mosaic Virus, *J. Gen. Virol.*, 1984, **65**, 253–279.

48. Y. Zhu, B. Carragher, D. J. Kriegman, R. A. Milligan and C. S. Potter, Automated Identification of Filaments in Cryoelectron Microscopy Images, *J. Struct. Biol.*, 2001, **135**, 302–312.

49. G. Stubbs, Tobacco mosiac virus particle structure and the initiation of disassembly, *Philos. Trans. R. Soc. Lond. B*, 1999, **354**, 551–557.

50. L. Specthrie, E. Bullitt, K. Horiuchi, P. Model, M. Russel and L. Makowski, Construction of a microphage variant of filamentous bacteriophage, *J. Mol. Biol.*, 1992, **228**, 720–724.

51. G. P. Martelli, M. J. Adams, J. F. Kreuze and V. V. Dolja, Family Flexiviridae: A Case Study in Virion and Genome Plasticity, *Ann. Rev. Phytopathol.*, 2007, **45**, 73–100.

52. O. Dolnik, L. Kolesnikova and S. Becker, Filoviruses: Interactions with the host cell, *Cellular and Molecular Life Sciences (CMLS)*, 2008, **65**, 756–776.

53. S. Tomar, M. M. Green and L. A. Day, DNA-Protein Interactions as the Source of Large-Length-Scale Chirality Evident in the Liquid Crystal Behaviour of Filamentous Bacteriophages, *J. Am. Chem. Soc.*, 2007, **129**, 3367–3375.

54. A. L. Olins and D. E. Olins, Spheroid Chromatin Units (ngr Bodies), *Science*, 1974, **183**, 330–332.

55. D. E. Olins and A. L. Olins, Chromatin history: our view from the bridge, *Nat. Rev. Mol. Cell Biol.*, 2003, **4**, 809–814.

56. D. E. Schones and K. Zhao, Genome-wide approaches to studying chromatin modifications, *Nat. Rev. Genet.*, 2008, **9**, 179–191.

57. C. A. Davey, D. F. Sargent, K. Luger, A. W. Maeder and T. J. Richmond, Solvent Mediated Interactions in the Structure of the Nucleosome Core Particle at 1.9 Å Resolution, *J. Mol. Biol.*, 2002, **319**, 1097–1113.

58. K. Luger, A. W. Mader, R. K. Richmond, D. F. Sargent and T. J. Richmond, Crystal structure of the nucleosome core particle at 2.8 Å resolution, *Nature*, 1997, **389**, 251–260.

59. R. D. Kornberg, Chromatin structure: a repeating unit of histones and DNA, *Science*, 1974, **184**, 868–871.

60. R. D. Kornberg and J. O. Thomas, Chromatin structure: oligomers of the histones, *Science*, 1974, **184**, 865–868.

61. F. Thoma, T. Koller and A. Klug, Involvement of histone H1 in the organisation of the nucleosome and of the salt-dependent superstructures of chromatin, *J. Cell Biol.*, 1979, **83**, 403–427.

62. S. C. R. Elgin, J. L. Workman (eds.), *Chromatin structure and gene expression*, Oxford University Press, New York, 2001.

63. N. J. Francis, R. E. Kingston and C. L. Woodcock, Chromatin Compaction by a Polycomb Group Protein Complex, *Science*, 2004, **306**, 1574–1577.

64. A. H. Hassan, K. E. Neely and J. L. Workman, Histone acetyltransferase complexes stabilize SWI/SNF binding to promoter nucleosomes, *Cell*, 2001, **104**, 817–827.

65. A. Alexeev, A. Mazin and S. C. Kowalczykowski, Rad54 protein possesses chromatin-remodelling activity stimulated by the Rad51-ssDNA nucleoprotein filament, *Nat. Struct. Mol. Biol.*, 2003, **10**, 182–186.

66. D. Reinberg and R. J. Sims, III. de FACTo nucleosome dynamics, *J. Biol. Chem.*, 2006, **281**, 23297–23301.

67. P. B. Becker and W. Horz, ATP-dependent nucleosome remodeling, *Ann. Rev. Biochem.*, 2002, **71**, 247–273.

68. S. Ercan, M. Carrozza and J. Workman, Global nucleosome distribution and the regulation of transcription in yeast, *Genome Biol.*, 2004, **5**, 243.

69. S. A. Spirin, The ribosome as an RNA-based molecular machine, *RNA Biol.*, 2004, **1**, 3–9.

70. J. H. Cate, M. M. Yusupov, G. Z. Yusupova, T. N. Earnest and H. F. Noller, X-ray Crystal Structures of 70S Ribosome Functional Complexes, *Science*, 1999, **285**, 2095–2104.

71. V. Berk, W. Zhang, R. D. Pai and J. H. D. Cate, Structural basis for mRNA and tRNA positioning on the ribosome, *Proc. Natl. Acad. Sci. USA*, 2006, **103**, 15830–15834.

72. K. H. Nierhaus, J. Wadzack, N. Burkhardt, R. Jünemann, W. Meerwinck, R. Willumeit and H. B. Stuhrmann, Structure of the elongating ribosome: arrangement of the two tRNAs before and after translocation, *Proc. Natl. Acad. Sci. USA*, 1998, **95**, 945–950.

73. P. Bieling, M. Beringer, S. Adio and M. V. Rodnina, Peptide bond formation does not involve acid-base catalysis by ribosomal residues, *Nat. Struct. Mol. Biol.*, 2006, **13**, 423–428.

74. V. Ramakrishnan, Ribosome Structure and the Mechanism of Translation, *Cell*, 2002, **108**, 557–572.

75. C. Weitzmann, P. Cunningham, K. Nurse and J. Ofengand, Chemical evidence for domain assembly of the Escherichia coli 30S ribosome, *FASEB J.*, 1993, **7**, 177–180.

76. R. Samaha, B. O'Brien, T. O'Brien and H. Noller, Independent in vitro Assembly of a Ribonucleoprotein Particle Containing the 3′ Domain of 16S rRNA, *Proc. Natl. Acad. Sci. USA*, 1994, **91**, 7884–7888.

77. S. C. Agalarov, E. N. Zheleznyakova, O. M. Selivanova, L. A. Zheleznaya, N. I. Matvienko, V. D. Vasiliev and A. S. Spirin, In vitro assembly of a ribonucleoprotein particle corresponding to the platform domain of the 30S ribosomal subunit, *Proc. Natl. Acad. Sci. USA*, 1998, **95**, 999–1003.

78. G. M. Culver and H. F. Noller, Efficient reconstitution of functional Escherichia coli 30S ribosomal subunits from a complete set of recombinant small subunit ribosomal proteins, *RNA*, 1999, **5**, 832–843.

79. M. Nomura, Assembly of Bacterial Ribosomes: In vitro reconstitution systems facilitate study of ribosome structure, function, and assembly, *Science*, 1973, **179**, 864–873.

80. R. Rohl and K. H. Nierhaus, Assembly map of the large subunit (50S) of Escherichia coli ribosomes, *Proc. Natl. Acad. Sci. USA*, 1982, **79**, 729–733.

81. M. Bochtler, L. Ditzel, M. Groll, C. Hartmann and R. Huber, The proteasome, *Ann. Rev. Biophys. Biomol. Struct.*, 1999, **28**, 295–317.

82. E. Lorentzen and E. Conti, The Exosome and the Proteasome: Nano-Compartments for Degradation, *Cell*, 2006, **125**, 651–654.

83. A. Grziwa, S. Maack, G. Pühler, G. Wiegand, W. Baumeister and R. Jaenicke, Dissociation and reconstitution of the Thermoplasma proteasome, *Eur. J. Biochem.*, 1994, **223**, 1061–1067.

84. J. Walz, A. Erdmann, M. Kania, D. Typke, A. J. Koster and W. Baumeister, 26S Proteasome Structure Revealed by Three-dimensional Electron Microscopy, *J. Struct. Biol.*, 1998, **121**, 19–29.

85. I. Nagy, T. Tamura, J. Vanderleyden, W. Baumeister and R. De Mot, The 20S Proteasome of Streptomyces coelicolor, *J. Bacteriol.*, 1998, **180**, 5448–5453.

86. M. Kaksonen, C. P. Toret and D. G. Drubin, Harnessing actin dynamics for clathrin-mediated endocytosis, *Nat. Rev. Mol. Cell Biol.*, 2006, **7**, 404–414.

87. Kirchhausen and T. Clathrin, *Ann. Rev. Biochem.*, 2000, **69**, 699–727.

88. M. A. Edeling, C. Smith and D. Owen, Life of a clathrin coat: insights from clathrin and AP structures, *Nat. Rev. Mol. Cell Biol.*, 2006, **7**, 32–44.

89. A. Fotin, Y. Cheng, P. Sliz, N. Grigorieff, S. C. Harrison, T. Kirchhausen and T. Walz, Molecular model for a complete clathrin lattice from electron cryomicroscopy, *Nature*, 2004, **432**, 573–579.

90. A. J. Jin and R. Nossal, Rigidity of triskelion arms and clathrin nets, *Biophys. J.*, 2000, **78**, 1183–1194.

91. A. Young, Structural insights into the clathrin coat, *Semin. Cell Dev. Biol.*, 2007, **18**, 448–458.

92. A. Fotin, Y. Cheng, N. Grigorieff, T. Walz, S. C. Harrison and T. Kirchhausen, Structure of an auxilin-bound clathrin coat and its implications for the mechanism of uncoating, *Nature*, 2004, **432**, 649–653.

93. T. Kanaseki and K. Kadota, The 'vesicle in a basket': a morphological study if the coated vesicle isolated from the nerve endings of the quinea pig brain, with special reference to the mechanism of membrane movements, *J. Cell Biol.*, 1969, **42**, 202–220.

94. S. M. Stagg, P. LaPointe and W. E. Balch, Structural design of cage and coat scaffolds that direct membrane traffic, *Curr. Opin. Struct. Biol.*, 2007, **17**, 221–228.

95. A. Fotin, T. Kirchhausen, N. Grigorieff, S. C. Harrison, T. Walz and Y. Cheng, Structure determination of clathrin coats to subnanometre resolution by single particle cryo-electron microscopy, *J. Struct. Biol.*, 2006, **156**, 453–460.

96. E. ter Haar, A. Musacchio, S. C. Harrison and T. Kirchhausen, Atomic structure of clathrin: A beta-propeller terminal domain joins an alpha-zigzag linker, *Cell*, 1998, **95**, 563–573.

97. B. M. F. Pearse, C. J. Smith and D. J. Owen, Clathrin coat construction in endocytosis, *Curr. Opin. Struct. Biol.*, 2000, **10**, 220–228.

98. M. G. Ryadnov, A Self-Assembling Peptide Polynanoreactor, *Angew. Chem. Int. Ed.*, 2007, **46**, 969–972.

99. S. Ghosh, M. Reches, E. Gazit and S. Verma, Bioinspired Design of Nanocages by Self-Assembling Triskelion Peptide Elements, *Angew. Chem. Int. Ed.*, 2007, **46**, 2002–2004.

100. C. A. Mirkin and C. M. Niemeyer (eds.), *Nanobiotechnology II: More Concepts and Applications*, Wiley-VCH, Weinheim, NY, 2007.

101. E. C. Theil, Ferritin: Structure, Gene Regulation and Cellular Function in Animals, Plants, and Microorganisms, *Ann. Rev. Biochem.*, 1987, **56**, 289–315.

102. E. Theil, M. Matzapetakis and X. Liu, Ferritins: iron/oxygen biominerals in protein nanocages, *J. Biol. Inorg Chem.*, 2006, **11**, 803–810.

103. P. Arosio and S. Levi, in Molecular and Cellular Iron Transport (ed. Templeton, D. M.) 125–154 (Marcel Dekker, Inc., New York, 2002).

104. T. Watabe and T. Hoshino, Observation of individual ferritin particles by means of scanning electron microscope, *J. Electron Microsc. (Tokyo)*, 1976, **25**, 31–33.

105. P. M. Harrison and P. Arosio, The ferritins: molecular properties, iron storage function and cellular regulation, *Bioch. Biophys. Acta*, 1996, **1275**, 161–203.

106. R. J. Hoare, P. M. Harrison and T. G. Hoy, Structure of horse-spleen apoferritin at 6 angstrom resolution, *Nature*, 1975, **255**, 653–654.

107. P. D. Hempstead, *et al.* Comparison of the three-dimensional structures of recombinant human H and horse L ferritins at high resolution, *J. Mol. Biol.*, 1997, **268**, 424–448.

108. A. E. Hamburger, J. A. P. West, Z. A. Hamburger, P. Hamburger and P. J. Bjorkman, Crystal Structure of a Secreted Insect Ferritin Reveals a Symmetrical Arrangement of Heavy and Light Chains, *J. Mol. Biol.*, 2005, **349**, 558–569.

109. T. Takahashi and S. Kuyucak, Functional properties of threefold and fourfold channels in ferritin deduced from electrostatic calculations, *Biophys. J.*, 2003, **84**, 2256–2263.

110. S. Mann, *Biomineralisation: Principles and Concepts in Bioinorganic Materials Chemistry*, Oxford University Press, Oxford, 2002.

111. M. Fischlechner and E. Donath, Viruses as Building Blocks for Materials and Devices, *Angew. Chem. Int. Ed.*, 2007, **46**, 3184–3193.

112. A. Treffry and P. M. Harrison, Spectroscopic studies on the binding of Iron, Terbium, and Zinc by Apoferritin, *J. Inorg. Biochem.*, 1984, **21**, 9–20.

113. J. Wardeska, B. Viglione and N. Chasteen, Metal ion complexes of apoferritin. Evidence for initial binding in the hydrophilic channels, *J. Biol. Chem.*, 1986, **261**, 6677–6683.

114. C. M. Barnes, E. C. Theil and K. N. Raymond, Iron uptake in ferritin is blocked by binding of $[Cr(TREN)(H_2O)(OH)]^{2+}$, a slow dissociating model for $[Fe(H_2O)6]^{2+}$, *PNAS*, 2002, **99**, 5195–5200.

115. X. Yang, P. Arosio and N. D. Chasteen, Molecular Diffusion into Ferritin: Pathways, Temperature Dependence, Incubation Time, and Concentration Effects, *Biophys. J.*, 2000, **78**, 2049–2059.

116. A. Treffry, E. R. Bauminger, D. Hechel, N. W. Hodson, I. Nowik, S. J. Yewdall and P. M. Harrison, Defining the roles of the threefold channels in iron uptake, iron oxidation and iron-core formation in ferritin: a study aided by site-directed mutagenesis, *Biochem. J.*, 1993, **296**, 721–728.

117. F. C. Meldrum, V. J. Wade, D. L. Nimmo, B. R. Heywood and S. Mann, Synthesis of inorganic nanophase materials in supramolecular protein cages, *Nature*, 1991, **349**, 684–687.

118. T. Douglas, D. P. E. Dickson, S. Betteridge, J. Charnock, C. D. Garner and S. Mann, Synthesis and Structure of an Iron(III) Sulfide-Ferritin Bioinorganic Nanocomposite, *Science*, 1995, **269**, 54–57.
119. F. C. Meldrum, T. Douglas, S. Levi, P. Arosio and S. Mann, Reconstitution of manganese oxide cores in horse spleen and recombinant ferritins, *J. Inorg. Biochem.*, 1995, **58**, 59–68.
120. S. Gider, D. Awschalom, T. Douglas, S. Mann and M. Chaparala, Classical and quantum magnetic phenomena in natural and artificial ferritin proteins, *Science*, 1995, **268**, 77–80.
121. F. C. Meldrum and S. Mann, Controlled synthesis of inorganic materials using supramolecular assemblies, *Adv. Mater.*, 1991, **30**, 316–318.
122. J. Polanams, A. D. Ray and R. K. Watt, Nanophase Iron Phosphate, Iron Arsenate, Iron Vanadate, and Iron Molybdate Minerals Synthesized within the Protein Cage of Ferritin, *Inorg. Chem.*, 2005, **44**, 3203–3209.
123. T. Douglas and V. T. Stark, Nanophase Cobalt Oxyhydroxide Mineral Synthesized within the Protein Cage of Ferritin, *Inorg. Chem.*, 2000, **39**, 1828–1830.
124. H.-A. Hosein, D. R. Strongin, M. Allen and T. Douglas, Iron and Cobalt Oxide and Metallic Nanoparticles Prepared from Ferritin, *Langmuir*, 2004, **20**, 10283–10287.
125. M. Okuda, K. Iwahori, I. Yamashita and H. Yoshimura, Fabrication of nickel and chromium nanoparticles using the protein cage of apoferritin, *Biotechnol. Bioeng.*, 2003, **84**, 187–194.
126. M. T. Klem, D. A. Resnick, K. Gilmore, M. Young, Y. U. Idzerda and T. Douglas, Synthetic Control over Magnetic Moment and Exchange Bias in All-Oxide Materials Encapsulated within a Spherical Protein Cage, *J. Am. Chem. Soc.*, 2007, **129**, 197–201.
127. F. Meldrum, B. Heywood and S. Mann, Magnetoferritin: in vitro synthesis of a novel magnetic protein, *Science*, 1992, **257**, 522–523.
128. K. K. W. Wong, T. Douglas, S. Gider, D. D. Awschalom and S. Mann, Biomimetic Synthesis and Characterisation of Magnetic Proteins (Magnetoferritin), *Chem. Mater.*, 1998, **10**, 279–285.
129. V. V. Nikandrov, C. K. Grätzel, J. E. Moser and M. Grätzel, Light induced redox reactions involving mammalian ferritin as photocatalyst, *J. Photochem. Photobiol. B.*, 1997, **41**, 83–89.
130. I. Kim, H.-A. Hosein, D. R. Strongin and T. Douglas, Photochemical Reactivity of Ferritin for Cr(VI) Reduction, *Chem. Mater.*, 2002, **14**, 4874–4879.
131. D. Ensign, M. Young and T. Douglas, Photocatalytic Synthesis of Copper Colloids from Cu(II) by the Ferrihydrite Core of Ferritin, *Inorg. Chem.*, 2004, **43**, 3441–3446.
132. M. T. Klem, J. Mosolf, M. Young and T. Douglas, Photochemical mineralisation of europium, titanium, and iron oxyhydroxide nanoparticles in the ferritin protein cage, *Inorg. Chem.*, 2008, **47**, 2237–2239.
133. T. Ueno, M. Suzuki, T. Goto, T. Matsumoto, K. Nagayama and Y. Watanabe, Size-selective olefin hydrogenation by a Pd nanocluster

provided in an apo-ferritin cage, *Angew. Chem. Int. Ed.*, 2004, **43**, 2527–2530.

134. C. Barnes, S. Petoud, S. Cohen and K. Raymond, Competition studies in horse spleen ferritin probed by a kinetically inert inhibitor, $[Cr(TREN)(H_2O)(OH)]^{2+}$, and a highly luminescent Tb(III) reagent, *J. Biol. Inorg. Chem.*, 2003, **8**, 195–205.

135. F. Bou-Abdallah, G. Biasiotto, P. Arosio and N. D. Chasteen, The Putative "Nucleation Site" in Human H-Chain Ferritin Is Not Required for Mineralisation of the Iron Core, *Biochemistry*, 2004, **43**, 4332–4337.

136. R. M. Kramer, C. Li, D. C. Carter, M. O. Stone and R. R. Naik, Engineered Protein Cages for Nanomaterial Synthesis, *J. Am. Chem. Soc.*, 2004, **126**, 13282–13286.

137. R. R. Naik, S. J. Stringer, G. Agarwal, S. E. Jones and M. O. Stone, Biomimetic synthesis and patterning of silver nanoparticles, *Nature Mater.*, 2002, **1**, 169–172.

138. S. R. Whaley, D. S. English, E. L. Hu, P. F. Barbara and A. M. Belcher, Selection of peptides with semiconductor binding specificity for directed nanocrystal assembly, *Nature*, 2000, **405**, 665–668.

139. K. I. Sano, K. Ajima, K. Iwahori, M. Yudasaka, S. Iijima, I. Yamashita and K. Shiba, Endowing a Ferritin-Like Cage Protein with High Affinity and Selectivity for Certain Inorganic Materials, *Small*, 2005, **1**, 826–832.

140. M. Li, K. K. W. Wong and S. Mann, Organisation of Inorganic Nanoparticles Using Biotin-Streptavidin Connectors, *Chem. Mater.*, 1999, **11**, 23–26.

141. D. Xu, G. D. Watt, J. N. Harb and R. C. Davis, Electrical Conductivity of Ferritin Proteins by Conductive AFM, *Nano Lett.*, 2005, **5**, 571–577.

142. K. M. Shin, G. D. Watt, B. Zhang, J. N. Harb, R. G. Harrison, S. I. Kim and S. J. Kim, Electrochemical analysis of the reduction of ferritin using oxidized methyl viologen, *J. Electroanal. Chem.*, 2006, **598**, 22–26.

143. C. Batya, D. Hagit, M. Gila, H. Alon and N. Michal, Ferritin as an Endogenous MRI Reporter for Noninvasive Imaging of Gene Expression in C6 Glioma Tumors, *Neoplasia*, 2005, **7**, 109–117.

144. M. L. Flenniken, D. A. Willits, S. Brumfield, M. J. Young and T. Douglas, The Small Heat Shock Protein Cage from Methanococcus jannaschii Is a Versatile Nanoscale Platform for Genetic and Chemical Modification, *Nano Lett.*, 2003, **3**, 1573–1576.

145. W. Shenton, S. Mann, H. Cölfen, A. Bacher and M. Fischer, Synthesis of Nanophase Iron Oxide in Lumazine Synthase Capsids, *Angew. Chem. Int. Ed.*, 2001, **40**, 442–445.

146. T. Douglas and A. Young, Viruses: making friends with old foes, *Science*, 2006, **312**, 873–875.

147. J. A. Speir, S. Munshi, G. Wang, T. S. Baker and J. E. Johnson, Structures of the native and swollen forms of cowpea chlorotic mottle virus determined by X-ray crystallography and cryo-electron microscopy, *Structure*, 1995, **3**, 63–78.

148. T. Lin, Z. Chen, R. Usha, C. V. Stauffacher, J.-B. Dai, T. Schmidt and J. E. Johnson, The Refined Crystal Structure of Cowpea Mosaic Virus at 2.8 Å Resolution, *Virology*, 1999, **265**, 20–34.

149. Q. Wang, T. Lin, L. Tang, J. E. Johnson and M. G. Finn, Icosahedral virus particles as addressable nanoscale building blocks, *Angew. Chem. Int. Ed.*, 2002, **41**, 459–462.

150. W. F. Ochoa, A. Chatterji, T. Lin and J. E. Johnson, Generation and structural analysis of reactive empty particles derived from an icosahedral virus, *Chem. Biol.*, 2006, **13**, 771–778.

151. T. Douglas and M. Young, Host-guest encapsulation of materials by assembled virus protein cages, *Nature*, 1998, **393**, 152–155.

152. T. Douglas, E. Strable, D. Willits, A. Aitouchen, M. Libera and M. Young, Protein engineering of a viral cage for constrained nano-materials synthesis, *Adv. Mater.*, 2002, **14**, 415–418.

153. N. F. Steinmetz and D. J. Evans, Utilisation of plant viruses in bio-nanotechnology, *Org. Biomol. Chem.*, 2007, **5**, 2891–2902.

154. M. T. Klem, D. Willits, M. Young and T. Douglas, 2-D Array formation of genetically engineered viral cages on Au surfaces and imaging by atomic force microscopy, *J. Am. Chem. Soc.*, 2003, **125**, 10806–10807.

155. M. Uchida, M. T. Klem, M. Allen, P. Suci, M. Flenniken, E. Gillitzer, Z. Varpness, L. O. Liepold, M. Young and T. Douglas, Biological containers: protein cages as multifunctional nanoplatforms, *Adv. Mater.*, 2007, **19**, 1025–1042.

156. Q. Wang, K. S. Raja, K. D. Janda, T. Lin and M. G. Finn, Blue Fluorescent Antibodies as Reporters of Steric Accessibility in Virus Conjugates, *Bioconjug. Chem.*, 2003, **14**, 38–43.

157. Q. Wang, E. Kaltgrad, T. Lin, J. E. Johnson and M. G. Finn, Natural Supramolecular Building Blocks: Wild-Type Cowpea Mosaic Virus, *Chem. Biol.*, 2002, **9**, 805–811.

158. A. Chatterji, W. F. Ochoa, M. Paine, B. R. Ratna, J. E. Johnson and T. Lin, New Addresses on an Addressable Virus Nanoblock: Uniquely Reactive Lys Residues on Cowpea Mosaic Virus, *Chem. Biol.*, 2004, **11**, 855–863.

159. C. M. Soto, A. S. Blum, G. J. Vora, N. Lebedev, C. E. Meador, A. P. Won, A. Chatterji, J. E. Johnson and B. R. Ratna, Fluorescent signal amplification of carbocyanine dyes using engineered viral nano-particles, *J. Am. Chem. Soc.*, 2006, **128**, 5184–5189.

160. A. S. Blum, C. M. Soto, C. D. Wilson, T. L. Brower, S. K. Pollack, T. L. Schull, A. Chatterji, T. Lin, J. E. Johnson, C. Amsinck, P. Franzon, R. Shashidhar and B. R. Ratna, An engineered virus as a scaffold for three-dimensional self-assembly on the nanoscale, *Small*, 2005, **1**, 702–706.

161. N. F. Steinmetz, G. Calder, G. P. Lomonossoff and D. J. Evans, Plant Viral Capsids as Nanobuilding Blocks: Construction of Arrays on Solid Supports, *Langmuir*, 2006, **22**, 10032–10037.

162. A. S. Blum, C. M. Soto, C. D. Wilson, J. D. Cole, M. Kim, B. Gnade, A. Chatterji, W. F. Ochoa, T. Lin, J. E. Johnson and B. R. Ratna,

Cowpea Mosaic Virus as a Scaffold for 3-D Patterning of Gold Nano-particles, *Nano Lett.*, 2004, **4**, 867–870.

163. C. L. Cheung, J. A. Camarero, B. W. Woods, T. Lin, J. E. Johnson and J. J. De Yoreo, Fabrication of Assembled Virus Nanostructures on Templates of Chemoselective Linkers Formed by Scanning Probe Nanolithography, *J. Am. Chem. Soc.*, 2003, **125**, 6848–6849.

164. A. Chatterji, L. L. Burns, S. S. Taylor, G. P. Lomonossoff, J. E. Johnson, T. Lin and C. Porta, Cowpea mosaic virus: from the presentation of antigenic peptides to the display of active biomaterials, *Intervirology*, 2002, **45**, 362–370.

165. S. S. Gupta, J. Kuzelka, P. Singh, W. G. Lewis, M. Manchester and M. G. Finn, Accelerated Bioorthogonal Conjugation: A Practical Method for the Ligation of Diverse Functional Molecules to a Polyvalent Virus Scaffold, *Bioconjug. Chem.*, 2005, **16**, 1572–1579.

166. K. S. Raja, Q. Wang, M. J. Gonzalez, M. Manchester, J. E. Johnson and M. G. Finn, Hybrid Virus-Polymer Materials. 1. Synthesis and Properties of PEG-Decorated Cowpea Mosaic Virus, *Biomacromolecules*, 2003, **4**, 472–476.

167. E. Strable, J. E. Johnson and M. G. Finn, Natural Nanochemical Building Blocks: Icosahedral Virus Particles Organized by Attached Oligonucleotides, *Nano Lett.*, 2004, **4**, 1385–1389.

168. C. Chen, M.-C. Daniel, Z. T. Quinkert, M. De, B. Stein, V. D. Bowman, P. R. Chipman, V. M. Rotello, C. C. Kao and B. Dragnea, Nanoparticle-Templated Assembly of Viral Protein Cages, *Nano Lett.*, 2006, **6**, 611–615.

169. R. W. Lucas, S. B. Larson and A. McPherson, The crystallographic structure of brome mosaic virus, *J. Mol. Biol.*, 2002, **317**, 95–108.

170. J. Sun, C. DuFort, M.-C. Daniel, A. Murali, C. Chen, K. Gopinath, B. Stein, M. De, V. M. Rotello, A. Holzenburg, C. C. Kao and B. Dragnea, Core-controlled polymorphism in virus-like particles, *PNAS*, 2007, **104**, 1354–1359.

171. K. J. Koudelka, C. S. Rae, M. J. Gonzalez and M. Manchester, Interaction between a 54-Kilodalton Mammalian Cell Surface Protein and Cowpea Mosaic Virus, *J. Virol.*, 2007, **81**, 1632–1640.

172. P. Singh, D. Prasuhn, R. M. Yeh, G. Destito, C. S. Rae, K. Osborn, M. G. Finn and M. Manchester, Bio-distribution, toxicity and pathology of cowpea mosaic virus nanoparticles in vivo, *J. Control Release*, 2007, **120**, 41–50.

173. M. Manchester and P. Singh, Virus-based nanoparticles (VNPs): Platform technologies for diagnostic imaging, *Adv. Drug Deliv. Rev.*, 2006, **58**, 1505–1522.

174. J. D. Lewis, G. Destito, A. Zijlstra, M. J. Gonzalez, J. P. Quigley, M. Manchester and H. Stuhlmann, Viral nanoparticles as tools for intravital vascular imaging, *Nat. Med.*, 2006, **12**, 354–360.

175. C. S. Rae, I. Wei Khor, Q. Wang, G. Destito, M. J. Gonzalez, P. Singh, D. M. Thomas, M. N. Estrada, E. Powell, M. G. Finn and M. Manchester,

Systemic trafficking of plant virus nanoparticles in mice via the oral route, *Virology*, 2005, **343**, 224–235.

176. J. M. Hooker, E. W. Kovacs and M. B. Francis, Interior surface modification of bacteriophage MS2, *J. Am. Chem. Soc.*, 2004, **126**, 3718–3719.

177. H. N. Barnhill, R. Reuther, P. L. Ferguson, T. Dreher and Q. Wang, Turnip yellow mosaic virus as a chemoaddressable bionanoparticle, *Bioconjugate Chem.*, 2007, **18**, 852–859.

178. R. Carbone, L. Giorgetti, A. Zanardi, I. Marangi, E. Chierici, G. Bongiorno, F. Fiorentini, M. Faretta, P. Piseri, P. G. Pelicci and P. Milani, Retroviral microarray-based platform on nanostructured TiO2 for functional genomics and drug discovery, *Biomaterials*, 2007, **28**, 2244–2253.

179. W. Shenton, T. Douglas, M. Young, G. Stubbs and S. Mann, Inorganic-Organic Nanotube Composites from Template Mineralisation of Tobacco Mosaic Virus, *Adv. Mater.*, 1999, **11**, 253–256.

180. H. Yi, S. Nisar, S.-Y. Lee, M. A. Powers, W. E. Bentley, G. F. Payne, R. Ghodssi, G. W. Rubloff, M. T. Harris and J. N. Culver, Patterned assembly of genetically modified viral nanotemplates via nucleic acid hybridization, *Nano Lett.*, 2005, **5**, 1931–1936.

181. T. L. Schlick, Z. Ding, E. W. Kovacs and M. B. Francis, Dual-Surface Modification of the Tobacco Mosaic Virus, *J. Am. Chem. Soc.*, 2005, **127**, 3718–3723.

182. E. Dujardin, C. Peet, G. Stubbs, J. N. Culver and S. Mann, Organisation of Metallic Nanoparticles Using Tobacco Mosaic Virus Templates, *Nano Lett.*, 2003, **3**, 413–417.

183. S. Y. Lee, J. Choi, E. Royston, D. B. Janesr, J. N. Culve and M. T. Harris, Deposition of platinum clusters on surface-modified Tobacco mosaic virus, *J. Nanosci. Nanotechnol.*, 2006, **6**, 974–981.

184. S. Fujikawa and T. Kunitake, Surface Fabrication of Hollow Nanoarchitectures of Ultrathin Titania Layers from Assembled Latex Particles and Tobacco Mosaic Viruses as Templates, *Langmuir*, 2003, **19**, 6545–6552.

185. R. A. Miller, A. D. Presley and M. B. Francis, Self-Assembling Light-Harvesting Systems from Synthetically Modified Tobacco Mosaic Virus Coat Proteins, *J. Am. Chem. Soc.*, 2007, **129**, 3104–3109.

186. H. Yi, G. W. Rubloff and J. N. Culver, TMV Microarrays: Hybridisation-Based Assembly of DNA-Programmed Viral Nanotemplates, *Langmuir*, 2007, **23**, 2663–2667.

187. M. Knez, A. M. Bittner, F. Boes, C. Wege, H. Jeske, E. Maib and K. Kern, Biotemplate Synthesis of 3-nm Nickel and Cobalt Nanowires, *Nano Lett.*, 2003, **3**, 1079–1082.

188. M. Knez, A. Kadri, C. Wege, U. Gosele, H. Jeske and K. Nielsch, Atomic Layer Deposition on Biological Macromolecules: Metal Oxide Coating of Tobacco Mosaic Virus and Ferritin, *Nano Lett.*, 2006, **6**, 1172–1177.

189. M. Law, J. Goldberger and P. Yang, Semiconductor nanowires and nanotubes, *Ann. Rev. Mater. Res.*, 2004, **34**, 83–122.

190. Z. Niu, M. A. Bruckman, S. Li, L. A. Lee, B. Lee, S. V. Pingali, P. Thiyagarajan and Q. Wang, Assembly of Tobacco Mosaic Virus into Fibrous and Macroscopic Bundled Arrays Mediated by Surface Aniline Polymerisation, *Langmuir*, 2007, **23**, 6719–6724.

191. E. Royston, S.-Y. Lee, J. N. Culver and M. T. Harris, Characterisation of silica-coated tobacco mosaic virus, *J. Colloid Interface Sci.*, 2006, **298**, 706–712.

192. A. Nedoluzhko and T. Douglas, Ordered association of tobacco mosaic virus in the presence of divalent metal ions, *J. Inorg. Biochem.*, 2001, **84**, 233–240.

193. U. Kriplani and B. K. Kay, Selecting peptides for use in nanoscale materials using phage-displayed combinatorial peptide libraries, *Curr. Opin. Biotechnol.*, 2005, **16**, 470–475.

194. J. W. Kehoe and B. K. Kay, Filamentous Phage Display in the New Millennium, *Chem. Rev.*, 2005, **105**, 4056–4072.

195. N. K. Petty, T. J. Evans, P. C. Fineran and G. P. C. Salmond, Biotechnological exploitation of bacteriophage research, *Trends Biotechnol.*, 2007, **25**, 7–15.

196. C. Mao, D. J. Solis, B. D. Reiss, S. T. Kottmann, R. Y. Sweeney, A. Hayhurst, G. Georgiou, B. Iverson and A. M. Belcher, Virus-Based Toolkit for the Directed Synthesis of Magnetic and Semiconducting Nanowires, *Science*, 2004, **303**, 213–217.

197. A. B. Sanghvi, K. P.-H. Miller, A. M. Belcher and C. E. Schmidt, Biomaterials functionalisation using a novel peptide that selectively binds to a conducting polymer, *Nature Mater.*, 2005, **4**, 496–502.

198. M. M. Tomczak, J. M. Slocik, M. O. Stone and R. R. Naik, Bio-based approaches to inorganic material synthesis, *Biochem. Soc. Transact.*, 2007, **035**, 512–515.

199. C. E. Flynn, C. Mao, A. Hayhurst, J. L. Williams, G. Georgiou, B. Iverson and A. M. Belcher, Synthesis and organisation of nanoscale II-VI semiconductor materials using evolved peptide specificity and viral capsid assembly, *J. Mater. Chem.*, 2003, **13**, 2414–2421.

200. B. D. Reiss, C. Mao, D. J. Solis, K. S. Ryan, T. Thomson and A. M. Belcher, Biological Routes to Metal Alloy Ferromagnetic Nanostructures, *Nano Lett.*, 2004, **4**, 1127–1132.

201. S.-K. Lee, D. S. Yun and A. M. Belcher, Cobalt Ion Mediated Self-Assembly of Genetically Engineered Bacteriophage for Biomimetic Co-Pt Hybrid Material, *Biomacromolecules*, 2006, **7**, 14–17.

202. S.-W. Lee, C. Mao, C. E. Flynn and A. M. Belcher, Ordering of Quantum Dots Using Genetically Engineered Viruses, *Science*, 2002, **296**, 892–895.

203. S.-W. Lee, B. M. Wood and A. M. Belcher, Chiral Smectic C Structures of Virus-Based Films, *Langmuir*, 2003, **19**, 1592–1598.

204. K. T. Nam, D.-W. Kim, P. J. Yoo, C.-Y. Chiang, N. Meethong, P. T. Hammond, Y.-M. Chiang and A. M. Belcher, Virus-Enabled Synthesis and Assembly of Nanowires for Lithium Ion Battery Electrodes, *Science*, 2006, **312**, 885–888.

205. C. Mao, C. E. Flynn, A. Hayhurst, R. Sweeney, J. Qi, G. Georgiou, B. Iverson and A. M. Belcher, Viral assembly of oriented quantum dot nanowires, *Proc. Natl. Acad. Sci. USA*, 2003, **100**, 6946–6951.

206. L. B. Giebel, R. Cass, D. L. Milligan, D. Young, R. Arze and C. Johnson, Screening of cyclic peptide phage libraries identifies ligands that bind streptavidin with high affinities, *Biochemistry*, 1995, **34**, 15430–15435.

207. S.-W. Lee, S. K. Lee and A. M. Belcher, Virus-Based Alignment of Inorganic, Organic, and Biological Nanosized Materials, *Adv. Mater.*, 2003, **15**, 689–692.

208. A. S. Khalil, J. M. Ferrer, R. R. Brau, S. T. Kottmann, C. J. Noren, M. J. Lang and A. M. Belcher, From the Cover: Single M13 bacteriophage tethering and stretching, *Proc. Natl. Acad. Sci. USA*, 2007, **104**, 4892–4897.

209. Y. Huang, C.-Y. Chiang, S. K. Lee, Y. Gao, E. L. Hu, J. D. Yoreo and A. M. Belcher, Programmable Assembly of Nanoarchitectures Using Genetically Engineered Viruses, *Nano Lett.*, 2005, **5**, 1429–1434.

210. K. T. Nam, B. R. Peelle, S.-W. Lee and A. M. Belcher, Genetically Driven Assembly of Nanorings Based on the M13 Virus, *Nano Lett.*, 2004, **4**, 23–27.

211. L. Yang, H. Liang, T. E. Angelini, J. Butler, R. Coridan, J. X. Tang and G. C. L. Wong, Self-assembled virus-membrane complexes, *Nature Mater.*, 2004, **3**, 615–619.

212. I. Koltover, K. Wagner and C. R. Safinya, DNA condensation in two dimensions, *Proc. Natl. Acad. Sci. USA*, 2000, **97**, 14046–14051.

213. P. J. Yoo, K. T. Nam, J. Qi, S.-K. Lee, J. Park, A. M. Belcher and P. T. Hammond, Spontaneous assembly of viruses on multilayered polymer surfaces, *Nature Mater.*, 2006, **5**, 234–240.

214. www.foresight.org/challenges/.

215. C. E. Thomas, A. Ehrhardt and M. A. Kay, Progress and problems with the use of viral vectors for gene therapy, *Nat. Rev. Genet.*, 2003, **4**, 346–358.

216. P. C. Hendrie and W. R. David, Gene Targeting with Viral Vectors, *Mol. Ther.*, 2005, **12**, 9–17.

217. A. L. Parker, C. Newman, S. Briggs, L. Seymour and P. J. Sheridan, Nonviral gene delivery: techniques and implications for molecular medicine, *Exp. Rev. Mol. Med.*, 2003, **5**, 1–15.

218. C. W. Pouton and L. W. Seymour, Key issues in non-viral gene delivery, *Adv. Drug Deliv. Rev.*, 2001, **46**, 187–203.

219. NIH report. *Hum. Gene. Ther.* **13**, 3–13.

220. M. L. Edelstein, M. R. Abedi and J. Wixon, Gene therapy clinical trials worldwide to 2007-an update, *J. Gene Med.*, 2007, **9**, 833–842.

221. K. Kostarelos and A. D. Miller, Synthetic, self-assembly ABCD nanoparticles; a structural paradigm for viable synthetic non-viral vectors, *Chem. Soc. Rev.*, 2005, **34**, 970–994.

222. B. Demeneix, Z. Hassani and J.-P. Behr, Towards Multifunctional Synthetic Vectors, *Curr. Gene Ther.*, 2004, **4**, 445–455.

223. G. Zuber, E. Dauty, M. Nothisen, P. Belguise and J.-P. Behr, Towards synthetic viruses, *Adv. Drug Deliv. Rev.*, 2001, **52**, 245–253.

224. I. A. Khalil, K. Kogure, H. Akita and H. Harashima, Uptake Pathways and Subsequent Intracellular Trafficking in Nonviral Gene Delivery, *Pharmacol. Rev.*, 2006, **58**, 32–45.

225. J. A. Wolff and D. B. Rozema, Breaking the Bonds: Non-viral Vectors Become Chemically Dynamic, *Mol. Ther.*, 2007, **16**, 8–15.

226. M. L. Read, A. Logan and L. W. Seymour, in *Adv. Genet*, Academic Press, 2005, 19–46.

227. J. A. Wolff and V. Budker, in *Adv. Genet*, Academic Press, 2005, 1–20.

228. D. L. Lewis and J. A. Wolff, Systemic siRNA delivery via hydrodynamic intravascular injection, *Adv. Drug Deliv. Rev.*, 2007, **59**, 115–123.

229. S. D. Li and L. Huang, Gene therapy progress and prospects: non-viral gene therapy by systemic delivery, *Gene Ther.* **13**, 1313–1319.

230. K. Anwer, in *Electroporation Protocols*, 2008, 77–89.

231. S. Mehier-Humbert and R. H. Guy, Physical methods for gene transfer: Improving the kinetics of gene delivery into cells, *Adv. Drug Deliv. Rev.*, 2005, **57**, 733–753.

232. J.-i. Miyazaki and H. Aihara, in *Gene Therapy Protocols*, 2002, 49–62.

233. R. Waehler, S. J. Russell and D. T. Curiel, Engineering targeted viral vectors for gene therapy, *Nat. Rev. Genet.*, 2007, **8**, 573–587.

234. D. J. Glover, H. J. Lipps and D. A. Jans, Towards safe, non-viral therapeutic gene expression in humans, *Nat. Rev. Genet.*, 2005, **6**, 299–310.

235. B. Demeneix and J.-P. Behr, Polyethylenimine, *Adv. Genet.*, 2005, **53**, 217–230.

236. I. Kopatz, J.-S. Remy and J.-P. Behr, A model for non-viral gene delivery: through syndecan adhesion molecules and powered by actin, *J. Gene Med.*, 2004, **6**, 769–776.

237. C. M. Ward, M. Pechar, D. Oupicky, K. Ulbrich and L. W. Seymour, Modification of pLL/DNA complexes with a multivalent hydrophilic polymer permits folate-mediated targeting in vitro and prolonged plasma circulation in vivo, *J. Gene Med.*, 2002, **4**, 536–547.

238. M. L. Read, S. Singh, Z. Ahmed, M. Stevenson, S. S. Briggs, D. Oupicky, L. B. Barrett, R. Spice, M. Kendall, M. Berry, J. A. Preece, A. Logan and L. W. Seymour, A versatile reducible polycation-based system for efficient delivery of a broad range of nucleic acids, *Nucl. Acids Res.*, 2005, **33**, e86.

239. R. C. Carlisle, T. Etrych, S. S. Briggs, J. A. Preece, K. Ulbrich and L. W. Seymour, Polymer-coated polyethylenimine/DNA complexes designed for triggered activation by intracellular reduction, *J. Gene Med.*, 2004, **6**, 337–344.

240. D. Oupicky, A. L. Parker and L. W. Seymour, Laterally Stabilized Complexes of DNA with Linear Reducible Polycations: Strategy for Triggered Intracellular Activation of DNA Delivery Vectors, *J. Am. Chem. Soc.*, 2002, **124**, 8–9.

241. D. L. McKenzie, K. Y. Kwok and K. G. Rice, A Potent New Class of Reductively Activated Peptide Gene Delivery Agents, *J. Biol. Chem.*, 2000, **275**, 9970–9977.

242. S. M. Albelda, R. Wiewrodt and J. B. Zuckerman, Gene Therapy for Lung Disease: Hype or Hope? *Ann. Intern. Med.*, 2000, **132**, 649–660.

243. A. Benigni, S. Tomasoni and G. Remuzzi, Impediments to successful gene transfer to the kidney in the context of transplantation and how to overcome them, *Kidney Inter.*, 2002, **61**, S115–S119.

244. A. El-Aneed, An overview of current delivery systems in cancer gene therapy, *J. Control Release*, 2004, **94**, 1–14.

245. H. Takeuchi, Y. Matsui, H. Sugihara, H. Yamamoto and Y. Kawashima, Effectiveness of submicrometre-sized chitosan-coated liposomes in oral administration of peptide drugs, *Int. J. Pharm.*, 2005, **303**, 160–170.

246. R. I. Mahato, Y. Takakura and M. Hashida, Nonviral vectors for in vivo gene delivery: physicochemical and pharmacokinetic considerations, *Crit. Rev. Ther. Drug Carrier Syst.*, 1997, **14**, 133–172.

247. R. Duncan, The dawning era of polymer therapeutics, *Nat. Rev. Drug Discov.*, 2003, **2**, 347–360.

248. V. N. Medvedkin, E. A. Permyakov, L. V. Klimenko, Y. V. Mitin, N. Matsushima, S. Nakayama and R. H. Kretsinger, Interactions of (Ala*Ala*Lys*Pro)n and (Lys*Lys*Ser*Pro)n with DNA. Proposed coiled-coil structure of AlgR3 and AlgP from Pseudomonas aeruginosa, *Protein Eng.*, 1995, **8**, 63–70.

249. M. Suzuki, SPKK, a new nucleic acid-binding unit of protein found in histone, *EMBO J.*, 1989, **8**, 797–804.

250. J. R. Khadake and M. R. S. Rao, Condensation of DNA and Chromatin by an SPKK-Containing Octapeptide Repeat Motif Present in the C-Terminus of Histone H1, *Biochemistry*, 1997, **36**, 1041–1051.

251. K. M. Wagstaff, D. J. Glover, D. J. Tremethick and D. A. Jans, Histone-mediated Transduction as an Efficient Means for Gene Delivery, *Mol. Ther.*, 2007, **15**, 721–731.

252. K. D. Murray, C. J. Etheridge, S. I. Shah, D. A. Matthews, W. Russell, H. M. D. Gurling and A. D. Miller, Enhanced cationic liposome-mediated transfection using the DNA-binding peptide μ from the adenovirus core, *Gene Ther.*, 2001, **8**, 453–460.

253. T. Tagawa, M. Manvell, N. Brown, M. Keller, E. Perouzel, K. D. Murray, R. P. Harbottle, M. Tecle, F. Booy, M. C. Brahimi-Horn, C. Coutelle, N. R. Lemoine, E. W. Alton and A. D. Miller, Characterisation of LMD virus-like nanoparticles self-assembled from cationic liposomes adeno-virus core peptide μ and plasmid DNA, *Gene Ther.*, 2002, **9**, 564–576.

254. M. Tecle, M. Preuss and A. D. Miller, Kinetic Study of DNA Condensa-tion by Cationic Peptides Used in Nonviral Gene Therapy: Analogy

of DNA Condensation to Protein Folding, *Biochemistry*, 2003, **42**, 10343–10347.

255. P. Erbacher, A. C. Roche, M. Monsigny and P. Midoux, Putative Role of Chloroquine in Gene Transfer into a Human Hepatoma Cell Line by DNA/Lactosylated Polylysine Complexes, *Exp. Cell Res.*, 1996, **225**, 186–194.

256. M. Ogris, R. C. Carlisle, T. Bettinger and L. W. Seymour, Melittin Enables Efficient Vesicular Escape and Enhanced Nuclear Access of Nonviral Gene Delivery Vectors, *J. Biol. Chem.*, 2001, **276**, 47550–47555.

257. W. S. Horne, C. M. Wiethoff, C. Cui, K. M. Wilcoxen, M. Amorin, M. R. Ghadiri and G. R. Nemerow, Antiviral cyclic d,l-α-peptides: Targeting a general biochemical pathway in virus infections, *Bioorg Med. Chem.*, 2005, **13**, 5145–5153.

258. M. Hansen, K. Kilk and Ü. Langel, Predicting cell-penetrating peptides, *Adv. Drug Deliv. Rev.*, 2008, **60**, 572–579.

259. R. Abes, A. Arzumanov, H. Moulton, S. Abes, G. Ivanova, M. J. Gait, P. Iversen and B. Lebleu, Arginine-rich cell penetrating peptides: Design, structure-activity, and applications to alter pre-mRNA splicing by steric-block oligonucleotides, *J. Pept. Sci.*, 2008, **14**, 455–460.

260. D. Derossi, A. H. Joliot, G. Chassaing and A. Prochiantz, The third helix of the Antennapedia homeodomain translocates through biological membranes, *J. Biol. Chem.*, 1994, **269**, 10444–10450.

261. E. Vives, P. Brodin and B. Lebleu, A Truncated HIV-1 Tat Protein Basic Domain Rapidly Translocates through the Plasma Membrane and Accumulates in the Cell Nucleus, *J. Biol. Chem.*, 1997, **272**, 16010–16017.

262. G. Elliott and P. O'Hare, Intercellular Trafficking and Protein Delivery by a Herpesvirus Structural Protein, *Cell*, 1997, **88**, 223–233.

263. L. E. Yandek, A. Pokorny, A. Floren, K. Knoelke, U. Langel and P. F. F. Almeida, Mechanism of the Cell-Penetrating Peptide Transportan 10 Permeation of Lipid Bilayers, *Biophys. J.*, 2007, **92**, 2434–2444.

264. S. Boeckle, J. Fahrmeir, W. Roedl, M. Ogris and E. Wagner, Melittin analogs with high lytic activity at endosomal pH enhance transfection with purified targeted PEI polyplexes, *J. Control Release*, 2006, **112**, 240–248.

265. E. Wagner, C. Plank, K. Zatloukal, M. Cotten and M. L. Birnstiel, Influenza virus hemagglutinin HA-2 N-terminal fusogenic peptides augment gene transfer by transferrin-polylysine-DNA complexes: toward a synthetic virus-like gene-transfer vehicle, *Proc. Natl. Acad. Sci. USA*, 1992, **89**, 7934–7938.

266. S. Gottschalk, J. T. Sparrow, J. Hauer, M. P. Mims, F. E. Leland, S. L. Woo and L. C. Smith, A novel DNA-peptide complex for efficient gene transfer and expression in mammalian cells, *Gene Ther.*, 1996, **3**, 448–457.

267. M. D. Pierschbacher and E. Ruoslahti, Cell attachment activity of fibronectin can be duplicated by small synthetic fragments of the molecule, *Nature*, 1984, **309**, 30–33.

268. J. Graf, Y. Iwamoto, M. Sasaki, G. R. Martin, H. K. Kleinman, F. R. Robey and Y. Yamada, Identification of an amino acid sequence in laminin mediating cell attachment chemotaxis, and receptor binding, *Cell*, 1987, **48**, 989–996.

269. R. G. Cooper, R. P. Harbottle, H. Schneider, C. Coutelle and A. D. Miller, Peptide Mini-Vectors for Gene Delivery, *Angew. Chem. Int. Ed.*, 1999, **38**, 1949–1952.

270. T. W. R. Lee, G. E. Blair and D. A. Matthews, Adenovirus core protein VII contains distinct sequences that mediate targeting to the nucleus and nucleolus, and colocalisation with human chromosomes, *J. Gen. Virol.*, 2003, **84**, 3423–3428.

271. A. D. Bangham, M. M. Standish and J. C. Watkins, Diffusion of univalent ions across the lamellae of swollen phospholipids, *J. Mol. Biol.*, 1965, **13**, 238–252.

272. P. L. Felgner, T. R. Gadek, M. Holm, R. Roman, H. W. Chan, M. Wenz, J. P. Northrop, G. M. Ringold and M. Danielsen, Lipofection: a highly efficient, lipid-mediated DNA–Transfection procedure, *Proc. Natl. Acad. Sci. USA*, 1987, **84**, 7413–7417.

273. N. S. Templeton, D. D. Lasic, P. M. Frederik, H. H. Strey, D. D. Roberts and G. N. Pavlakis, Improved DNA: liposome complexes for increased systemic delivery and gene expression, *Nature Biotech.*, 1997, **15**, 647–652.

274. H.-J. Butt, K. Graf and M. Kappl, *Physics and Chemistry of Interfaces*, Wiley-VCH, Weinheim, 2003.

275. T. Segura and L. D. Shea, Materials for non-viral gene delivery, *Ann. Rev. Mater. Res.*, 2001, **31**, 25–46.

276. L. Wasungu and D. Hoekstra, Cationic lipids, lipoplexes and intracellular delivery of genes, *J. Control Release*, 2006, **116**, 255–264.

277. M. X. Tang, W. Li and F. C. Szoka, Toroid formation in charge neutralized flexible or semi-flexible biopolymers: potential pathway for assembly of DNA carriers, *J. Gene Med.*, 2005, **7**, 334–342.

278. J. O. Radler, I. Koltover, T. Salditt and C. R. Safinya, Structure of DNA-Cationic Liposome Complexes: DNA Intercalation in Multilamellar Membranes in Distinct Interhelical Packing Regimes, *Science*, 1997, **275**, 810–814.

279. I. Koltover, T. Salditt, J. O. Radler and C. R. Safinya, An Inverted Hexagonal Phase of Cationic Liposome-DNA Complexes Related to DNA Release and Delivery, *Science*, 1998, **281**, 78–81.

280. T. Blessing, J.-S. Remy and J.-P. Behr, Monomolecular collapse of plasmid DNA into stable virus-like particles, *Proc. Natl. Acad. Sci. USA*, 1998, **95**, 1427–1431.

281. C. Chittimalla, L. Zammut-Italiano, G. Zuber and J. P. Behr, Monomolecular DNA Nanoparticles for Intravenous Delivery of Genes, *J. Am. Chem. Soc.*, 2005, **127**, 11436–11441.

282. E. Dauty, J. S. Remy, T. Blessing and J. P. Behr, Dimerisable Cationic Detergents with a Low cmc Condense Plasmid DNA into Nanometric

Particles and Transfect Cells in Culture, *J. Am. Chem. Soc.*, 2001, **123**, 9227–9234.

283. G. Zuber, L. Zammut-Italiano, E. Dauty and J.-P. Behr, Targeted Gene Delivery to Cancer Cells: Directed Assembly of Nanometric DNA Particles Coated with Folic Acid, *Angew. Chem. Int. Ed.*, 2003, **42**, 2666–2669.

284. T. Nakai, T. Kanamori, S. Sando and Y. Aoyama, Remarkably Size-Regulated Cell Invasion by Artificial Viruses. Saccharide-Dependent Self-Aggregation of Glycoviruses and Its Consequences in Glycoviral Gene Delivery, *J. Am. Chem. Soc.*, 2003, **125**, 8465–8475.

285. A. Yasuhiro, Macrocyclic Glycoclusters: From Amphiphiles through Nanoparticles to Glycoviruses, *Chemistry*, 2004, **10**, 588–593.

286. Y. Aoyama, T. Kanamori, T. Nakai, T. Sasaki, S. Horiuchi, S. Sando and T. Niidome, Artificial Viruses and Their Application to Gene Delivery. Size-Controlled Gene Coating with Glycocluster Nanoparticles, *J. Am. Chem. Soc.*, 2003, **125**, 3455–3457.

287. J. E. Padilla, C. Colovos and T. O. Yeates, Nanohedra: Using symmetry to design self assembling protein cages, layers, crystals, and filaments, *Proc. Natl. Acad. Sci. USA*, 2001, **98**, 2217–2221.

288. S. Raman, G. Machaidze, A. Lustig, U. Aebi and P. Burkhard, Structure-based design of peptides that self-assemble into regular polyhedral nanoparticles, *Nanomedicine*, 2006, **2**, 95–102.

289. C.-J. Tsai, J. Zheng, D. Zanuy, N. Haspel, H. Wolfson, C. Aleman and R. Nussinov, Principles of nanostructure design with protein building blocks, *Proteins: Structure, Function, and Bioinformatics*, 2007, **68**, 1–12.

290. S. C. Zimmerman, F. Zeng, D. E. C. Reichert and S. V. Kolotuchin, Self-Assembling Dendrimers, *Science*, 1996, **271**, 1095–1098.

291. V. Percec, A. E. Dulcey, V. S. K. Balagurusamy, Y. Miura, J. Smidrkal, M. Peterca, S. Nummelin, U. Edlund, S. D. Hudson, P. A. Heiney, H. Duan, S. N. Magonov and S. A. Vinogradov, Self-assembly of amphiphilic dendritic dipeptides into helical pores, *Nature*, 2004, **430**, 764–768.

292. D. A. Tomalia, H. M. Brothers, L. T. Piehler, H. D. Durst and D. R. Swanson, Partial shell-filled core-shell tecto(dendrimers): A strategy to surface differentiated nano-clefts and cusps, *Proc. Natl. Acad. Sci. USA*, 2002, **99**, 5081–5087.

293. S. Hecht and J. M. J. Fréchet, Dendritic Encapsulation of Function: Applying Nature's Site Isolation Principle from Biomimetics to Materials Science, *Angew. Chem. Int. Ed.*, 2001, **40**, 74–91.

294. R. M. Versteegen, D. J. M. van Beek, R. P. Sijbesma, D. Vlassopoulos, G. Fytas and E. W. Meijer, Dendrimer-Based Transient Supramolecular Networks, *J. Am. Chem. Soc.*, 2005, **127**, 13862–13868.

295. J. C. T. Carlson, S. S. Jena, M. Flenniken, T. F. Chou, R. A. Siegel and C. R. Wagner, Chemically Controlled Self-Assembly of Protein Nanorings, *J. Am. Chem. Soc.*, 2006, **128**, 7630–7638.

296. A. J. Olson, Y. H. E. Hu and E. Keinan, Chemical mimicry of viral capsid self-assembly, *Proc. Natl. Acad. Sci. USA*, 2007, **104**, 20731–20736.

CHAPTER 4
Reassembling Multiple

Acquiring an ability to self-regenerate is an everlasting dream of human-kind.[1] Throughout the history of medicine there has hardly been another subject that would cause such an explosive mixture of stimulating excitement and holding scepticism.[2,3] Yet, however remote may its therapeutic feasibility seem, regeneration is ubiquitous to all multicellular organisms.[1]

The main question regeneration poses and that remains unanswered is why some forms of life, even within the same family, have such an ability to restore lost or damaged body parts and others do not. Even more puzzling, regen-eration, which would appeal to the rational mind as an obvious solution to many biomedical problems, does not appear to be evolutionarily advantageous.[4] In contrast, albeit better expressed in lower forms of life, it is characteristic of very few.[5]

For example, although possible it is not so common for plants to fully regenerate from individual cells, for a small number of invertebrates to reborn from body parts, and for some (again not all) amphibians to restore lost limbs. Nonetheless, restrictions in exhibiting the ability go further to be defined by degrees with which different species regenerate. This is what stimulates the investigation of the phenomenon and inspires attempts to "tame" its mechanisms for medical applications.[6,7]

The purpose of regeneration can thus be formulated as to maintain the integrity of tissues and organs.[1] This is consistent and equally vital for all organisms including those with lower regenerative capabilities – us.[2] Typical relevant examples are skin and blood that undergo complete cellular renewal an innumerable number of times during a single life span. Biologists studying regeneration call this "maintenance regeneration".[8]

In this process reserves of nondifferentiated cells (stem cells) maintain con-stant contact with environmental signals that activate their differentiation. This is opposed to a less common and more conditional process, which occurs as a result of tissue damage and is usually referred to as injury-induced

RSC Nanoscience and Nanotechnology No. 7
Bionanodesign
By M Ryadnov
© Maxim Ryadnov 2009
Published by the Royal Society of Chemistry, www.rsc.org

regeneration. Injured tissues require much faster turnover for regeneration and therefore "mobilise" stem cells *via* a partial reversal of their original embryonic development, which involves regression of specialised cells (differentiated) into unspecialised cells (embryonic or stem). Therefore, the process is recognised as dedifferentiation and accentuates the formation of an unspecialised tissue at an injury site to serve as a regrowth bud.[9]

Dedifferentiation shows remarkable efficiency in plants and invertebrates in repairing damaged or missing tissues and organs. Low vertebrates such as urodele amphibians can renew lost or severed body parts relying on similar mechanisms.[10] Mammalian cells do not regenerate *via* dedifferentiation, an ability believed to have been long lost as a result of evolution.[4,5] However, wound repair and tissue regeneration are inherent to mammals including humans. Healing is provided by another mechanism – fibrosis.[11]

Fibrosis occurs as a result of an inflammatory response accompanied by the formation of large amounts of fibrous tissues that undergo remodelling and maturation into scars. It can also be triggered in healing extensive wounds, the size of which is beyond the regenerative capacity of regeneration-competent tissues.[11] Strictly speaking, fibrosis is not a regeneration mechanism as it does not restore native tissues. Moreover, due to growing evidence that many tissues typically regarded as nonpermissive to regeneration contain regeneration-competent stem cells, scarring is often seen as the result of either an energy-efficient abortion or survival-favoured suppression of preinitiated regeneration.[12] Although our understanding of how regeneration can be differed from fibrosis is quite limited, both phenomena underlie one of the main strategies in regenerative medicine. This stems from the excessive production of extracellular matrix (ECM), which provides a cell-supporting scaffold for tissue growth and accompanies fibrosis.[13] Given that ECM is a "must" for the growth and development of most cells, developing synthetic ECM analogues can be conceived as a main strategy in the quest for alternative solutions to tissue regeneration.[6–7,14] Indeed, the last decade has witnessed significant progress in this direction, the review of which forms the core of this chapter.

4.1 Keeping All in Touch

Most cells of multicellular organisms require and produce ECM, which serves as a scaffolding platform for their attachment and subsequent assembly into tissues,[15,16] Figure 4.1(a). It is a collagenous mesh that embeds matrix glyco-proteins, growth factors, metalloproteinases and other macromolecular mediators of matrix–cell interactions.

The role of ECM in tissue development is two-fold.[17] Firstly, ECM provides tensile strength to a tissue, and defines its 3D architecture through the formation of complex supramolecular structures. Secondly, such structures promote cell adhesion and migration, provide a storage depot of growth factors and signalling for cellular homeostasis (morphogenesis and differentiation). This makes ECM an essential tissue component, a dynamic or even living part

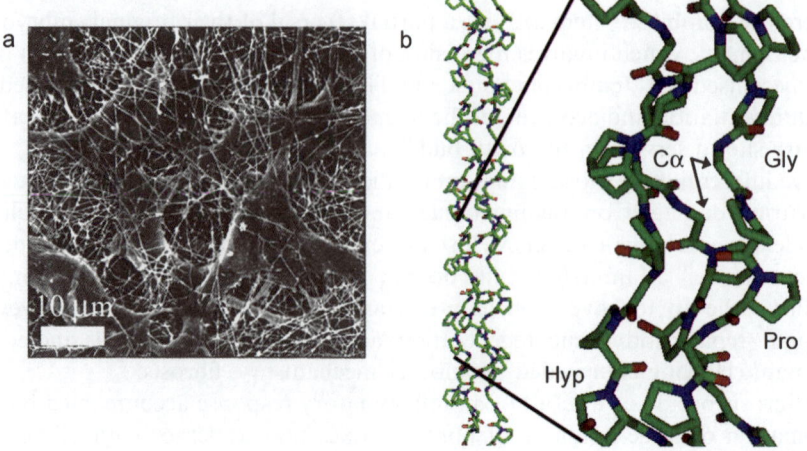

Figure 4.1   Collagen structure, (a) Electron micrograph of a native extracellular
matrix composed of collagen fibrils, (reprinted from Kim, B.-S. and
Mooney, D. J. Development of biocompatible synthetic extracellular
matrices for tissue engineering. *Trends Biotechnol.*, **16**, 224–230, Copyright
1998, with permission from Elsevier), (b) The collagen triple helix built
from three (Pro-Hyp-Gly)n chains. Arrows indicate the C_α atoms of
glycines (PDB 2D3F).

of a tissue. Furthermore, ECM is very flexible in the compilation of its
domains, which allows it to be structurally heterogeneous or specialist as
required by specific tissues.[18] For example, in kidneys highly porous fibrillar
networks serve as molecular filters, narrowly sized and regularly tilted orthogonal lattices sustain the convex curvature of cornea necessary for optical
transparency, and mechanical resilience of ECM is responsible for elasticity in
cartilage and provides fracture resistance in bone. This functional diversity
implies a variability of extracellular matrices that compose different ECM
families. The composition and functions of an ECM family are regulated by the
constitution of ECM or a relative ratio of its components. Matrices of rapidly
growing connective tissues such as cartilage, bone and skin require extensive
extracellular connectors and are largely shaped by polymolecular assemblies of
proteoglycans and collagens I and II. Lesser amounts of the scaffolding
assemblies and growth factors are compensated by the enrichment of adhesion
proteins, entactin and amorphous collagen IV in basement membranes to
endow boundaries between different tissue types and the controlled masking of
enzymatically targeted cryptic sites. However, irrespective of the family it
belongs to the impact of any given ECM on tissue and organ development
originates from the highly repetitive nature of its modular morphology.[19]

   The ECM hosts and is formed by many individual components that selectively communicate with cells at the molecular level. It acts as a multiplicator of
signals that pass between both sides – messenger molecules and cells. Multiple
signals are transmitted from extended surfaces of ECM via cell receptors to

intracellular pathways that lead to cellular responses. The part ECM takes in signal transduction and broader in the regulation of cell behaviour renders it indispensable for torrents of biological functions ranging from cell adhesion and apoptosis to vascularisation and organ morphogenesis.[16]

The property of multiplying extra- and intracellular interactions is unique, and it is not surprising that mimicking it has been a major target of regeneration research.[20,21] Several approaches have been described in the literature on how and by what means to achieve this. Most of them address either challenges of replacing ECM with synthetic fibrillar matrices or finding bulk materials able to reproduce biomechanical properties of the ECM mainly attributable to its gel-type viscoelasticity. These two major paths are mutually related and are often pursued concomitantly, with the rationale behind both routes being in the compositional nature of the native ECM.

4.1.1 Unravelling the Essential

4.1.1.1 Winding Three in One

ECM is a complex fibrillar framework assembled with nanoscale precision. Collagen is a protein that makes up the bulk of ECM and accounts for a good quarter of the total protein content of the body.[22] It is also one of the most conserved and recognisable protein folds.[22–26] A tripeptide Gly-Xaa-Yaa, in which Xaa and Yaa can be any amino acids, but often, and usually together, are proline (Pro) and hydroxyproline (Hyp), is known as the collagen motif,[22] Figure 4.1(b). The tandems of Gly-Pro-Hyp with marginal insertions of other amino acids make up 300-nm long contiguous polypeptide chains that interdigitate into rod-like bundles of 1.5 nm diameter. One bundle constitutes a right-handed helix shaped by three individual but almost identical left-handed helices. The fold is called the triple helix and is frequently referred to as tropocollagen. Side-chain-free Cα-atoms of glycine residues placed at every third position of the motif are oriented towards the interior of the fold, which has no space for any side groups, Figure 4.1(b). This is induced and held by the outward orientation of the proline rings. Polyproline helices are very common in proteins and are characterised by the absence of intrachain hydrogen bonding necessary for other oligomerising elements. Left-handed helices are lowest in energy for polyprolines and are stretched and open to accommodate local conformational inversions caused by glycines. Inclusion of glycines in polyprolines may also be required for derivatisation (*e.g.* glycosylation, crosslinking) of collagen fibres *via* modified amino acids (hydroxyproline, hydroxylysine, allysine) as these were found to be specifically oriented relative to glycines in the triple helix. Once folded, tropocollagen conforms the building block of collagen fibres.[27]

4.1.1.2 Aligning Stagger

There exist about 30 genetically encoded collagen types. However, only a few (I, II, III, V, IX and XI) are directly involved in fibrillogenesis.[28–30] As judged by electron microscopy, the fibres appear as morphologically uniform cylinders with a striated surface pattern of light and dark bands that repeat every 67 nm,[15,30] Figure 4.2.

The striated pattern is described as an axial D-period and is used as a measure unit of the fibre length.[31] The length of one tropocollagen (300 nm) is approximately 4.5 D. Tropocollagens aligned side-by-side in a parallel and regular stagger by multiples of D are believed to spontaneously assemble into five-stranded "early fibrils" of about 90 periods (~ 7 micrometres),[15,31] Figure 4.2. These are intermediate or seed structures, whose further coiled-like associations lead to mature fibres hundreds of micrometres in length. Type I and II collagens are major fibrillar components and are the ones that are synthesised in response to injury.

However, these fibrils are not functional without copolymerising with collagen molecules of other types. For instance, types III and V coassemble with type I into heterotypic fibrils, whereas type XI determines the structure and assembly of type II fibrils.[28] Type IX, also known as a FACIT (fibril-associated collagens with interrupted triple helices), decorates the surfaces of fibrils.[32] FACIT is a small collagen subfamily of 4 types, none of which can form fibrils but instead each associates with fibril-forming collagens. All FACIT collagens share one common feature, namely their collagen triple-helix domains are separated by noncollagen (NC) domains. NC domains are glycosylated and contain cysteine, methionine or hydroxylated amino acid residues that covalently crosslink and stabilise fibrillar structures. Therefore, one of the main functions of FACIT collagens is to modify fibres. Another is to mediate their interactions with cells.[28] For example, IX collagen binds oligomeric matrix proteins such as fibronectin and vitronectin that interact with cell-surface integrins, which in this notation act as responsive mediators to mechanical changes in the matrix. FACIT collagens are also thought to regulate lateral associations of the early fibrils into mature fibres. Similarly, the function of a FACIT-related collagen IV lies in the generation of sheet-like networks in basement membranes. The networks are arranged by hexameric hubs of two trimeric NC domains that are assembled through head-to-head interactions and are stabilised by covalent crosslinking of hydroxylysine residues.[33] Another type of collagen superstructure forming a dominant part of

---

Figure 4.2   Collagen fibril assembly, (a) Schematic of tropocollagens assembling into five-stranded "early fibrils" with an axial D-periodicity of 67 nm, (b) Electron micrographs of a collagen fibril with the characteristic periodic pattern of light and dark bands, (c) Electron micrographs of an *N,N*-bipolar collagen fibril with tapered ends. The higher magnification box indicates the 8D-long region of polarity transition, (a–c, reprinted from Starborg, T. *et al*. Electron microscopy in cell–matrix research. *Methods*, **45**, 53–64 (2008), and Graham, H. K. *et al*. Identification of collagen fibril fusion during vertebrate tendon morphogenesis. The process relies on unipolar fibrils and is regulated by collagen–proteoglycan interaction. *J. Mol. Biol.*, **295**, 891–902, Copyright 2008 and 2000, with permission from Elsevier), (d) Scanning electron and atomic force micrographs (inset) of an individual tapered collagen fibre showing the physical nature of the banding pattern (image courtesy of Andrzej Fertala).

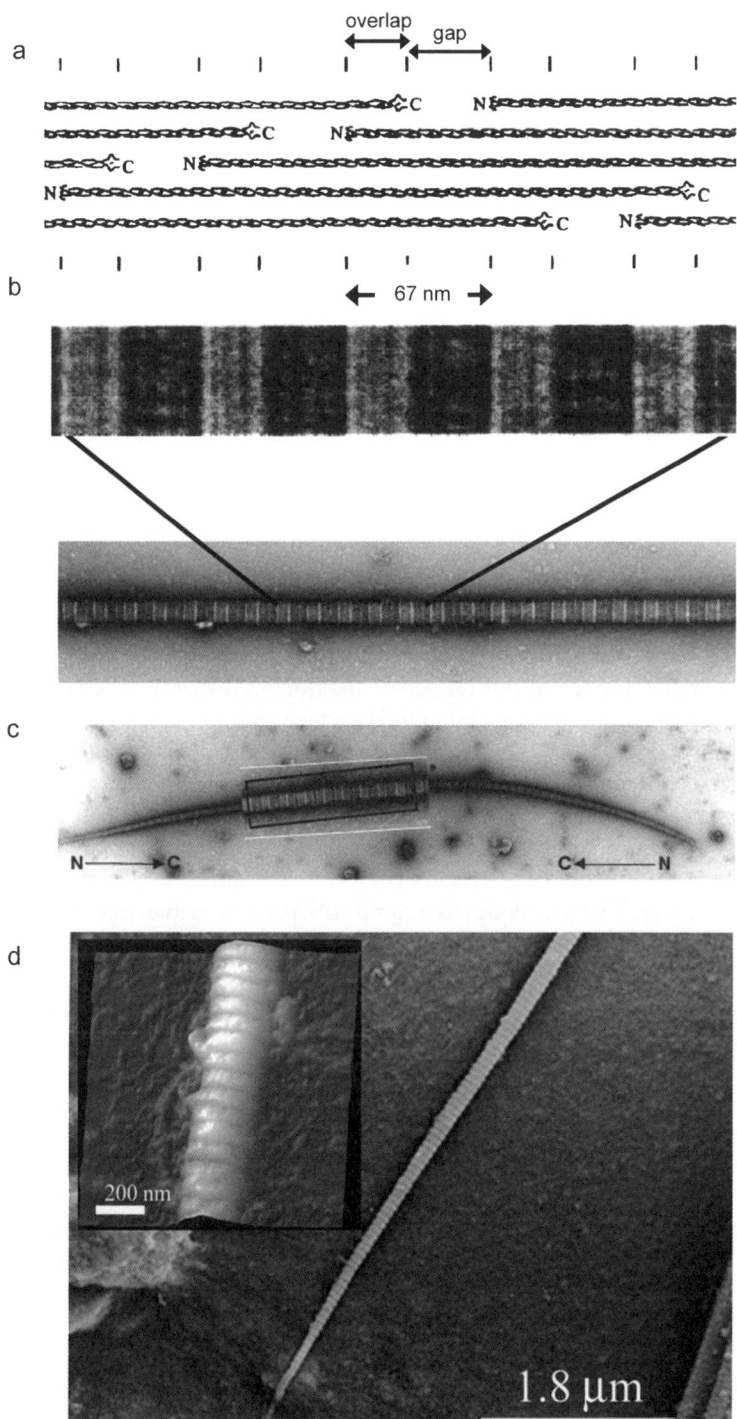

vascular subendothelial matrices is assembled as 3D polygonal lattices by collagen VIII. Other collagens that occur in various tissues are equally indispensably contributive to the innate functioning of the matrix.[28]

As a rule, impaired or failing coassembly of the "minor" collagens with major fibrillar types is a normal cause of tissue disorders. For example, collagen IX and XI gene defects develop functionally anomalous matrices that lead to degenerative diseases such as osteoarthritis and chondrodisplasia syndromes. Basement membranes lacking collagen IV disintegrate with time to result in kidney failures (*e.g.* Alport syndrom(e)). Reduced productions of collagens III and V are dominant factors in the development of intracranial and arterial aneurysms and connective tissues degenerations. The spectrum of diseases caused by collagen mutations is in fact very broad, which strongly advocates the importance of preserving connatural collagen assembly.[34]

4.1.1.3 Tapering Polar

As for many other scleroproteins (proteins that form fibrillar or filamentous assemblies), the assembly of collagen fibres is entropy driven and is forced by the hydrophobic effect. The exclusion of bulk water from the surfaces of individual collagen molecules allows them to multiply in millions into a highly ordered paracrystalline phase.

The cylindrical shape of this phase, *i.e.* the fibre, is a thermodynamically stable and preferable form to provide minimised surface area to volume ratio. A number of protein filamentous structures ranging from fibrin to intermediate filaments have been characterised in biological systems.[35] All these structures are cylindrical paracrystals with nanometre-precise periodicities similar to that observed in collagen fibres.[35] Fibrin fibres, microtubule-associated fibres, cyto- and nucleoskeletal filaments exhibit cross-striated patterns on their surfaces indicative of paracrystalline order,[36] which alludes to general structural principles that drive protein-based fibrillar assemblies regardless of the nature of their building blocks.[37]

These are (1) polar assembly that suggests that protein building blocks in periodic fibres should point in the same direction, and (2) lateral associations or bundling, which implies a bottom-up approach in the assembly. For instance, microtubules and microfilaments arise from a directional assembly of globular proteins tubulin and actin, respectively.[38,39] Analogously, the spontaneous polymerisation of fibrinogen molecules into fibrin fibrils is initiated at one end by thrombin. Intermediate filaments such as vimentins and lamins present rope-like structures of long contiguous α-helices bundled together.[35] Bundling is also thermodynamically favoured as long single helices are not stable in aqueous media. In intermediate filaments, for instance, helices first coil around each other with formation of 2-nm wide rod-like dimers that then associate in a head-to-tail fashion to form filaments of 10 nm in diameter. These principles are universal in fibrillogenesis and equally apply to collagen fibres.

Indeed, the observation of D-periodicity in collagen fibres as well as other continuous nanometre periodicities in similar systems evidently stands for the unipolar mode of assembly. Because proteins are normally depicted as polypeptide chains running in the N-to-C (from amino to carboxyl termini) direction a periodic fibre should also have distinguished N- and C-ends. Consistent with this, collagen fibres were found to be tapered with carboxyl and amino termini closest to the opposite tips, Figures 4.2(c) and (d). Interestingly, fibrils with the same ends were also reported.

Kadler with colleagues established that some of fibrils isolated from chick embryo leg tendons had N-terminal paraboloidal tips.[15,40] These fibres are characterised by a switch in molecular orientation that occurs in a region of axial extent of approximately 8 D-periods, Figure 4.2(c). Furthermore, it was shown that collagen fibres of some invertebrates are exclusively *N,N*-bipolar and symmetrically bipolar. The latter is particularly intriguing as bipolar fibres are of uneven diameters, with their thickest parts located precisely at the switch regions, Figure 4.2(c). The lengths of such spindle-shaped fibrils are limited to a few micrometres. This can be indicative of the fact that the fibrils are seeding or nucleating structures. Indeed, both unipolar and bipolar fibrils were shown to elongate *via* tip fusions into longer fibrils.[41] C-ends appear to be the main requirement in fusion reactions suggesting a mechanism of fibre assembly involving unipolar fibrils as main building blocks and bipolar fibrils as capping elements. Consequently, the process ultimately yields longer bipolar structures with depleting unipolar reactants. Thus, the ratio of unipolar/bipolar fibrils determines the final composition and the size of a matrix. For instance, while bipolar fibrils are not capable of fusing together, homogenous mixtures of unipolar fibrils were shown to associate at their *C*-ends into star-like aggregates, with longer fibrils formed by 4D tip overlaps generating the thinnest and mechanically most fragile regions.[42] Subsequent thickening of the fusion sites that stabilises the structure occurs following an unknown mechanism, but might involve lateral associations with tropocollagens. In contrast, side-to-side fusions along fibrillar surfaces do not occur and were revealed to be inhibited by small proteoglycans that are conjugated to the fibrillar surfaces.[40] Furthermore, as collagen fibres mature they branch and form networks. One mechanism proposed for branching involves tip-to-shaft fusions of collagen fibres that result from the interactions of C-tips with the molecular switch regions.[40]

Branching is also a universal event in protein fibrillogenesis, evidence of which is abundantly found in various scleroproteins.[27,35,43–45] Fibre branching renders ECM three-dimensional and defines its viscoelastic properties. Individual insoluble fibres when branched gel and thus become soluble and compatible with local cell environments they are designed to support. Therefore, understanding the mechanisms of branching is extremely important for designing biomimetic matrices.

4.1.1.4   Branching and Stretching

In this regard, the main principles of mechanical properties and assembly of fibrin matrices or clots drawn from empirical observations by Weisel and coworkers present a vivid picture of immediate relevance to the problem.[44–46] Fibrin fibres assemble from twisted fibrinogen-converted protofibrils that

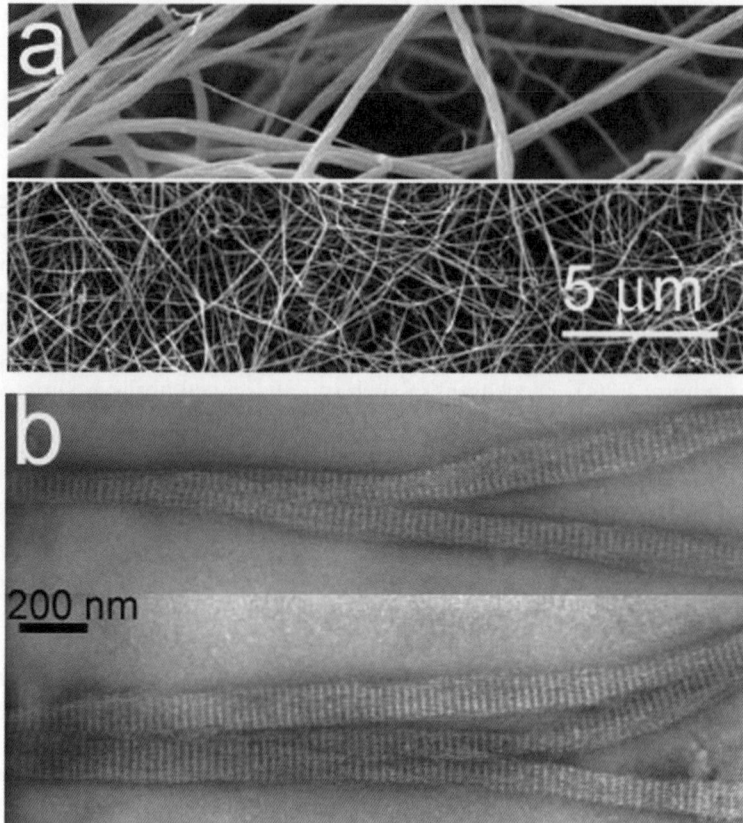

Figure 4.3   Fibrin; electron micrographs of (a) fibrin clots of thick fibres with few
branching points (top) and thin fibres with many (bottom), (b) of branched fibres with the characteristic band pattern of 22.5 nm, (reprinted with
permission from Weisel, J. W. The mechanical properties of fibrin for
basic scientists and clinicians. *Biophys. Chem.*, **112**, 267–276, Copyright
2004, with permission from Elsevier).

laterally associate by wrapping around each other.[47] Protofibrils are periodic
half-staggered structures.[48] A periodicity of 22.5 nm is precisely preserved in
matured fibres, suggesting surface tension as a limiting factor for the aggregation of protofibrils,[47] Figure 4.3. As a result, most of the fibres forming native
clots are straight, displaying no or little curvature, Figure 4.3(a). However,
depending on the conditions of fibrin assembly curved and branched fibres can
be observed, Figure 4.3(b).

In such fibres most branch or junction points were established to be built up
from three fibres and can be separated at various distances depending on
polymerisation conditions of fibrinogen. Interestingly, the band patterns of
diverging fibers were found to be perfectly aligned, Figure 4.3(b). Although
it is difficult to judge about geometric parameters of branch points based

on two-dimensional electron micrographs it typically appeared that two diverging fibres are joint under a small angle at almost all branching points. This may stand for splitting of a growing fibre into two others at an early polymerisation stage or/and the addition of other growing protofibrils. Subsequent studies confirmed the proposed model of branching, according to which the lateral association of preformed protofibrils is randomly intervened by divergence or splitting generating branching.[49] Noteworthy, the bundling and branching of protofibrils appear to be antagonistic events in fibrin and the preference for one or the other is dictated by conditions of fibrin assembly. For example, high concentrations of thrombin – initiator of fibrinogen polymerisation into fibrin – favour branching that gives dense networks, whereas low concentrations produce much thicker fibres with very few branch points.[45]

While crucial in the assembly of collagen and fibrin matrices, branching is not the only option available for network formations. Other polymerising proteins cocomposing the ECM of certain tissues also form intricate networks.

The elastic properties of connective tissues of lung, aorta and skin are mainly attributed to ECM proteins elastin and fibrillin.[50] These are the core building components of elastic fibres.[51] Both proteins polymerise independently in a stepwise manner and merge at a final stage resulting in crosslinked networks.

Fibrillin first assembles into microfibrils with untensioned periodicities of 56–60 nm. Untensioned regions that have a bead-like appearance and are separated by stretched "string" regions are able to uncoil and coil back, thus endowing the unique elastic properties of fibrils,[52] Figure 4.4.

Elastin assembles into fibrils as tropoelastin, which has two distinctive and alternating domains – insoluble hydrophobic and crosslinking. Concentrated by their hydrophobic domains tropoelastin molecules are aligned on the preformed microfibrills that serve as templates or scaffolds for directing the deposition of tropoelastins and their subsequent crosslinking.[52–54] The final

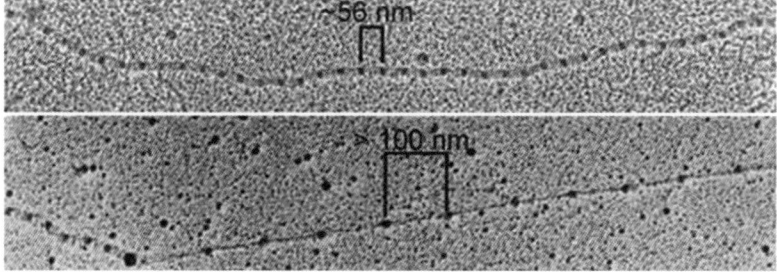

Figure 4.4 Elastin; electron micrographs of fibrillin-rich beaded structures in closed (top) and stretched (bottom) forms with characteristic periodicities, (adapted with kind permission from Springer Science: Kielty, C. *et al. J. Mus. Res. Cell Motil.*, **23**, 581–596 (2002)).

product of this concomitant assembly is rubber-like matrices capable of responding to an externally applied force by stretching and retracting.

4.1.2 Replicating Apparent

Whether it is fibrillar branching or covalent crosslinking, the physicochemical properties of resulting materials are alike and are largely attributed to gelation. Virtually all variants of ECM by their biomechanical parameters are hydrogels yet at the ultrastructural level can differ to significant degrees. This specifies approaches towards the design of gelation for biomedical applications (*e.g.* tissue engineering). For instance, collagenous and low-branched fibrin matrices can be mixed with cells upon gelation. Dry protein contents in hydrated collagen and fibrin fibres are astonishingly low, less than 1%, with the other 99% reserved for water. This creates large spaces within fibrillar networks, *i.e.* between fibres, without compromising the viscoelastic properties of matrices. Although narrow-pored elastic matrices or highly branched fibrin networks gel, their pore sizes are much smaller than the diameter of the smallest cell. Therefore, these may provide better performance as gel precursors or when mixed with cells prior to crosslinking or branching.

Elucidating the rationale behind gelation alongside the understanding of the molecular organisation of cell-supporting matrices should form the cornerstone of attempts aiming at designing ECM substituents. Mimicking these properties is guiding in developing instructive regeneration. However, reproducing the whole set of native properties in one separate artificial system presents almost a formidable task. Hence, it does not come as a surprise that several approaches are being accessed if not to fill the gap then to approach the problem from different angles.

One of the most straightforward paths to be taken is a partial reconstitution of the ECM. However, this can be regarded as a design approach only if put into the strict brackets of morphogenetic adaptation of the ECM for advancing tissue-engineering methodology.

In most related cases ECM analogues derived from biological scraps of the native ECM.

4.1.2.1 Scraping Refusal

Collagen-rich tissues such as placental, dermis, pericardial, vascular, bone and connective have been major sources of collagen. Collagen isolated and purified using various methods including acid extraction, high-speed homogenisation at high salts or thermal and vacuum dehydration can be reconstituted and resuspended to form gels. The obtained materials are of noncrosslinked collagens. The mesh sizes of the formed networks are on the cellular length scales, but somewhat larger than in the native state, which proved to be notably useful in studying cell motility and tissue mechanics. For example, native dermal bovine collagen helped to reveal that cell migration and further metastasis can proceed through mechanisms independent of matrix metallo-proteinases (MMPs).[55] Long thought to involve proteolytic interactions with the ECM in an elongated morphology, metastasising cells were shown to be equally capable of escaping their location in a nonproteolytic fashion.[55] In this study drugs that inhibit MMPs and so designed to prevent malignant cells from their detachment from surrounding ECM failed to stop the migration of cells when used in tissue cultures.[56] The blockage of MMPs was very efficient, but did stimulate cells to follow another route involving acquiring an amoeba-like morphology, which allowed the cells to find gaps in the matrix. Migrating through the matrix cells would leave self-closing tunnel tracks without disrupting the fibrous networks. The time needed for such traction zones to close varied depending on the degree to which collagen gels are constrained at their edges.[56]

These findings can be used as an indicator of the mechanical strength of the gels in response to cell-induced contractions. The role of tissue contraction is very important in regenerative and healing processes. An inability of a given matrix to defy contraction forces may lead to pathological contractures.[57] Therefore, to be used for *in situ* tissue engineering the materials must have substantially higher levels of mechanical strength.

In part to address this, adjustments can be made through chemical (glyco-sylation), enzymatic (transglutaminase), and physical (elevated temperatures) crosslinking or biomineralisation that can yield gels that are more stress resistant.[58–60]

In this regard, the preparation of polyanionic elastin–collagen hybrids *via* hydrolysis of raw collagen extracts is perhaps the most efficient route to gels with enhanced elastic as well as piezoelectric and dielectric properties.[61] The latter can find use in engineering coatings for cardiovascular prostheses[62] or in mimicking and monitoring piezoelectric effects intrinsic for bone remodelling.[63] Indeed, having polar uniaxial orientation of molecular dipoles fibrillar collagen can be used for the development of bio-electret materials.[64,65] As discussed above, collagen fibres exhibit polar assembly, with longitudinally aligned polymeric chains. Such a directionally preferential assembly leads to the paracrystalline structuring of polymeric chains and creates a polarisation-responsive architecture. As a result, pyroelectricity and piezoelectricity in the axial direction is observed and can be further induced.

For instance, following externally enforced polarisation collagen fibres can be used in the construction of biomedically important polymeric materials such as antithrombogenic surfaces and coatings, artificial membranes or mineralised films. Thus, by employing the collagen polarity it is possible to generate magnetically oriented gels or to achieve the preferentially nonmechanical alignment of fibrillar matrices on nonbiological surfaces.[66] Collagen materials of enhanced alignment are beneficial for and are in particular demand in applications requiring directed tissue growth, *i.e.* neurite and vascular out-growth.[67,68] A number of drawbacks for collagen extracts have been reported, mainly related to immunogenicity for those from animal sources and to disease transfection for those derived from human tissues. This situation is, however, rapidly changing with the emergence of recombinant and chemically synthesised collagens that are free of such risks.[69,70]

Fibrin and elastin can also be made available from biological sources. Traditionally, fibrin can be obtained by cryoprecipitation of blood plasma, whereas elastin can be reconstituted from human or animal tissues by acidic or hot alkali extractions.[71,72] Materials based on both proteins have been equally successful in tissue-regeneration approaches. Owing to the remarkable adhesive properties autologous fibrin has been used in the production of tissue adhesives and glues for a diverse range of applications in cosmetic and plastic surgery. Sutureless fixation of skin grafts and replacements, knee arthroplasty, vascular, intestinal and colonic anastomosis and seroma prevention[73,74] are amongst the most popular uses of fibrin, so are fibrin sealants in craniomaxillofacial and orthopaedic surgeries as carriers for slow release of demineralised bone, bone morphogenetic proteins and growth factors.[75,76] The rubber-like properties elastin endows materials with are of particular use for the regeneration of tissues that are critically dependent on resilience and elasticity such as arteries, lung, ligament and skin. The deposition of elastin sheets on crosslinked hyaluronan gels advances the construction of functional composites that can mimic complex multilayered structures of aortic valves, the elastin-rich ventricularis or viscoelastic spongiosa.[77]

From the diversity of tissue-engineering approaches now available in both clinical and laboratory settings it is clear that elastin, fibrin and collagen have matured to traditional materials with the range of properties that are of universal demand across all tissue types. Although all three predominantly contribute to the tensile strength of the ECM constituting its mechanical and structural nature, their sustained responsiveness to cell support and tissue growth also implies a contribution to the compressive qualities of the matrix.

Most tissues, however, allocate the mission to other types of biopolymers. One example is a mucopolysaccharide hyaluronic acid, a major element of connective, epithelial and neural tissues and synovial fluid, which provides viscous, elastic and lubricant properties. Mimicking these two dominant functions has been a focus of research of several groups. The key driver behind the intention is to produce elastic gel-alike materials without concentrating on ultrastructural features of the native fibrillar matrices that remain difficult to reproduce.

This becomes unambiguously topical in the context of three-dimensional cell cultivation in which intimate contact of cells with a supporting matrix is a prerequisite. A major hurdle for the synthesis or prefabrication of artificial fibrous matrices here is the mesh size of fibrillar networks that is generally much smaller than the diameters of most cells. Materials with the capacities may not be readily accommodatable to cellular distribution and growth.

As discussed above, the limitation can be addressed with the help of large-pore gels reconstituted from raw collagen-rich extracts, but such gels have poor mechanical resilience and are associated with drawbacks not characteristic of synthetic materials. An alternative to both is the use of related gel materials from nonanimal sources.

Alginates derived from seaweeds have been thoroughly explored in this regard.[78,79]

4.1.2.2 Tempting Compatible

Alginates are polysaccharides belonging to the vast class of natural gums – materials that are capable of generating substantial increases of viscosity in a concentration-disproportional manner.[17] Alginates make up the bulk of algae, with brown seaweed being a major source of the material. These are random anionic composites of β-D-mannuronic (M) acid and its epimer, α-L-guluronic (G) acid that are linked by 1-4 glycosidic linkage into linear copolymeric chains, Figure 4.5(a).

M and G blocks are coupled in varying ratios but typically three types of monomeric blocks, namely MM, GG and GM, dominate the structure of alginates.[80] The distribution of the blocks over the lengths of the polymers is the major determinant of physicochemical properties of alginates. According to the limiting viscosity number of dilute aqueous solutions of alginates, the

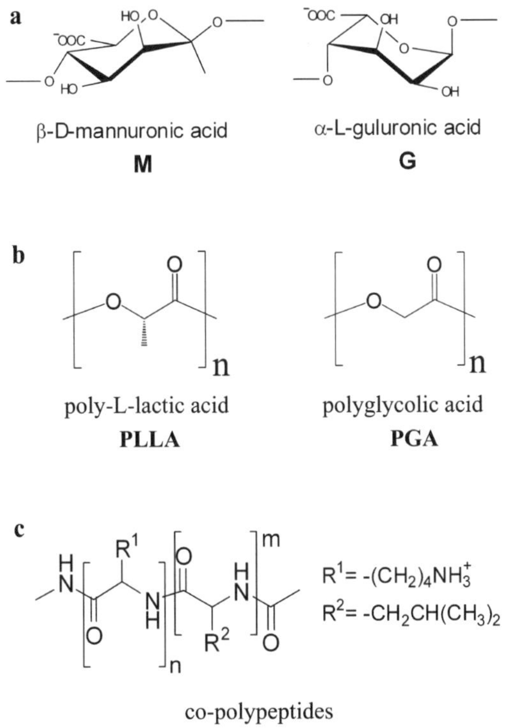

Figure 4.5 Chemical structures of hydrogel-forming polymers.

average square end-to-end distance per uronate residue is approximately two times larger for G than that for M. Generally, it implies that G endows alginates with greater stiffness as compared to M. This in turn suggests that the properties of alginates can be rationally controlled. Indeed, it has been found that MM and GG blocks are conformationally different and can be disorderly linear and helically constrained, respectively.[81,82] PolyMM prefer linear or 3-fold left-handed helices characterised by single-type intramolecular hydrogen bonding between the hydroxyl in 3 position and the subsequent ring oxygen, whereas polyGG fold into stiffer 2-fold screw-type helices stabilised by intramolecular hydrogen bonds between and within GG blocks involving carboxylic function and 2 and 3 hydroxyl groups of the prior and subsequent G residues, respectively. Consistent with this, GG blocks rigidify the polymer and G-rich alginates present denser materials than MM-rich ones that tend to be structurally and physically amorphous. Furthermore, alternating MG blocks makes the molecule more flexible to yield less dense gels with superior elasticity. PolyMG adopt conformations with hydrogen bonds between the carboxyl group of M and 2 and 3 hydroxyls of the following G. The character of the binding gives, however, more conformational freedom to either residue, resulting in higher flexibility. The overall flexibility in alginate polymers increases in the order GG < MM < MG. As a result, the conversion of MM or GM to MG improves the elasticity of the polymer.[83] The reason for this is that otherwise more elastic M-rich segments do not contribute to crosslinking – the indispensable requirement for gelation. Regardless of their composition and length alginate polymers gel only upon contact with divalent cations, *e.g.* Ca^{2+}, Ba^{2+} or Sr^{2+}. Aqueous solutions of alginates even at < 2wt% are instantly forming gels once treated with divalent cations.[83] Stronger gels are those that reach the equilibrium of intrinsic stiffness and elasticity allowing additional increases in flexibility, which directly depends on the increased number of intermolecular crosslinks of divalent cations. Due to the exhibited conformational preferences GG blocks bind divalent cations with the carboxylic groups by forming confined egg-box-like clusters between adjacent polymer chains. This leads to gel crosslinked networks. By preferring a less-gelling ribbon conformation MM blocks use carboxylic groups on sequential residues to chelate multivalent cations intramolecularly, which gives rise to more amorphous liquid materials.[84]

This tuneable behaviour of alginates can be utilised to suit different applications. The possibility for controlling the phenomenon is supported by natural process of alginate synthesis. The ratio of guluronate that is produced from mannuronate can be programmed during the alginate biosynthesis. In this way, the strength and elasticity of the required gel are made responsive to different parameters such as seaweed species, sea conditions and age through the regulation of the molecular weight, content and distribution of G/M blocks in the alginate. For example, rapidly gelating products demanded by some species growing in raw seas use high alginate concentrations that are most efficiently achieved with a high GG content of low molecular weight polymers.

Similar reasoning supported empirically and applied to practical needs in regenerative medicine has led to a range of alginate-based materials with specific functions.[85-87] Alginate gels prove to be not only perfectly biocompatible being easily generated at physiological conditions but also as exhibiting unique physicomechanical properties promising better control over key material parameters such as drug release, cell binding, biodegradation and erosion. The studies on relationship between the structure of alginate gels and these and other parameters have been widely carried out in the context of biomedical compatibility.[77,87-89] Understandably, the closest and preliminary focus on alginate gels have been with regard to their mechanical properties as being predisposed to and independent on the incorporation of cells or *in situ* tissue growth.[88] This is of a direct and major concern with achieving the desirable form of gels as materials for regenerative medicine – tissue implants.[90,91] In order to be used in tissue implantation gels have to be both sufficiently durable in maintaining physical integrity and responsive in supporting assigned functions over a given time, *i.e.* the time necessary for the formation of a new tissue.[88,92,93] A key factor to meet these requirements lies in the controlled degradation of alginate gels. Due to its polysaccharide nature and nonanimal origin the dissolution of alginate proceeds slowly *in vivo*. The degradation is also uncontrollable as opposed to that in the native ECM whose proteins contain specific signal sequences recognised and targeted by matrix-metallo-proteinases thus allowing matrix remodelling.[93] The complete lack of such features is the weakest point of alginates and a major hurdle for their systematic use in regenerative medicine. Therefore, devising methods to control or at least monitor alginate gel degradation solicits for a main strategy in the development of alginate-based implants.

Several groups have undertaken substantial efforts to solve this problem, with Mooney and coworkers leading the quest.[93]

In a series of studies the team investigated partial oxidation as means of promoting the hydrolysis of alginates in water. Periodate-mediated cleavage of the carbon–carbon bonds of the *cis*-diol group in uronate residues allowed degradation of the polymer within nine days.[94] The degradation rate was found to depend on the pH and temperature of the alginate preparations. Alginate oxidised to a low extent ($< 5\%$) retained its ability to bind calcium ions and to crosslink forming gels. Further studies revealed that degradation rates can be tuned by the combination of oxidation and an appropriate ratio of high-to-low molecular weight alginates.[95] Besides, adjusting the molecular weight distribution of partially hydrolysed but gel-competent alginates can be used to retain elasticity. Increased fractions of low molecular weights maintained similar rates of elasticity compared to that for the high molecular-weight gels. Interestingly, this gave more rapidly degrading hydrogels that were more receptive to transplanted bone-marrow cells in the formation of new bone tissues.[96] The oxidation did not compromise biocompatibility either as judged by the adhesion of myoblasts, with proliferation and differentiation rates being comparable to those exhibited by cells cultured on untreated gels.

Alternative approaches involving other modifications can also be employed to control the degradation of alginates. It was shown, for example, that interactions between cells and cell-adhesion motifs have an enhancing effect on the mechanical properties of alginate gels. The assessment of alginates modified with the motifs and mixed with myoblasts using compression and tensile tests suggests that additional mechanical integrity acquired by hydrogels through stronger interactions between myoblasts and polymer-anchored peptides should occur above critical peptide and cell densities.[97]

Other examples may include the generation of surface-reinforced alginates by "secondary" polymers such as polycations (polyethyleneimine),[98] the use of slow-gelling ionic crosslinkages in producing structurally uniform gels,[99] the rapid crosslinking of oxidised alginates and gelatins triggered by tetraborates[100] or polyion complexation of alginate and chitosan.[101]

The number and variety of crosslinking methods is potentially significant. However, the ultimate limitation for all is alginate itself. In this regard, many argued that purely synthetic polymers possessing properties analogous to those displayed by alginate are more technologically sound and can offer a straightforward access to biodegradable gels.[7,93,102]

4.1.2.3 Likening Synthetic

Considerable attention is paid to synthetic polymers as biomimetic scaffolds. Synthetic scaffolds can be produced and reproduced practically on any scale and according to strict specifications.[103] Their mechanical, physicochemical and degradation properties are relatively easy to control and tailor. Polymeric matrices are free of immunogenic response and infection transmission notoriously characteristic for biological extracts and can be made adjustably biodegradable to erode in the body.[7,14] In principle, any nontoxic water-soluble polymer can be used to produce cell-supporting gels.[13,20,93] Synthetic polymers currently used as regenerative templates range from polydioxanones to ceramics. Best studied and medically approved are polyesters, polyglycolic (PGA), poly L-lactic (PLLA) acids and their copolymers (PLGA), Figure 4.5(b).

These materials are being extensively used in clinical practice as well as in tissue-engineering applications.[20,104] One of their most attractive features is that *in vivo* they degrade to natural metabolites, eventually converting to carbon dioxide and water. The degradation of the polymers undergoes via nonenzymatic fashion as the ester bonds are intrinsically susceptible to hydrolysis, the rate of which can be predefined by specifying the ratio of the monomer types.

The polymers are also rigid thermoplastics with high crystallinity ($\sim 50\%$) and can be manufactured into different shapes. However, their high sensitivity to hydrolytic degradation requires careful control of processing conditions and implies the prefabrication of implants. As a consequence, the types of materials produced from the polymers are porous solids and foams, but never gels.

Critically, materials able to give stable gels in water are popular across nearly all biomedical branches, which is mainly attributed to hydrogels possessing properties necessarily desirable for implants and not typical of other types of materials.[93]

Hydrogels are relatively easy to construct as biocompatible and biodegradable. Their degradation modes, including finding reliable ways for the elimination of degradation by-products, can be rationally assigned through the selection of safe building precursors. Physically, once formed they exhibit a low interfacial free energy with body fluids, which prevents unspecific adhesion and adsorption of cells, bacteria and biopolymers. Hydrogels are elastic materials, which makes them capable of minimising tension and irritation to surrounding tissue. Elasticity as the key property of hydrogels remains.

Elastic hydrogels can be generated from water-soluble polymers with lower or upper critical solution temperatures. Block-copolymers of polyethylene and polypropylene glycols are common examples. Analogously to alginates, gelation of such polymers is a result of a liquid–gel phase transition. While alginates require divalent cations to form gels, polymeric glycols undergo thermo-reversible gelation.[105] In elastic gels at physiological temperatures the polymers are also prone to diffusion into the liquid phase. This poses significant stability issues of potential implants based on synthetic gels due to their almost instant and complete dissolution in fluidically open and dynamic body environments. Crosslinking strategies that are successfully applied to stabilising biological extracts of natural elastomers are also efficient in the case of the polymers in shifting the equilibrium towards gelation.[106] Toxicological requirements for crosslinked synthetic polymers are somehow stricter because of their artificial organic origin. This is particularly true for approaches requiring mixing cells in the precursor components of the gel.[107] Apart from pH and temperature ranges that are usually limited to physiological ones, use of organic media and toxicity of molecular groups generated as a result of crosslinking are amongst the main side effects to be avoided. This primarily defines the choice of polymeric candidates together with the methods of their polymerisation and stabilisation. Methods used to identify safe and efficient crosslinking sum up into a mainstream trend in gel design. Photopolymerisation,[108] transglutaminase-catalysed condensation[109] and biointeractive Michael-type addition reactions[110] give best results with preserving cell viability. These and other related strategies have advanced live vaccine delivery,[111] cell encapsulation,[112] cartilage and bone repair,[110,113,114] wound healing[115] and the like.

Unlike matrices reconstituted from biological extracts, synthetic polymers can be assembled in tissue culture from scratch (from low molecular weight precursors) and thus not only can encapsulate cells upon gelation but can also be made environmentally responsive in the context of matrix development.[116] Crosslinking in this regard may be used as an instrument in mimicking an inherent and very important characteristic of the native ECM – remodelling. This mechanism of matrix degradation and resynthesis is a complex process that has been largely ignored in most matrix-mimicking designs. The complexity of biological interactions between cells and the ECM involves matrix restructuring that enables cell invasion and migration and accommodates subsequent cell proliferation. This altogether assumes the existence of efficient

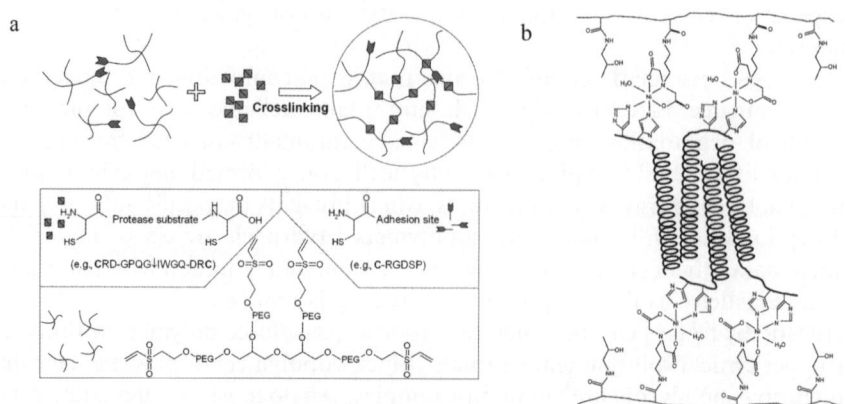

Figure 4.6   "Smart" hydrogels; (a) Hybrid matrices capable of proteolytic remodelling, (reprinted by permission from Macmillan Publishers Ltd: (Lutolf,
M. P. *et al.* Repair of bone defects using synthetic mimetics of collagenous
extracellular matrices. *Nature Biotech.*, **21**, 513–518, Copyright (2003)), (b)
Reversible gelation of self-assembling artificial proteins, (120) (c) Assembly of protein–polymer hydrogels mediated by metal complexation, (reprinted by permission from Macmillan Publishers Ltd: Wang, C. *et al.*
Hybrid hydrogels assembled from synthetic polymers and coiled-coil
protein domains. *Nature*, **397**, 417–420, Copyright (1999)).

processes that can allow such matrix turnover in a controllable and molecularly
reproducible way. The native ECM is susceptible to cell-triggered proteolysis;
that is, partial and targeted degradation of ECM components mediated by
metalloproteinases (MMP) and controlled by cells themselves. This is what
makes the ECM a dynamic and "alive" system.

Hubbell with colleagues aimed to address this in engineered synthetic
hydrogels containing a combination of crosslinking MMP substrates and
pendant adhesion sites,[110,116] Figure 4.6(a).

The substrates assembled into branched polymeric networks interacting with
cells via cell-adhesion RGD-based peptide motifs were shown to follow cell-
mediated proteolytic degradation in further remodelling into cell-secreted bone
matrices at injury sites.[110] The mimetic was proposed as a model biomaterial
for clinical *in situ* applications.

The synthetic scheme developed included addition reactions between conjugated unsaturated ends of polyethylene glycol chains and thiol-containing
peptides in yielding bioactive PEG networks upon mixing the components
under physiological conditions. Swelling measurements of the obtained material revealed that the network mesh sizes, taken as the distance between consecutive crosslinks, were considerably below (<50 nm) the cell sizes and
dimensions of cellular processes. This was used in assessing the sensitivity of the
gels to cell-secreted MMP as cell migration in this case is strictly limited to
proteolytic degradation. *In vitro* fibroblast-based cell-invasion models were

applied. Fibroblast 3D outgrowth was hindered for gels lacking MMP substrates or when MMP inhibitors were used. In contrast, radial migration of fibroblast clusters pre-entrapped in fibrin matrix was observed into the surrounding PEG gels at a good invasion rate. The migration was found to be also dependent of the inclusion of cell-adhesion motifs (RGD peptide) into gel networks, with no invasion occurring within networks lacking the RGD ligand or containing an adhesion-inactive RDG sequence. The ability of a given matrix to remodel following its partial degradation on the molecular level is mirrored by that in recovering its physicomechanical strength.

This thinking led to the emergence of the so-called "smart" hydrogels.

As opposed to "smart" matrices, as may be exemplified by the above design, these display self-control in maintaining their physical qualities under external rather than cell-induced stimuli such as pH, temperature or electric field.[117] This necessity of introducing the notion of recovering gels originates from a number of factors that continue putting restriction brackets on broader uses of hydrogels. For example, one of the main limitations to overcome consists in the relatively long time needed for hydrogels to swell or to turn "on" after being exposed to an external stimulus and "off" again once the stimulus is reversed.[118]

The importance of tackling this lies in all characteristic properties of a hydrogel – elasticity, degradation, stability and mechanical strength are all affected and can be corrected by how quickly the hydrogel can recover.[118] It is accepted that such limitations are the direct consequences of the pathways taken to synthesise or produce synthetic hydrogels. Traditional polymerisation methods often do not leave room for specifying sequence, length, composition and most of all three-dimensional preferences of resulting materials. Polydispersity and structural heterogeneity of hydrogels are logical outcomes. Concentrating on how to avoid or circumvent these problems has led to new routes in engineering hydrogels.[119]

4.1.2.4 Recovering Intelligent

The main feature here is to merge the predictability of molecular recognition patterns with highly controlled synthetic methods capable of providing monodisperse products. Several groups proposed different models of practically the same approach – self-assembly of block-copolymers. Early examples of attempting the rationale include the construction of stimuli-sensitive and reversible polymer hydrogels.

Petka *et al.*[120] described how protein engineering can be applied to producing hydrogels based on triblock-copolymers of synthetic and biological polymers, Figure 4.6(b). The polymers contained self-assembling coiled-coil protein sequences incorporated to provide interchain polymer interactions to permit crisscross points in polymolecular networks. The protein domains once assembled were allowed to swell; that is, to remain in solution, owing to the linked polyelectrolyte domains of PEG–peptide conjugates. The peptide portion was designed as unstructured solely to enforce solubility provided by PEG and reversibility that was additionally strengthened by oxidising cysteine residues. Such a combination allowed for the system to be capable of reversible swelling rather than precipitation in water and facilitated in the tuneable gel–solution transition of the hydrogels. Further studies proved this to be particularly critical in controlling the erosion rate of the hydrogels. It was established that hydrogels more resistant to surface dissolution can be produced by tuning the strength of protein–protein interactions and by changing the overall network topology.[121]

A similar design of reversible hydrogels were proposed by Wang *et al.*,[122] Figure 4.6(c). These were constructed as branched noncovalent copolymers of methacrylamides as primary chains and His-tagged coiled-coil domains as pendant groups. The two components coassemble by crosslinking upon complexation with Ni^{2+} ions to give reversible hydrogels. Hydrogels, for example, could be assembled by mixing chelating methacrylamide copolymers charged with Ni^{2+} and protein domains and then dried on Teflon sheets to give homogeneous films. The films when rehydrated were made to remain physically intact from 2 to 24 h before disintegrating due to excess swelling. Several distinct features were proposed with such hydrogels: metal complexation of recombinant proteins containing affinity tags, the use of divalent ions for protein-mediated stabilisation of hydrogels, and *in vivo* monitoring of the hydrogels by magnetic resonance imaging.

The concept of "smart" hydrogels essentially introduced by the two designs was elaborated further in engineering systems with rapid on-off responses. For instance, Nowak *et al.*[123] reported diblock co-polypeptide amphiphiles composed of poly (L-lysine) or poly(L-glutamic acid) and poly(L-leucine), poly(L-valine) or poly(D/L-leucine) as the hydrophilic and hydrophobic components respectively, Figure 4.5(c). The chosen model is interesting in that co-polypeptides of identical compositions, albeit of statistically random sequences, were never found to yield hydrogels. In contrast, synthesised as narrowly distributed low-molecular-weight polymers the reported amphiphiles associated into hydrogels at very low polymer concentrations. Furthermore, the hydrogels maintain their mechanical strength at high temperatures and were able to rapidly recover after applied stress. The latter correlated with structural parameters of the polymeric chains and in particular with their ability to fold under gelation conditions. Given that hydrophobic domains can adopt regular conformations insoluble in water – rod-like α-helices for poly(L-leucine) and crystalline β-sheets for poly(L-valine)[124] – the authors capitalised on the differences to reveal that helical segments have a better tendency to gel then β-sheets or random coils. Analogously to protein folding, the found correlation indicated that the gelation process is conformation specific. Using another analogy that concerns the dependence of gelation on the porosity of macromolecular meshes, namely the more porous the mesh the faster it gels, the observed abnormally fast response can be attributed to increased porosity as the amphiphiles were found to concentrate exclusively within a gel matrix leaving liquid domains polymer-free. The combination of the two features (rapid gel recovery and sustained large porosity) holds particular promise for those applications that require the bulk load of macromolecules or nanosized elements and their subsequent release to be physiologically triggered and repetitive.

Undoubtedly, biodegradable gels, so attractive for many applications, would benefit further from having controllable lability and "intelligent" responsiveness rationally introduced at the molecular level. Indeed, cleavable bonds can be made to be present either in the polymer backbone, within the gel-stabilising crosslinks or as reversibly folding protein or co-polypeptide domains.[125] Such cementing bonding is to be broken under physiological conditions in a "smart" fashion, which is better and selectively done in a response to a specific stimulus or with the help of enzymes or even, as in some cases, via simple hydrolysis.

In this way, manipulating degradation and recovery kinetics of hydrogels can become possible to be used in monitoring, for example, tissue growth or controlled release and delivery of therapeutic proteins, growth factors or nanoscale composites. Nonetheless, the development of novel hydrogels is intrinsically and inevitably associated with biocompatibility of their degradation products, hence, the stressed requirement for their low toxicity and immunogenicity. The gratifying lead for by-products is safe metabolism or traceless excretion. This, however, is by no means taken as a holding and disadvantageous limitation in pursuing and extending the development of hydrogels as materials for regenerative medicine. The range of unique properties hydrogels offer and the relative ease with which these can be achieved and guaranteed will continue to stimulate new designs whose constantly growing number amply support or rather generate this statement.

One argument that places hydrogel designs into a more critical perspective as ECM substituents is the poor resemblance and inconsistency of their ultrastructure with the fibrillar architecture of the native ECM. As conceived, most nonfibrous hydrogels in this regard only emulate the physical properties of the ECM – gelation, viscoelasticity, mechanical and tensile strength, responsiveness to external stimuli. Principle means by which such bulk physical information is implemented in the ECM as a material are found on the cellular and more primarily on the subcellular levels. This is where concurrence of molecular recognition and protein-folding patterns, biomolecular self-assembly and spatial adaptation of surface features, local cell-ECM biomechanics and intermolecular dynamics are expressed as fundamental and governing cues to be communicated to cells and tissues via and by the hierarchical organisation of the fibrillar ECM. This is the subject of the following section, in which recent and leading directions in the development of fibrillar ECM substituents are discussed.

4.2   Restoring Available

To this end, all highlighted routes taken in the pursuit for best-performing ECM alternatives have been focused on reproducing or achieving its final, testable but not necessarily sustainable physical form – elastic gels. Strictly speaking, provided that all the main requirements concerning biosafety are met, an ideal hydrogel may arise from any chemical archetype, *i.e.* synthetically it is not restricted.

This certainly creates a generous start for novel chemistries and fabrication methods and may give a material of exceptional quality and efficacy. But this might be true only in the short term as the overall success gets less apparent when a need for modifications of once-recommended protocols takes place.

4.2.1 Prompting Longitudinal

The lack of robust prediction between the physical properties and chemical nature of ECM substituents in such approaches, any search for which can go as far down as the subnanoscale, is exactly that missing link that makes attempts to mimic the fibrillar nature of the native ECM inescapable. Unsurprisingly, this vision is being shared by many and is beginning to prevail in the field. Many examples available in the literature bring the idea of an artificial matrix into increasingly more technical rather than purely design terms.[126]

This trend is further influenced by the availability and steady improvement of high-resolution techniques such as electron microscopy and X-ray crystallography that allow better hands-on experience tobe gained with the morphology and architecture of the ECM with precision and reproducibility set on the nanoscale. The possibility of applying the techniques and the degree of their involvement in the characterisation and hence the monitoring of a given development is often taken as recommendation for the choice of suitable candidates.

Electron microscopy, for example, is utilised to analyse the morphological features of both bulk and specific regions of nano-to-meso fibrous materials irrespective of their fabrication paths. Whilst manufacturing methodologies used in the production of mesoscopic fibres can vary, the morphologies and shapes of their final products are better observed and directly assessed.

Concomitantly, X-ray fibre diffraction has been the technique of overwhelming preference in elucidating paracrystalline and near-crystalline characteristics of hierarchical protein aggregates. Furthermore, the latter has been developed to meet specific requirements imposed by the analysis of patterns uniquely characteristic for fibres to which traditional crystallographic methods used for crystals cannot be applied. The detailed structural information fibre diffraction offers has a long story of success and reliability for different macromolecular systems taking an early start from main biopolymer folds such as protein α-helices and β-structures and DNA double helix.

In relation to synthetic matrices, the applicability of both techniques, likewise tested in the structural elucidation of ECM fibres, provides an original support in stimulating novel designs. The analogous design is often the very first step in chasing the idea of an artificial ECM and is traditionally linked to the process of the ECM assembly.

4.2.1.1 Invoking Granted

As previously discussed, the native ECM results from the self-assembly of specific folding elements that are common for all biological matrices. Following this established assembly model would seem to give an obvious and success-granting access to the right ECM analogue.

Yet surprisingly, very few attempts have been made as regards synthetic collagen-based assemblies even given that the current state of chemical synthesis of proline- and glycine-rich sequences does not permit collagen-size building blocks.

Nevertheless, with none taking the idea into the level of applications, still positively promising and very interesting designs independently described by several laboratories may be just the first steps towards fully functional collagen mimetics.

The notion of aligning staggered or sticky ended[127] peptide chains extensively exploited over the last decade by different groups served as a central paradigm in designing fibrillar collagen mimetics.

Koide *et al.*[128] employed the principle into building long triple helices of collagen-like peptides. Because in natural fibrillar collagen,[129] as well as in some earlier described artificial systems,[130,131] lengths of the formed triple helices were largely defined by those of the peptide chains, the authors hypothesised that an alternative approach to generate extended helices could be a poly-molecular elongation of shorter peptides provided strong intermolecular binding. To achieve this, the combination of regular staggering and forced intermolecular interactions was designed into heterotrimeric self-com-plementary peptides. The peptides were synthesised as classical (Gly-Pro-Hyp) tandem repeats that become laterally linked upon oxidation of cysteine residues incorporated into the opposite ends of the chains to ensure the stagger. The linked peptides had 12- or 13-residue complementary collagen overhangs each. These assembled into larger aggregates as revealed by circular dichroism spectroscopy and laser diffraction particle analysis. However, no data made available on morphological and folding characteristics of such an assembly questions the formation of ordered collagen-like fibres.

In another collagen triple-helix design Kotch and Raines[70] used similar assumptions. The researchers rightly argued that unlike DNA the noncovalent association of collagen strands is not mediated by a "code". Therefore, the design had to rely upon an additional capacity that would be able to con-siderably limit the promiscuity of potential interactions. To realise this, cova-lent bonds formed by a pair of disulfide bonds were introduced to tether the strands of collagen fragments and also to offset the strands in setting their register. The design rationale is featured by the formation of an intramolecular triple helix between $(Pro-Pro-Gly)_3$ segments of two identical strands and a core strand promoting other, $(Pro-Hyp-Gly)_5$, segments to overhang and assemble intermolecularly yielding collagen-alike fibrillar structures, Figure 4.7(a).

The resulting assemblies were examined by atomic force microscopy (AFM) to reveal 20–120 nm long fibrils. Further analyses of the AFM data gave fibril

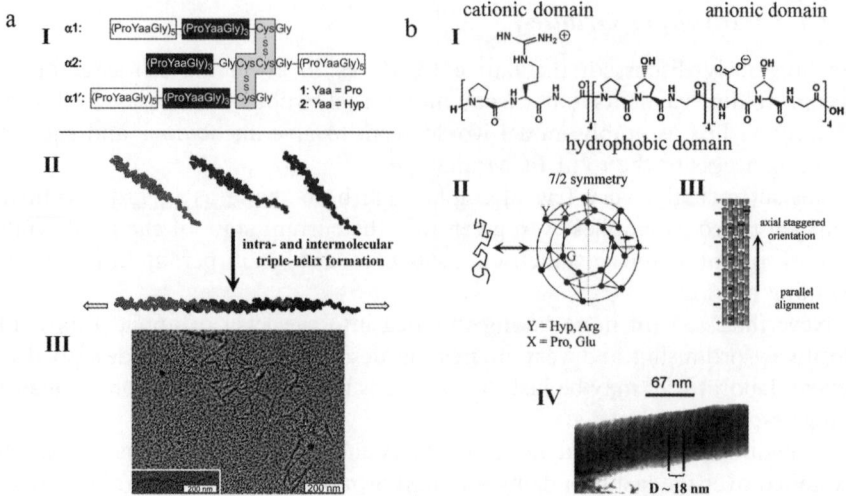

Figure 4.7   Fibrillar collagen mimetics; (a) (I) design of a three-stranded staggered array held by disulfide bonds, (II) schematic of the self-assembly of the array, (III) electron micrographs of assembled fibrils, (reprinted with permission from Kotch, F. W. and Raines, R. T. Self-assembly of synthetic collagen triple helices. *Proc. Natl. Acad. Sci. USA*, **103**, 3028–3033. Copyright (2006) National Academy of Sciences, USA), (b) (I) amino acid sequence of a triple-helical protomer composed of flanking hydrophilic domains, cationic and anionic, and a central hydrophobic domain, (II) helical wheel diagram of the protomer assuming 7/2 superhelical symmetry (only one XYG triad is shown), (III) schematic of axially staggered assembly of the protomer, (IV) electron micrograph of the assembled collagen fibre with the characteristic D banding pattern, (reprinted with permission from Rele, S. *et al.* D-Periodic Collagen-Mimetic Microfibers. *J. Am. Chem. Soc.*, **129**, 14780–14787. Copyright (2007) American Chemical Society).

thicknesses of 0.5–1.0 nm – the diameter range of naturalfibrillar collagen,[129] Figure 4.7(a). TEM complemented the AFM data in providing additional data on fibre morphology. The morphology and diameter of such assemblies were also found to strongly resemble natural collagens.[132,133] However, a broad distribution of fibre lengths from 30 nm to >400 nm was observed. The tendency proved to be common in collagen designs and can be attributed to the impaired stability of designed assemblies compared to native collagen. This is a serious issue as reproducing the native morphology of collagen fibrils underlies the functional mimicry of ECM. The problem can be partly addressed through chemical modifications of polypeptide backbones that in some cases lead to the stabilisation of the triple helix and stronger interhelical interactions.[134–137] However, designing more efficient polypeptide assemblies that at a later stage can be adapted to recombinant technology would be more attractive in this regard.

Leading this path, Conticello and coworkers[138] designed triple-helix protomers that assembled into D-periodic microfibres – structures strikingly characteristic of the native fibrillar collagen, Figure 4.7(b). Protomers were engineered as peptide sequences of three different (Xaa-Yaa-Gly) domains. A sequence of Pro-Hyp-Gly repeats was used as a central or core hydrophobic domain coupled to two oppositely charged sequences of similar sizes. Electrostatic interactions between arginine residues of one charged domain (cationic) and glutamate residues of the other (anionic) ensured the linearly aligned and axially staggered assembly of triple helices, Figure 4.7(b). The main assumption of the design is that such electrostatic interactions that are made complementary are reinforced by hydrogen bonding. The hydrophobic segment in this respect serves to maintain the thermodynamically favoured network of hydrogen bonds.

As is common to this type of systems,[139,140] an external trigger was necessary for the formation of ordered structures. Indeed, at benign conditions and at lower salt morphologically, amorphous and structurally undefined fibres were observed. In contrast, thermally annealed protomers yielded smooth fibrils of hundreds of nanometres in length and tens of nanometres in diameter. As early as the first hours of their assembly the fibrils revealed tactoidal morphology.[141] Longer post-melted incubations gave uniform, several micrometres long and 70-nm thick fibres with a characteristic transverse banding, Figure 4.7(b). The observed axial D-period was found to be approximately 18 nm and to be formed by 63 residues. This is in a remarkably but somehow unexpectedly (for a 10-nm long peptide) good agreement with the 67 nm periodicity of over a thousand residues tropocollagens. In this respect, it was hypothesised that the apparent discrepancy between the peptide length and the found banding distance suggests that the nucleation or functional unit of the assembly may be a lateral oligomer of protomers whose staggered bundling is defined by regions, albeit periodic, but of varied packing and alternating charge density. This observation is very important as it is applicable to the general area of designing self-assembling protein fibres.

4.3 Reposing Modular

The combination of hydrophobic cores or interfacial hydrophobic interactions together with alternating oppositely charged domains or spatial clusters has been viewed as a main driving force in the supramolecular polymerisation of designed peptides into fibres. A growing variety of systems assembled from sequences based on different protein structural elements including β–pleated sheets,[142–145] swapping domains[146,147] and the aforementioned α-helical coiled coils[127,148] has been reported. All such examples obey a number of common structural restrictions that contribute to mainly two key requirements found necessary for fibrillogenesis.

First, to assemble a sequence has to be complementary, whether to itself or to a partner sequence. This property is inherent to protein-folding motifs and can easily be reproduced in any design. Complementarity ensures homo- or heterogenic oligomerisation, but does not guarantee the directional polymolecular extension of an oligomer.

The latter constitutes the second requirement that is met by another key principle of fibrillar assembly – alternation of charges. This has been found critically contributive to achieving monomer propagation in a number of designs. As first employed by Zhang *et al.*[142,144] for the fabrication of β-structured fibrillar membranes and matrices such an ionic application has proved to be universally amendable for many other fibrillar designs.

More specifically, in the original design by Zhang and coworkers self-complementary oligopeptides composed of alternating hydrophobic (H) and polar (P) residues were utilised thus following the classical β-structure pattern – $(HP)_n$. The pattern provides the segregation of the amino acid residues into two surfaces – hydrophobic and polar or charged. The hydrophobic surface was made up of moderately hydrophobic alanines. The polar face consisted of alternating basic arginines and acidic aspartates.[144] Altogether, this generates a peptide repeat, RAD or RADA, Figure 4.8. An alternative version used a reverse of lysines and glutamates correspondingly; that is, EAK. Both peptide types were termed "modulus two" reflecting the arrangement of charged residues on the ionic side of the folded β-sheet – when folded two positively charged residues would be followed by two negatively charged aspartates, and *vice versa*. Other ionic moduli can also be produced. These are classified

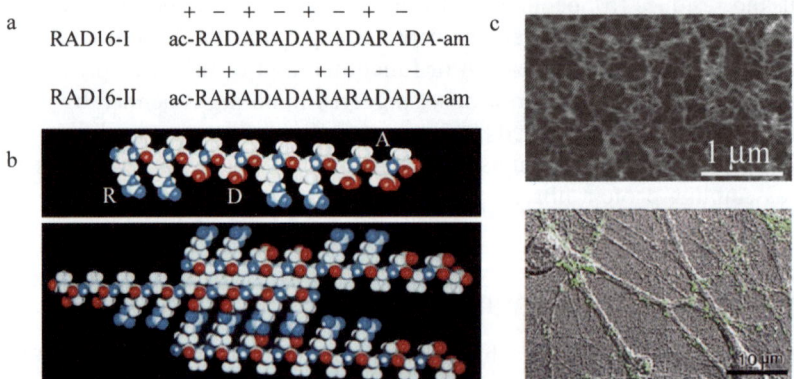

Figure 4.8   Design of self-complementary polypeptide matrices; (a) Modulus one and modulus two RAD sequences, (b) Molecular models of RAD16-II as a β-strand (top) and when assembled in a staggered fashion (bottom). Two distinct faces, polar (arginines and aspartates) and nonpolar (alanines) are distinguished, (c) Microscopy images of the assembled RAD16II matrix (top), and neurons forming active connections (green) on the matrix, (bottom), (reprinted with permission from Holmes, T. C. *et al.* Extensive neurite outgrowth and active synapse formation on self-assembling peptide scaffolds. *Proc. Natl. Acad. Sci. USA*, **97**, 6728–6733. Copyright (2000) National Academy of Sciences, USA).

numerically according to a modulus number that is correlated with the number of alternating basic and acidic amino acid residues. For example, $- - + \ + -$ $- + +$ consistent with the described modulus would be numbered II, whereas I would stand for $- + - + - +$ and $- - - - + + + + \ - - - - + + + +$ would be IV. A number of different modulus versions have been described.[149,150]

The format allows for an assembly model that involves the formation of β–sheets via intermolecular hydrogen bonding, alanine and charged residues shaping hydrophobic interfaces and intermolecular ionic bonds, and the staggered association of individual peptides, Figure 4.8(b). Peptide monomers in this respect are to form staggered β-sheets that act as nucleators in the assembly of extended protofilaments that in turn by undergoing side-by-side aggregation yield macroscopic fibrous membranes and tapes.

The model is consistent with spectroscopic data supporting the strong preference for β-sheet formation, an almost quantitative estimation in the β-content for both designs. Further support was provided by electron microscopy that revealed the strictly hierarchical assembly of folded peptides. Under various conditions the peptides were found to produce distinctive microscopic morphologies including membranous structures, thread- and tape-like assemblies. Protofilaments of 10–20 nm in diameter were observed as constituents of fibrous meshworks with relatively narrow distribution of pores of 50–100 nm, Figure 4.8(c). The macroscopic or bulk dimensions of the materials could be made to be subject to peptide concentrations, ionic strength or pH of buffer solutions, processing protocols or different devices used in the study. This may be unexpected given that the designed peptides were just 16-mers. The sequences may be too short for appreciable folding and are significantly shorter as compared to the components of native matrices. However, once formed, all of the generated materials were extremely stable.[151] This finding strongly supports the model as proposed and strengthens the position of the design from the perspective of natural mimicry.

Indeed, architectures featured by alternately spaced alanines of axially shifted sequences are characteristic of a number of native fibrous proteins. Notably, silk proteins underpinned by such sequential constitutions have fully extended backbones forming β-sheets with staggered alanines. By virtue of having the smallest side chains alanines facilitate in perhaps the densest or best-fitted packing of hydrophobic faces for creating contiguous threads of β-sheets to spontaneously and irreversibly form macroscopic protofillaments.[152]

Very high proportions of small amino acids in fibrous proteins prove to represent a universal answer of sequence arrangements to fibrous packing. In β-structured proteins such as keratins or fibroins as well as in others represented by different folding elements (*e.g.* collagens and elastins) inclusions of glycines or alanines are made periodic and faithfully reproducible over the entire sequence lengths.[152,153] This is believed to considerably relieve steric hindrance in the formation of hydrogen bonds between neighbouring polypeptide chains that can otherwise be imposed by larger side chains thus preventing their lateral association and alignment. Similarly, in collagens and elastins steric hindrance is undesirable due to limited interior space within

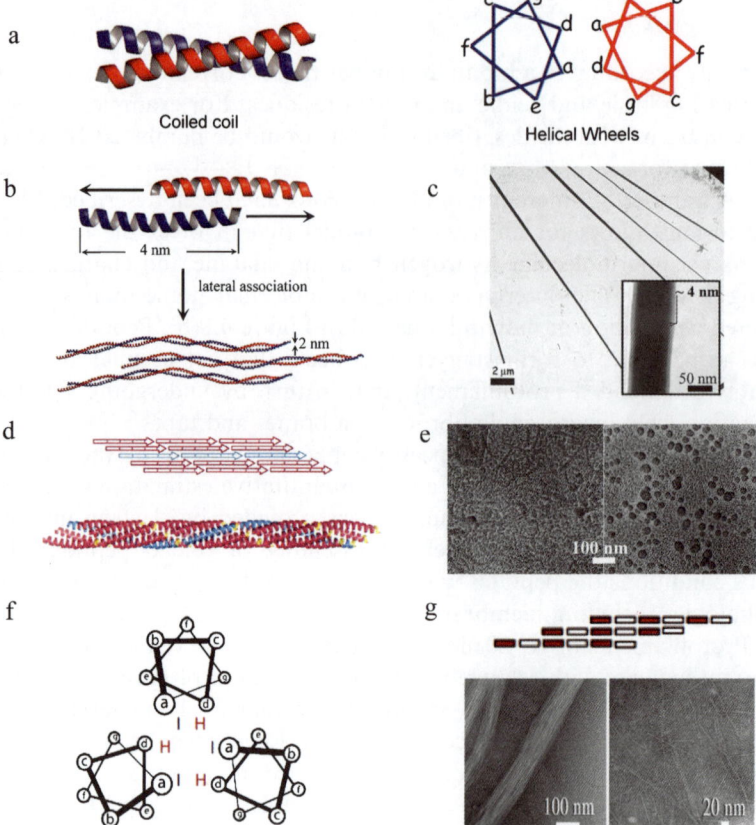

Figure 4.9 Staggered helical assemblies; (a) Two helices are coiled together forming a rod-like dimer (left) (PDB 2ZT(A) Helices configured into regular 3.5 residues per turn helical wheels (right). The wheels reflect the arrangement of canonical coiled coils characterised by a heptad repeat (HPPHPPP)$_n$ labelled *a–g*, with *a* and *d* positions being hydrophobic, (b) Axially shifted assembly of two helices into a sticky-ended dimer. The dimer propagates longitudinally in both axial directions with forming coiled-coil protofilaments that laterally associate into mature fibres, (c) Electron micrographs of resulting self-assembling fibres. The inset features the periodic striated pattern of 4 nm corresponding to the length of an individual peptide, (d) Schematic and model representations of a pentameric coiled coil design. Staggered helices are shown as arrows (top) and ribbons (bottom). One contiguous stand is highlighted in blue. The N-termini are shown in yellow to indicate the abutting ends of aligned helices, (e) Electron micrographs of assembled fibrils (left) and spherical particles (right), ((d) and (e), reprinted from Potekhin, S. A. *et al.* De novo design of fibrils made of short alpha-helical coiled coil peptides. *Chem. Biol.*, **8**, 1025–1032, Copyright 2001, with permission from Elsevier), (f) Helical wheel diagram of a trimeric helical bundle with core residues highlighted in red (histidines) and blue (isoleucines), (g) Schematic of staggered packing of helices resulted from an axial displacement of two heptads between adjacent helices (top), with heptads shown as white (containing His) and red (without His) bricks. Electron micrographs of assembled fibrils, bundled (left) and individual of ∼3 nm in diameter (bottom), ((f) and (g), reprinted with permission from Zimenkov, Y. *et al.* Rational Design of a Reversible pH-Responsive Switch for Peptide Self-Assembly. *J. Am. Chem. Soc.*, **128**, 6770–6771. Copyright (2006) American Chemical Society).

helical filaments. Such prevalence for small and chemically inert amino acids can serve as a structural signature for proteins whose preferences over chemical specificity lie in the close possible packing of associating building blocks. Partly, this is also the reason for the incorporation of specific functional domains within purely scaffolding sequences of structural proteins or the existence of specialist proteins within separate structural families. Interestingly, the "modulus two" design combines both functions in one sequence providing an example of astonishing space economy.

The assembly of either modulus peptide is both spontaneous and instantaneous in a range of aqueous buffers including cell culture media. Essentially directional in the development of cell-supportive matrices this is further fostered by observations that the RAD motif used in one of the peptides may mediate integrin-dependent cell adhesion analogously to the more common fibronectin triplet RGD.[154,155] In In this way, the RAD-based peptide appears to express at least two functions necessary for artificial extracellular matrices – formation of microscopic networks and cell adhesion, Figure 4.8(c). A series of studies based on the design comprehensively confirmed this.[126,156,157]

4.3.1 Displacing Coil

The described principle of ionic complementarity in conjunction with the staggered assembly of hydrogen-bonded β-pleated structures can be adapted by other patterns. In this regard, particularly interesting is the similarity between α-helices and β-strands. Strictly speaking, β-strands are helices, or rather stretched helices, as the number of residues per turn they have is lower than that of α-helices, 2 against 3.6. Also, a rise per residue is higher than that of an α-helix, 3.3 Å against 1.5 Å. Thermodynamically, it is unfavourable for a structure with such parameters to form autonomously stabilised structures and therefore β-strands exhibit strong preferences for intermolecular hydrogen bonding. In other words, singular β-strands are rare in Nature and, as a rule, group into β-sheets that by having a slight twist may further interdigitate around each other to shape helical fibres.

In marked contrast, α-helices are fully autonomous structures. Their folding does not depend on oligomerisation, nor does it prevent helices from associating laterally as would be required for a staggered assembly.[158] Moreover, the very notion of a staggered assembly implies the formation of at least a dimeric structure with two helices axially shifted with respect to each other.[159,160] α-helices that are employed in the context of oligomerising elements and able to meet the criterion are referred to as coiled coils – bundle-like oligomers of α-helices.[161–165]

Interestingly, examples of staggered coiled coils are not normally presented in biosystems and are limited to very few.[160] Some may include tropomyosins[166] or intermediate filaments such as vimentins and lamins[35,167] that use abrupt modes at different assembly levels. Nevertheless, most known coiled coils prefer lateral association and rely upon oligomerisation by stacking rather than on arranging axially shifted helices. The preference is not entirely clear as any staggered assembly would seem to be more effective in generating extended fibrillar forms of the same dimensions as compared with their laterally shaped counterparts.

One explanation may come from an original sticky-ended design by Pandya *et al.*[127] The design initially conceived as a test for an ability of short helical peptides to assemble longitudinally into an infinite coiled coil led to the generation of a unique α-helical fibre, Figure 4.9.

The fibre, referred to as SAF (self-assembling fibre), arises from the coassembly of two complementary peptides that form a staggered heterodimer with oppositely charged "sticky ends", Figure 4.9(b). The concept paid off in assembling extremely long coiled-coil-based structures, Figure 4.9(c). However, these were unexpectedly thicker and relatively heterogeneous. The diameter of a coiled-coil dimer is approximately 2 nm, whereas the thickness of observed fibres exceeded 50 nm. Unambiguously, this indicates lateral association of folded coiled coils, Figure 4.9(b).

More recent studies from the group suggest that the mechanism of the assembly should involve the tight bundling-up of the extended coiled coils that can be initially formed as 2-nm thick protofibrils of varied lengths.[168] The principle and key feature of the design – longitudinal alignment of individual peptide blocks into helical strands – was strongly supported by (1) an X-ray fibre diffraction pattern showing a 4-nm periodicity, which equates to the length of one folded peptide block;[168] and (2) the chemoselective ligation of peptide blocks within one noncovalent strand into contiguous micrometre-long polypeptide chains estimated at ≥ 3 MDa in mass.[169] In providing further evidence, the most recent studies revealed that the thickness of the assembled fibres can be controlled by introducing extended networks of electrostatic interactions.[170] Programming multidentate intra- and intercoiled-coil electrostatic interactions to the surfaces of the folded helical dimers produced thinner and bent or more flexible fibres.

This is curious as similar strategies can be found in naturally occurring assemblies. For example, vimentins, desmins and some keratins devote highly conserved arginine-containing clusters in their sequences to presumably control the stiffness and stability of the formed filaments by regulating inter- and intrahelical electrostatic interactions.[35] Furthermore, the conspicuous thinning observed in the designed fibre was found to directly correlate with decreases in its stiffness and stability resulting in partially branched or disintegrated (shorter) structures. The lowest observable cut-off in thickness, however, did not fall beyond 10 nm, which suggests there may be a concomitant mechanism involving both lateral and axial control over the assembly in producing uniformly sized 2-nm filaments.[171,172] The design primarily alludes to the fact that ultimately this property can be replicated in building blocks themselves making designing primary peptide sequences a major instrument in addressing this and related questions.

Interestingly, and somewhat conversely to this, all of more than sixty different sequences belonging to the family of intermediate filament proteins assemble into 10-nm thick filaments[35] – the same resulting morphology irrespective of the type of the protein, its function and origin. Furthermore, point mutations in these sequences may lead to severe and inheritable diseases that are often associated with the loss of the native filament structure,[173,174] Figure 4.10(c). This both directly and indirectly indicates that the careful assignment of primary sequences is of prime importance in constructing α-helical fibrillar structures.

The remaining question, however, concerns the critical thickness of filamentous coiled-coil assemblies. In this regard, Kajava and coworkers[175] hypothesised that filaments with the thickness of a single coiled coil can be rationally designed. Unlike the described SAF and most intermediate filaments that are based on helical dimers, the team proposed a five-stranded coiled-coil designed to assemble into 2-nm wide fibrils.[175] The design can serve as an introduction to a general model in constructing *n*-stranded coiled-coil ropes. The main rationale is based on an assumption that the repetition of identical heptad repeats in a given sequence makes the occurrence of mismatched peptide blocks highly probable. Helical blocks built of identical repeats would have the same interhelical interactions in the parallel coiled coils and would be shifted by multiples of the heptad leading thus to an axial stagger, Figure 4.9(d).

To realise this, the stagger or shift between adjacent helices should be made equivalent to an integral number of the repeats or a value divisible by one heptad. Importantly, the *i* and (*i*+*n*) helices of an *n*-stranded rope should not overlap upon the completion of one turn. The shift in this case would not be more than one residue – an addition in the sequence to provide an adequate space for head-to-tail packing of α-helices. Larger shifts would lead to larger gaps between abutting ends of the peptide blocks and consequently to unrectifiable irregularities in the assembly. Thus, most favourable helical arrangements to form *n*-stranded ropes can be reflected in a proposed general equation ($n \times 7 - 1$); that is, helices composed of 20, 27 and 34 residues for three-, four- and five-stranded ropes, respectively, would trigger and promote multimeric longitudinal propagations of the corresponding coiled-coil oligomers.[175]

The simplest coiled coil obeying the equation would be a parallel dimer assembled from a 13-mer, which is just under two heptads. However, the length may not be sufficient not only to form nucleating stable structures to promote fibrillar associations but also to sustain the stoichiometry of a coiled-coil dimer. Having taken this into consideration and given the minimum requirement of three contiguous repeats to form stable assemblies,[176,177] the authors chose to exemplify the concept using a 34-residue α-helical fibril forming peptide (αFFP) that was designed to assemble as a five-stranded coiled coil architecturally similarly to five-stranded α-helical bundles.[177,178] As revealed by TEM and X-ray fibre diffraction, the peptide assembled into 2.5-nm wide fibrillar ropes with helical blocks longitudinally aligned and staggered along the rope axis, Figure 4.9(d). No lateral association ensued in the assembly, which is perfectly consistent with the design and matches the predicted arrangement of a fixed lateral dimension.

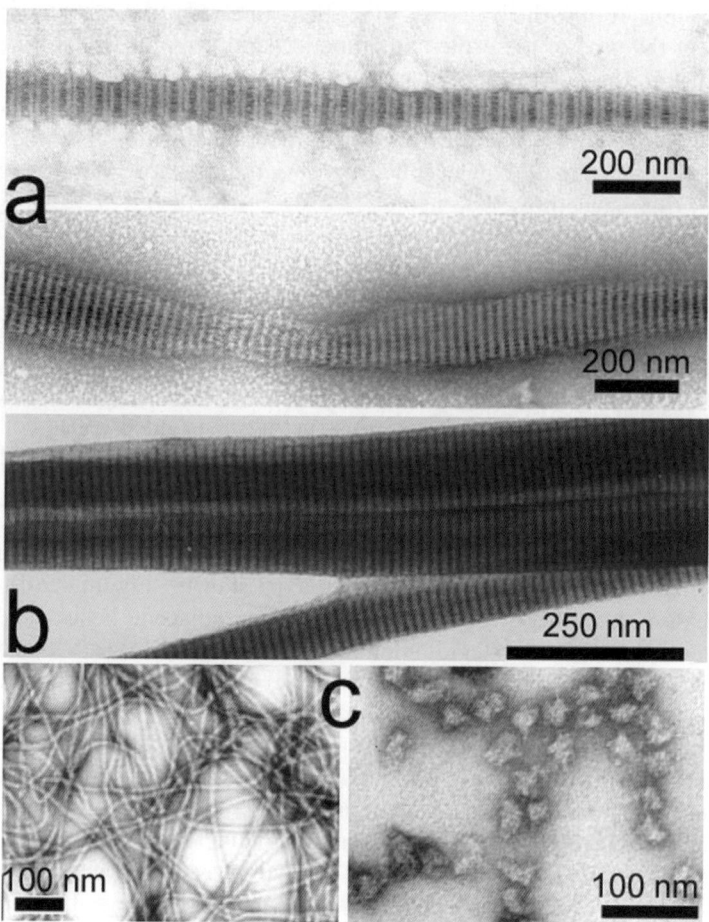

Figure 4.10   Natural helical filamentous assemblies; (a) Lamin-derived intermediate
filaments, (reprinted from Karabinos, A. *et al.* The Single Nuclear Lamin
of Caenorhabditis elegans Forms *In vitro* Stable Intermediate Filaments
and Paracrystals with a Reduced Axial Periodicity. *J. Mol. Biol.*, **325**,
241–247, Copyright 2003, with permission from Elsevier), (b) Microtubule-associated fibres assembled from SF-assemblin, (reprinted from
Lechtreck, K.-F. Analysis of striated fiber formation by recombinant SF-
assemblin *in vitro*. *J. Mol. Biol.*, **279**, 423–438, Copyright 1998, with
permission from Elsevier), and (c) Assembly products of wild (left) and
point-mutated (right) vimentin. Characteristic banding patterns are
apparent in all filaments, (reproduced with permission from Herrmann,
H. and Aebi, U. Intermediate filaments: Molecular Structure, Assembly
Mechanism, and Integration Into Functionally Distinct Intracellular
Scaffolds. *Annu. Rev. Biochem.*, **73**, 749–789 (2004)).

This appears to be the only example of the successful prescription of restricting the lateral association of coiled-coil protofilaments and sets a promising precedent for future follow-up designs.

However, the formation of fibrillar structures was not exclusive. Spherical assemblies of 10–15 nm diameter were also observed as formed preferentially at neutral pH, Figure 4.9(e). Such spheroids present certain interest for a number of applications including those highlighted in Chapter 3. Yet, one striking lesson from the finding is the manifestation of the highly unstable nature of formed fibrils.

This may constitute circumstantial evidence for the rational, *i.e.* stabilising, thickening observed in natural and in most designed α-helix-based fibres. Independent reports from several laboratories seem to support this.

4.3.2 Settling Lateral

4.3.2.1 Bundling Exclusive

A dual-component design strategy using hydrophobic interactions was proposed to aim at favouring axial propagation and electrostatic forces to regulate lateral assembly in the formation of nanoscale rope-like structures.[179]

A helical homodimer, CpA, assembling to nanoscale filaments and ropes was designed to comprise two identical sequence elements from the transcription factor GCN4 separated by a short linker. The linker was introduced to give a two-residue phase shift in the heptad repeat analogously to the one-residue shift in the αFFP design. In CpA, however, the two-residue spacer is believed to generate two hydrophobic surfaces, C- and N-terminal, that become oriented at ∼200° with respect to each other. Such a configuration leads to a third level of supercoiling and makes blunt-end dimers of the native GCN4 no longer possible. Instead, dimers with exposed hydrophobic "sticky ends" are formed as opposed to charged "sticky ends" in SAF and alternating ionic and hydrophobic faces in the β-sheet moduli. In this way, an open-assembly polymerisation is to be initiated by the enhanced hydrophobic effect. Additional assembly enhancers including cysteine residue inserts to provide covalent crosslinking and the activators of the hydrophobic effect (*e.g.* glycerol) were employed to foster the polymerisation. Consistent with the design, the peptide folded as an α-helix and assembled into mesoscopic filamentous structures. However, the assemblies appeared to be shorter, sporadically branched and more structurally irregular than fibres of other designs. This is most likely related to poor and unspecific intercoiled-coil associations within the filaments. In addition, oxidised cysteine residues may be contributive to promiscuity in polymerisation[180] and partial misfolding[181,182] that can lead to fractal patterns of assembly similarly to those ascertained in order-impaired aggregates derived from different peptide self-assembling motifs.[180,182,183]

Importantly, the CpA design proves that applying the enhanced hydrophobic effect to self-assembling systems under thermodynamic control as a means of morphological tuning does not provide sufficient specificity and requires subsidiary subtler forces to adapt its intensive or bulk solute-rejecting function for extensive variables operating in localised systems. Within the dimensions of fibrillar systems the need for lateral association is coherent with the extremely rare occurrence of bends and kinks in matured thickened fibres that becomes more frequent as fibre widths decrease. As a result, curved and in some cases irregularly split fibrils are formed, which proved to be characteristic of all the described designs. For example, subsequent analysis of α-FFP fibrils that disclosed the tendency of five-stranded protofilaments to double into 4–5-nm fibrils was particularly intriguing.[184] Once formed, the fibrils were able to associate further, forming structures with a range of thicknesses. The formation of thicker fibres was shown to be promoted by mutations in peptide sequences, namely through the replacement of charged residues, arginines and glutamates, with neutral glutamines.[185] Consistent with the outlined structural features of designed helical fibres and intermediate filaments, this finding builds in a probable rationale for adjusting the stability and morphology of fibres by controlling lateral associations *via* specific sequence assignments.

In this vein, Zimenkov *et al.*[186] took a reversed approach to controlling fibrillogenesis and set out to promote rather than discourage thickening. With this in mind, a coiled-coil homodimer was designed to undergo staggered assembly into long helical fibrils in a fashion that was almost identical to the SAF assembly. The design also served as an experimental verification of the concept of mismatched staggering, thus far exemplified only by αFFP design. Yet, lateral extension, an assembly condition deemed to be rectifiable in the aforedescribed designs, in this case was prompted by electrostatic interactions between charge-complementary residues, lysines and glutamates, incorporated into outer or solvent-exposed faces of folded coiled coils. In essence, this probes an assumption that lateral association results from multiple interactions between residues at noncore positions in adjacent fibrils. The observation of high aspect ratio fibrillar structures of diameters comparable to thickened assemblies may be in good agreement with the hypothesis.[186] However, the lack of comparative analysis of the model and in relation to other designs, particularly with regards to surface positions primarily responsible for the said effect, could not lead to an explicit conclusion.

Indeed, as recently shown by Papapostolou *et al.*,[170] electrostatic interactions provided by noncore residues can either stabilise and destabilise fibres or can cement and discourage their formation. Finding the right balance between intra- and intercoiled-coil interactions, which is inevitably set in primary sequences, is what appears to be persistent in predefining the fibre maturation. The sought relationship between such interactions and thickening, on the one hand, and thickening and stability and morphological perseverance, on the other, remains unveiled and can only be tackled by comparing various designs.

For instance, subsequent designs by Conticello and coworkers focused on modulating the ability of fibrillar systems to respond to incremental changes caused by external stimuli. An indispensable and gratifying requirement here is to find a responsive mechanism that would be compatible with the chosen assembly mode. A peptide sequence adopting a trimeric coiled-coil fold was designed to stagger into fibres in a fully reversible and pH-dependent manner. Otherwise similar to that of SAF and αFFP, the reversible pH dependence of this assembly is regulated by a conformational switch facilitated by histidine residues incorporated into the core of the coiled-coil bundle, Figure 4.9(f). The mechanism involves the protonation of imidazole side chains of histidines upon which random coil-helix conformational transition occurs.[187] The switch can be performed within a pH range narrowed by the pK_a of imidazole. As expected, no fibres formed at acidic pH at which the peptide is in a random coil conformation. At the limits of helical folding (up to 8.2) the coil–helix switch occurs and fibrils of 2.5–5 nm in diameter become clearly visible, Figure 4.9(g). Interestingly, the fibrils, tended to bundle into 40–100 nm associations. These, however, were not of fused, homogeneous fibres as the individual fibrils could be distinguished within single bundles. The aggregated bundles of similar fibrils were observed when assembled from the same peptide in the presence of silver ions, which represents an example of a metal-ion switch.[188] Upon folding of the peptide into the coiled-coil trimer histidine residues shape trigonal coordination sites constituting perfect settings for metal ions. As confirmed by CD spectroscopy and isothermal titration calorimetry, a coil–helix switch transpires upon the equimolar binding of the peptide exclusively by silver (I) ions, supporting the proposed model in which each peptide block contributes one net silver(I) coordination site.

Intriguingly, the bundling behaviour of switching fibrils developed from three-stranded coiled coils matches that of five-stranded αFFP designs and might be attributed to increased oligomerisation states as compared to designs based on coiled-coil dimers. The stronger disposition of double-stranded coiled coils to pack into larger structures may be a result of their surface areas being buried to a lesser extent that is also reflected in much lower stabilities compared to those of higher oligomers. The strikingly similar thickening characteristics of three- and five-stranded coiled coils allude to the fact that three-stranded ropes may present a minimal requirement or a starting point for designing laterally regular fibrils. Speculatively, this can also be in good accord with the Nature's choice for the triple helix assembly in building dimensionally well-defined collagen fibrils.

Elucidating structural principles underlying the phenomenon of fibre thickening in α-helix-based fibres is gaining momentum and beyond doubt will be under close scrutiny in future studies using different model systems. An excellent acid test for designer systems in respect to engineering predictable and highly ordered nanoscale systems, with particular relevance to fibrillar matrices, it is also fundamentally interesting as helical fibres are abundant in natural biosystems and any insight into their organisation would advance our knowledge of macromolecular architecture in general. Specifically within this chapter, the main factor of developing helical fibres is the level of prediction and control over assembly that helical designs may offer on the nanoscale.

4.3.2.2   Permitting Distinctive

As shown above, this is true in engineering nanoscale order into fibrous systems, which can be seen as a *de novo* approach to emulating internal para-crystalline structures characteristic of natural fibrils. Striated patterns typically found in fibrin and collagen fibres and intermediate filaments can be reproduced by design in helical fibres, as demonstrated by Papapostolou *et al.*[168] Specifically, the matured fibres of the SAF design described above are featured by a readily recognisable periodic banding pattern of light and dark striations that run across the long fibre axis, Figure 4.9(c). As experimentally characterised, the SAF fibres derive from two random coil peptides that when mixed together assemble into two parallel α-helical continuous strands. The axes of the strands therefore align within the mature fibre along its axis, which allows relating the observed banding pattern to repeating structural features in the folded peptides.

The spacing between striations measured as ~4.2 nm closely matches the distance covered by either SAF peptide of 28 residues in a fully folded coiled-coil conformation (*i.e.* 28 × 0.148 nm = 4.144 nm), Figure 4.9(c). This, in conjunction with the uninterrupted coverage of individual striations over the entire fibre width highlights a longitudinal repeat of the peptides along the fibre axis, and also stands for the fact that the folded peptides have to be in register across the widths regardless of their sizes. Consistent with this, extended peptides comprising 35 amino acids and are thus designed to span ~5.2 nm assembled into fibres with an identical striated pattern, the experimental analysis of which returned a spacing of ~5.2 nm.[168]

The design clearly demonstrates the feasibility of controlling structural features within a few amino acids residues. Such structural responsiveness of helical fibres can be tuned by peptides of varied lengths and different topologies that would considerably facilitate the programming of 3D polygonal fibrillar networks and matrices derivatised with oriented and surface-adjustable bio-functional elements – a promising and logical route to synthetic ECM analogues.

In exploring this, Ryadnov and Woolfson showed that matured fibres, seemingly unsusceptible to morphological alterations due to their mechanical rigidity, can be reprogrammed from the very bottom up.[189–192] Using the SAF assembly it was demonstrated that the final appearance, form and shape of the resultant fibres can be determined by specifically designed peptides. The researchers proposed an empirical algorithm according to which the SAF system can be supplemented with special peptide constructs, *specialists*, which by being complementary to standard fibre-forming peptides, *standards*, would guide the SAF assembly by redirecting it longitudinally.[189,192] In this case, fibres themselves may be viewed as building blocks for more complex super-structures – matrices.[192] The main driving force here is the fast binding of *standards* by fibre-shaping *specialists* and their kinetically controlled concentration in localised sites within a growing fibre that enable the physical expression of distinguishable morphologies, Figure 4.11.

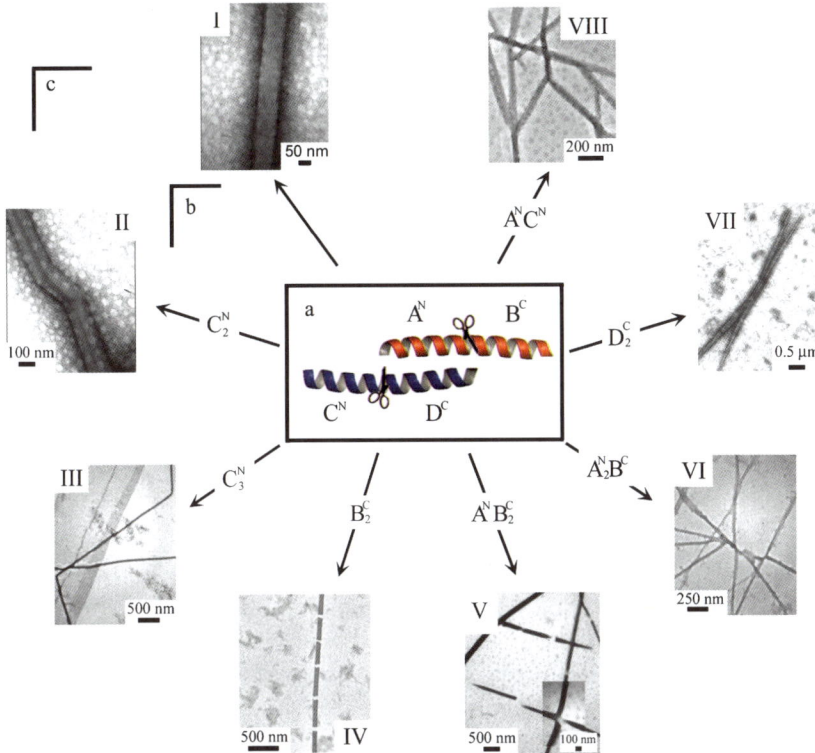

Figure 4.11 Programming the morphology of helical fibres; (a) Schematic of splitting of the two standard peptides into four subunits, labelled A–D, used in the construction of specialist conjugates. Superscripts denote the nature of the free termini, (b) Specialists coassembling with the standards, with numerical subscripts indicating the number of subunits (higher than 1) used in one conjugate, (c) Electron micrographs of various fibrous assemblies: (I) standard, (II) close and (III) distant kinks, (IV) linear segmented and (V) branching segmented fibres, (VI) multiple branches, (VII) splits, (VIII) polygonal networks, (reprinted by permission from Macmillan Publishers Ltd: Ryadnov, M. G. and Woolfson, D. N. Engineering the morphology of a self-assembling protein fibre. *Nature Mater.*, **2**, 329–332, Copyright (2003), and from Ryadnov, M. G. and Woolfson, D. N. MaP Peptides: Programming the Self-Assembly of Peptide-Based Mesoscopic Matrices. *J. Am. Chem. Soc.*, **127**, 12407–12415. Copyright (2005) American Chemical Society).

In this case, fibre shapers have to be sequences-complementary to and topologically distinctive from linear *standards*. The most rational way to do this is to split each *standard* containing two oppositely charged fractions of identical sizes into two building blocks, such that four unique units are generated. The units can then be recoupled in various combinations and numbers as orthogonal constructs in head-to-head or tail-to-tail fashions, depending on the orientation of the original sequences.[189] Although permitted to some extent, no amino acids mutations are required by this format. The *specialists*, upon combining with *standards*, then direct convergent or divergent modes of the SAF assembly leading to nonliner fibres. Depending on the nature of combined units and their cooperative functioning one preferential, or as in many cases, exclusive microscopic feature is introduced. Morphological expressions such as kinking, splitting, branching, segmentation or interconnection were shown to be selectively assigned to an individual fibre by a *specialist*, Figure 4.11(b).

The algorithm also allows rational extensions within the main assigned architectures. For instance, it is possible to adjust or tune the distances between kinks or vary branching density by incorporating subtle changes into the *specialists*,[192] Figure 4.11(c).

Because these assemblies are intrinsically biocompatible and can be made reversible – the approach is particularly attractive for engineering matrices capable of remodelling in cellular environments, the feasibility of which, however, remains to be experimentally demonstrated.

4.3.2.3   Inviting Captive

Nonetheless, even with the availability of such restructuring or "smart" scaffolds an outstanding caveat in designing extracellular matrices concentrates on all biomolecular designs having to associate with biologically relevant functionalities. A concept to mirror this innate quality of the ECM in synthetic matrices was formulated as fibre recruiting[191] or supramolecular decoration of fibrillar surfaces with biologically active molecules, Figure 4.12.

In this design, fibres were assembled from peptides modified with small molecules or peptide antigens. Such appendages act as baits by capturing their partner proteins from the solution phase and recruiting them on the fibre surfaces.[191] In this way the fibres become decorated with folded proteins that can mimic the interactions that occur between the native ECM and the integrin receptors of cellular surfaces.

The interactions can also be directly modelled as shown in several other designs based on nonhelical fibrillar systems. For example, Stupp and coworkers[193,194] reported a peptide amphiphile that assembles into tubular micelles with the hydrophobic core formed by aggregated aliphatic alkyl chains and the solvent-exposed surface shaped by cell-adhesion motifs at nearly van der Waals packing densities, Figure 4.13(a).

The assembled scaffold shared similar properties with the native ECM including nanoscale dimensions, fibrous morphology and cell adhesion. It was successfully tested in encapsulating neural progenitor cells.[195] Engineered to present neurite-sprouting laminin epitopes on its surfaces the scaffold was assembled in cell suspensions to permit the cell-governing formation of a highly hydrated matrices (> 99.5 wt% water). Fibres of about 8 nm in diameter and a few micrometres in length formed dense 3D fibrillar networks, Figure 4.13(b), that rapidly converted to gels embedding cells. Given the high surface areas of the formed matrices cell-adhesion epitopes were presented at densities considerably higher than those typical for the native ECM. Although it is evident

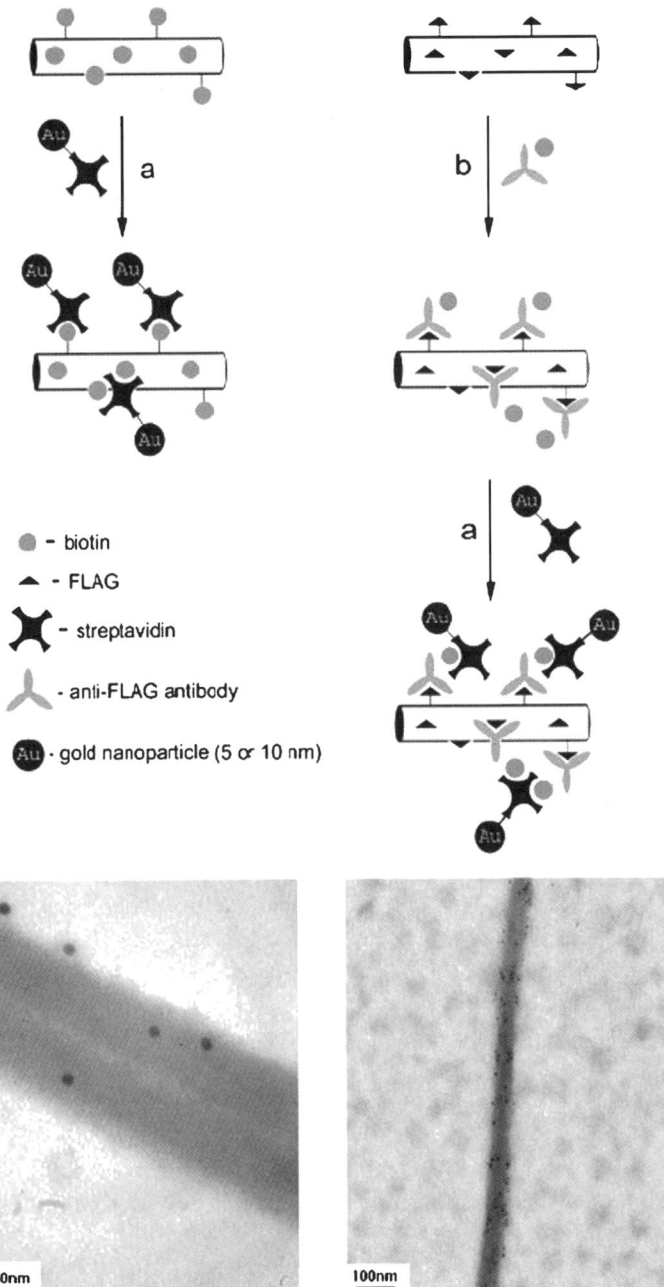

● - biotin

▲ - FLAG

✦ - streptavidin

⅄ - anti-FLAG antibody

🔵 - gold nanoparticle (5 or 10 nm)

Figure 4.12 Fibre recruiting; fibres depicted as cylinders are assembled from peptides carrying baits that after assembly decorate fibre surfaces and become solvent exposed to interact with partner proteins. Two routes (left and right) follow two different molecular-recognition patterns: protein–ligand (biotin–streptavidin) and antigen–antibody (FLAG peptide–antiFLAG antibody). Corresponding electron micrographs showing the recruitment of active proteins labelled with gold nanoparticles, (reprinted with permission from Ryadnov, M. G. and Woolfson, D. N. Fiber Recruiting Peptides: Noncovalent Decoration of an Engineered Protein Scaffold. *J. Am. Chem. Soc.*, **126**, 7454–7455. Copyright (2004) American Chemical Society).

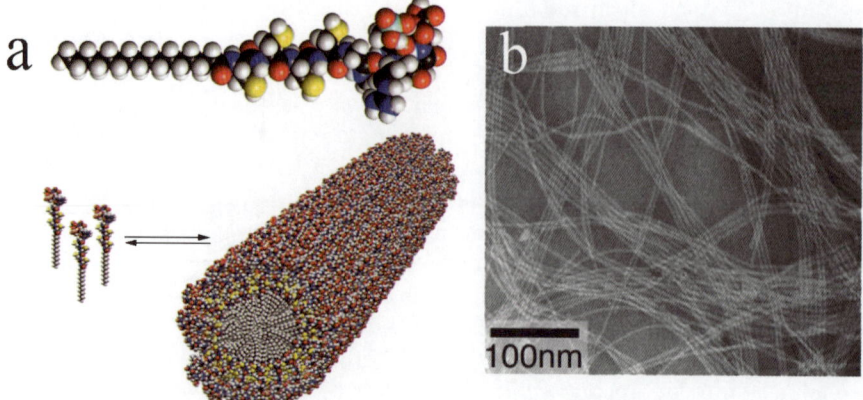

Figure 4.13 Self-assembling peptide amphiphile; (a) Molecular models of the
amphiphile and its assembly into a tubular micelle highlighting the key
structural features of the design. The amphiphile comprises a hydrophobic tail (grey) forming the core of the micelle, a crosslinking region
containing cysteine residues to cement the assembly (yellow), and a polar
head displaying cell adhesion motifs on fibre surfaces, (from Hartgerink,
J. D. *et al*. Self-Assembly and Mineralisation of Peptide-Amphiphile
Nanofibers. *Science*, **294**, 1684–1688 (2001). Reprinted with permission
from AAAS), (b) An electron micrograph of the assembly showing
nanoscale fibrillar networks (reprinted with permission from Hartgerink,
J. D. *et al*. Peptide-amphiphile nanofibers: A versatile scaffold for the
preparation of self-assembling materials. *Proc. Natl. Acad. Sci. USA*, **99**,
5133–5138. Copyright (2002) National Academy of Sciences, USA).

that not all of the motifs available for cell binding can be targeted by epitope-
binding integrins, this was sufficient to support cell signaling. Usefully, the
combination of amphiphiles that by carrying different functional moieties can
decorate the micellar corona with broadly distributed and statistically spaced
interacting motifs can be employed. However, the main investment of this
design into the goal of engineering artificial matrices is the notion of a peptide
assembly that occurs concomitantly with cell growth at extremely low concentrations. Providing mechanical and adhesion support sufficient for the
grouping of cells the assembly directs cell migration and stimulates cell differentiation and growth.

The concept of matrix decoration has been extended to other self-assembling
fibrillar systems,[196,197] but mainly in relation to functionalisation with integrin-
binding motifs aiming at promoting cell adhesion.[198–200]

Covalent grafting of preassembled scaffolds or fibrous extracts has also been
proposed, but this may be less promising due to toxicities associated with the
used coupling reagents and the uncontrollable character of their reactions.[201,202] Clearly, other ways of fibre coating ought to be sought.

A strategy reported by Yu and coworkers[203] appears to be of particular
relevance in this respect. The strategy is presented as a "physical" modification

as opposed to chemical modification and relies on using collagen mimetic peptides (CMPs) with strong affinity to type I collagen under controlled thermal conditions.

CMPs typically of less than 30 amino acids were designed as multimers of the collagen triads. Since such sequences are by nature prone to form collagen-alike triple helices CMPs were envisaged and subsequently found to bind to partially denatured collagen fibres by associating with their thermally labile and disentangled domains. The incorporation of the peptides into native and denatured collagens, be it effective, may provide an elegant route to mimicking nonfibrous FACIT collagens that could function as individual decorating units for the collagen fibrils. For example, by inserting into collagen fibres CMPs can bring or introduce specific functionalities and patterning (*e.g.*, MRI contrasting, cell attachment) or therapeutic entities (*e.g.*, growth factors, drugs) on their surfaces that may be either directly used in living tissues[204] or applied to engineering responsive biomaterial scaffolds.[205] Such a combination of site-specific unfolding and following decoration-targeted intervention may open principally novel routes to nanoprecise matrix functionalisation. The technique is, however, of a self-assembly method that strongly depends on the preassembling conditions of collagen fibres.

4.3.3 Clearing Limiting

The native ECM is a quintessence of biological nanodesign. With this, fabricating various fibrillar architectures in an attempt to replicate if not the very chemistry and assembly of the ECM but its morphological and physical characteristics, certainly eliminates all restrictions on chemistries and molecular hierarchies. This quite literally accepts any ideas that are progressive enough to approach the problem as intimately as practical.

In this regard, the highlighted concepts of ionic alternation, longitudinal staggering and tubular amphiphile micellisation successfully applied to the self-assembly of fibrillar matrices are to be brought forward into the next generations of designs and are being challenged by other ideas, with many already borne fruit.

4.3.3.1 Equilibrating Transitional

The exploitation of how subtle tailoring in the assembly of fibrillar designs may affect the key properties of resulting materials can be exemplified by an approach reported by Messersmith and coworkers.[206]

In this work, the assembly of the complementary β-structured moduli described above was shown to follow a salt-dependent mechanism that can be substantially sensitised by redesigning the hydrophobic face of the moduli with phenylalanines replacing small alanine residues. Seemingly simple, the design creates a compelling answer to readily executable properties of fibrous materials by maintaining physicochemical tuning in a highly specific manner. In this design, the assembly of a FEK16 sequence into highly homogenous hydrogels was controlled by stimuli-responsive liposomes constructed to release divalent metal ions at set temperatures or in response to near-infrared light exposure. At low concentrations the peptide, folded as an α-helix, was forced to undergo an α-to-β transition upon the release of the cations leading to the gelation of β-sheet fibrillar networks.

A helical version of a fibrillogenesis-accompanying transition – coil-helix – was reported by Ghosh and coworkers.[207] The design presented an overlapping version of the complementary staggering assembly in which two coiled-coil sequences, each displayed on a tetravalent dendritic hub, coassembled into rapidly bundling fibrils. As in the other described helix-based fibre designs, the preadopted helical conformation programmed in these sequences was preserved upon aggregation. Importantly, this is consistent with the fibre-shaping concept described above, suggesting no circumscriptions being imposed by a particular topology on the chosen folding type.

In unison with this, a topological format promoting a sticky-ended assembly was demonstrated by Ogihara *et al.*,[146] based on 3D domain swapping – a natural oligomerisation mechanism employed by two or more proteins. The mechanism involves exchanging identical structural domains between protein monomers.[208] During the process, folded monomers intertwine, with most of their intramolecular interactions preserved in a resulting swapped structure. To emulate the mechanism in constructing extended fibrillar structures an up-down-down topology of *de novo* helices[209] was proposed compiled into a three-stranded bundle monomer such that helical regions I and II of one copy of the monomer and a region III' of another copy would associate into one structural subunit correspondingly leaving I'/II' and III domains to overhang as sticky ends to incite a lengthwise assembly, Figure 4.14(a). Consistent with the design, microscopic filaments were formed, with the units oriented along the main filaments axis, Figure 4.14(b).

Interestingly, as in the three- and five-stranded designs by the Conticello and Kajava groups, respectively, the filaments tended to bundle rather than fuse or mature into thickened fibres.

In further extrapolation of domain swapping into transitional fibrillation, Schneider and colleagues programmed a thermally switchable formation of mechanically rigid hydrogels.[147] The design followed a series of studies by the group on the conformationally controlled assembly of β-structured fibrils induced by reversible β-hairpin folding.[210,211]

A strand-swapping peptide (SSP) containing an exchangeable β-strand domain appended to a nonexchangeable β-hairpin domain was designed as an assembly unit. At temperatures up to physiological ones, the peptide is predictably a random coil and is fully soluble in water. As the temperature increases SSP switches to a two-domain structure, with the folded β-hairpin acting as a scaffold on which the exchangeable strand is displayed. The latter thus becomes available to exchange with its counterpart of another SSP copy, giving rise to a bilayer of strand-swapped dimers. The dimer in turn propagates laterally into β-sheet-stacked fibrils on forming a hydrogel, Figure 4.14(c). By

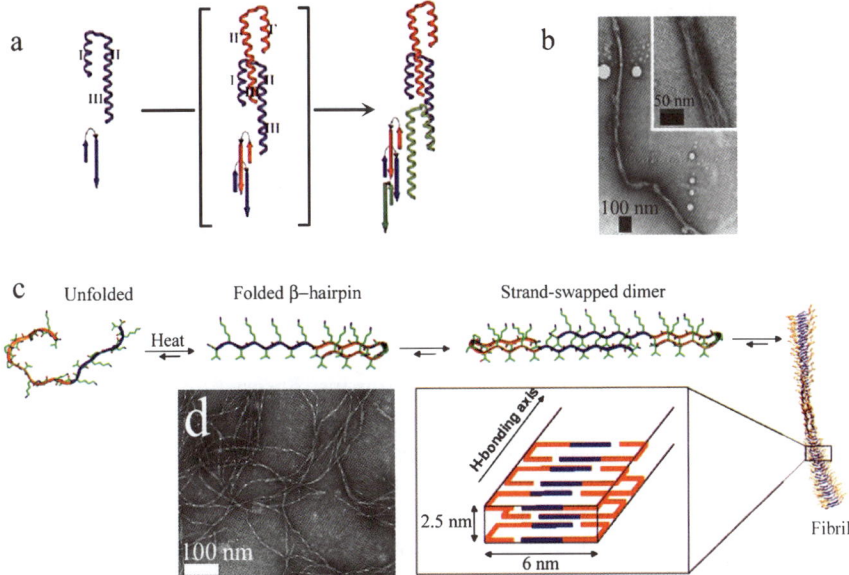

Figure 4.14 Domain-swapping designs; (a) A three-helix bundle assembling with up-down-down topology. Copies of the bundle are coloured differently, highlighting the lengthwise character of the assembly, (b) An electron micrograph showing the resulting fibrous oligomers, ((a) and (b), reprinted with permission from Ogihara, N. L. *et al.* Design of three-dimensional domain-swapped dimers and fibrous oligomers. *Proc. Natl. Acad. Sci. USA*, **98**, 1404–1409. Copyright (2001) National Academy of Sciences, USA), (c) The temperature-dependent assembly of a strand-swapping peptide into β-structured fibrils. Exchangeable domains are shown in blue and nonexchangeable domains in orange, (d) An electron micrograph of assembled fibrils, ((c) and (d), reprinted with permission from Nagarkar, R. P. *et al.* De Novo Design of Strand-Swapped beta-Hairpin Hydrogels. *J. Am. Chem. Soc.*, **130**, 4466–4474. Copyright (2008) American Chemical Society).

varying the length of the exchangeable strand it was possible to determine the mechanical properties of hydrogels that were made to relate to the specific morphologies of assembled fibrils. For example, shorter domains disposed to yield nontwisted fibrils were found to give hydrogels nearly twice as mechanically rigid as those composed of twisted ribbons produced by domains of doubled lengths.

4.3.3.2 *Extracting Minimal*

Biomimetic designs constitute the dominant force behind the replication of the ECM. This appears to be reasonable given the inherent property of biomimicry of applying selecting brackets of positive (towards) and negative (away) design strategies to construction approaches. In return, this impacts on preferring more problem-focused and feature-based approaches. However, inspiration can also come from comparatively unrelated research giving the problem a fresh start.

One notable example originating from a series of studies by Gazit with colleagues[212] on the structure of amyloid fibrils has led to the discovery of a minimal or "core" sequence capable of nucleating amyloidogenesis.[213] The identified sequence – phenylalanyl-phenylalanine (FF) – was shown to have a tendency to self-assemble into fibrillar nanotubes.[214] The formation of hollow nanofibre structures by FF was earlier predicted by Görbitz who also showed that the tendency is pronounced for many other hydrophobic dipeptides.[215–217]

The ability of FF to form nanostructures has been further explored by other groups.

Xu and coworkers[218] made a serendipitous observation that dipeptides derivatised with *N*-(fluorenyl-9-methoxycarbonyl) (Fmoc) group become highly efficient hydrogelators. The hydrogels exhibited fibrillar morphology as established by electron microscopy[218] and prompted the development of responsive biomaterials for a number of biological applications.

Ulijn and coworkers made use of Fmoc-dipeptides for producing gelating scaffolds to support the proliferation of chondrocytes in both two- and three-dimensional cell cultures.[219] Furthermore, the group investigated the ultra-structure of the formed gels.[220]

As originally proposed,[214,215] individually assembled needle-like FF-based fibrillar structures derive from the antiparallel arrangement of hydrogen-bonded β-sheets and π–π stacking interactions of aromatic phenylalanine rings. In the case of Fmoc-FF, the incorporation of Fmoc groups contributing to gelation otherwise untypical for dipeptides was suggested to be due to zippering-up of Fmoc groups into π-stacked alternate pairs with interleaved phenyl rings.[218,220] Because β-pleated structures are slightly twisted, associating β-sheets become persistently rotated around each other, thus furnishing a cylindrical structure.

Importantly, the model implies that the resultant morphology can be made subject to the nature of the hydrophobic N-terminal moiety, which can constitute a minimalistic approach in engineering fibrillar self-assembling systems. Indeed, subsequent experiments demonstrated this to be the case. A range of fibrillar and tubular structures were generated using a variety of hydrophobic dipeptide motifs, Figure 4.15.

Understandably, not all fibrillar systems reported to date have been covered here. With the pace the chemistry of fibrillar materials is developing this would be principally impossible to achieve. But critically, the design of fibrillar systems and gelating materials that has been relatively successful thus far or at least certainly rising provides only a partial answer to artificial matrices.

Other issues such as (i) what biology to build into a designed system; (ii) what molecular features are to aid in integrating into dynamic cellular environments; (iii) what means are to be introduced to respond to spatial and surface cues and patterns, and many others are progressively important in nanodesign. These and other points are accentuated upon in the following section.

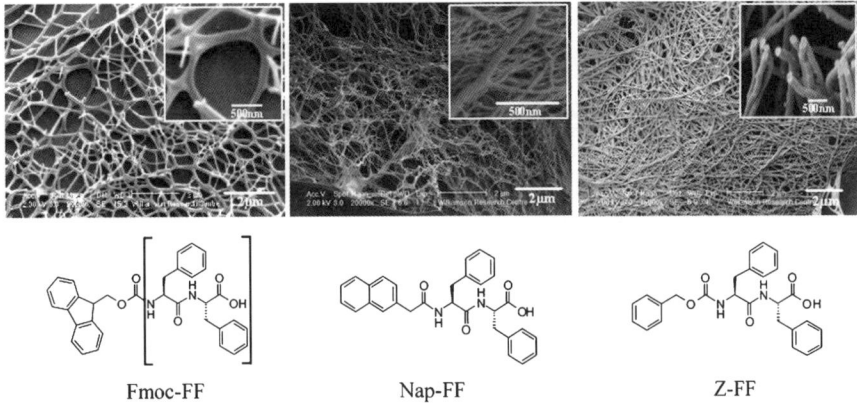

Fmoc-FF   Nap-FF   Z-FF

Figure 4.15 Minimalist design of fibrillar hydrogels; a variety of fibrillar structures as defined by different hydrophobic moieties linked to the FF motif (in brackets) (image courtesy of Rein Ulijn).

4.4 Gambling Beyond

Over the last decade the term "regenerative medicine" has been applied liberally to embrace tissue-engineering (TE) approaches that aim at the acceleration of natural healing processes. The central paradigm of TE is the use of ECM analogues in the native 3D cellular environment.[6,221,222]

In technical terms, TE is the use of fibrous supramolecular scaffolds for seeding tissue-specific cells that become fixed in 3D to further develop into living tissues. The logic underlying most designs is thus relatively straightforward – to grow in 3D, cells need to be embedded in a structure that mimics the native extracellular matrix (ECM) – ideally, a 3D nanostructured fibrillar protein scaffold. Naturally, derived fibrillar materials extensively tested as ECM substituents prove to have serious limitations, such as handling problems, morphological incongruity, and difficulties concerned with engineering specific properties and diseases transmissions, which eventually prompted the concept of designed fibrous systems presenting a very sound alternative.

In its classical sense TE is the use of tissue-specific cells for seeding a scaffold *ex vivo*. This is a cell-based definition that is normally distinguished from guided tissue regeneration in which designed scaffolds direct and maintain the regeneration of a tissue only from cells residing at the site of transplantation.[223]

However, in practical terms these details are apparent and have gradually evolved to a more general definition of the approach referred to as the strategy of "bioartificial tissues".[224,225]

4.4.1 Guiding Proliferative

Principally, the construction of bioartificial tissues aims at devising an appropriate replacement for tissue through the use of scaffolds that either are shaped

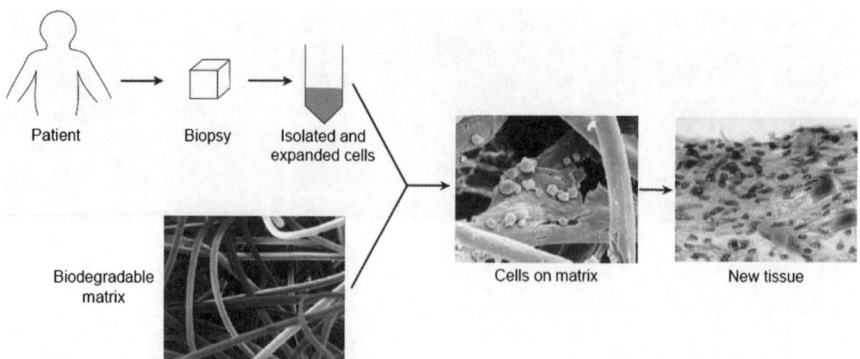

Figure 4.16 *Ex vivo* tissue-engineering approach; tissue-specific cells expanded *in vitro* are seeded onto an exogenous biodegradable matrix to allow the growth of a tissue transplant, (reprinted from Kim, B.-S. and Mooney, D. J. Development of biocompatible synthetic extracellular matrices for tissue engineering. *Trends Biotechnol.*, **16**, 224–230, Copyright 1998, with permission from Elsevier).

like the damaged or missing tissue or organ (*ex vivo*) or once deposited into the defect site are able to adapt in the tissue environment (*in vivo*). The former presents the concept of "spare body parts" and lends itself to a practice of replacing injured or lost organs. In this approach, cells are expanded *in vitro* and seeded onto the scaffold often of a preformed shape to mature into a tissue or organ prior to implantation into the patient, Figure 4.16.

The latter solicits for a continuous restoration of self-renewing tissues and implies the use of injectable and responsive materials that by filling the injury site are arranged into a 3D supporting matrix to encourage the surrounding cells to develop into living tissues *in situ*. In both cases the scaffolds would ideally mimic the ECM. Both strategies are perfectly adaptable to any cell or tissue type making the potential of their experimental designs inexhaustible.[222] Indeed, the number of successful examples showing steady growth include the *ex vivo* constructions of cardiac valve substitutes, corneal transplants, urinary bladders and *in vivo* nerve, bone and cartilage regeneration.[223–226]

Yet, the future perspective of a particular TE strategy is eventually selected by its clinical feasibility.[12,227]

4.4.1.1 Feeding Proximate

Apart from the key properties to be provided by ECM analogues at the cellular level, such as maximising cell adhesion and migration, sequestering and releasing biological signals to support cell proliferation and differentiation, there are other equally important criteria that are imposed by the compatibility of the strategies with clinical practice.

Poor cell diffusion, risk of infection or disease transmission, mechanical and physical incompatibility of preformed scaffolds, immunorejection of the implanted material and invasive surgery are typical drawbacks that can lead to treatment complications and increased costs.[12,221]

In this respect, the use of living or "smart" materials that can be assembled inside the body from directly introduced precursors allowing them to fully integrate within an injury site or a tissue lesion is of preferential interest.[6,7] Furthermore, the strategy of onsite forming scaffolds entails the simultaneous use of scaffold components and cells that can be distributed homogenously within a hosting tissue cavity regardless of its irregularities in shape or size. Moreover, the given cell-scaffold format makes invasive surgery, otherwise necessary for other TE methods, practically excessive.[228]

Within the approach, bioartificial tissues can be constructed as open and closed.[229]

Open are those that become vascularised by the host tissue and rely on temporary scaffolds. These provide initial support for the cell growth and should be made biodegradable in order to be replaced by the native ECM produced by cells. The time periods to allow this vary from several days to weeks depending on the tissue type and application.[12] Common examples of open systems are fibrous matrices and reversible hydrogels described previously.

Closed tissues are substantially less "leaky" and arise from the slow diffusion of cells encapsulated into a scaffold resistant to degradation. Scaffolds used for closed tissues are typically porous microcapsules or membranes of < 500 μm diameter and < 5 μm thickness. These are hydrogels (alginates) or copolymeric shells (acrylonitriles, methylacrylates). A typical microcapsule design would comprise an inner cell-holding endoskeleton (matrix) and an outer exoskeleton (matrix-surrounding shell), with a charged biopolymer making up the inner layer (matrix component) and an oppositely charged synthetic polyelectrolyte shaping the outer layer of the capsule. The combination of the two gives a macroporous membrane. The porosity renders such a microcapsule permeable to nutrients for normal metabolic functioning of cells and to toxins released by the cells.

Both open and closed designs are being shown to give comparably efficient extracellular systems and share many similarities in supporting the normal cell development. Having both approaches available and continually developed gives additional flexibility in choosing appropriate strategies in experimental tissue designs or regenerative therapies. However, in some areas the preferential use of one type over the other can be well justified.

For example, foreign xenogenic or allogeneic cell materials widely used in tissue engineering can elicit immunorejection and therefore should be expected to be better used in combination with closed constructions. The nature of closed systems permits immunoprotection which is not characteristic of open scaffolds. On the other hand, open bioartificial tissues make a better use of stem cells that need only a short-term boost for differentiation.

4.4.1.2 Rooting Renewal

Open bioartificial tissues attract widespread attention stimulated by the view of stem cells,[230] in particular embryonic stem cells (ESC),[231] presenting an ideal source for transplantable cells. Due to their continuous and potentially extensive self-renewal ESC can be expanded and maintained *in vitro* indefinitely while retaining their ability for directed differentiation into any cell type in the body.[232]

The potential of ESC has been shown in repairing ischemic tissues, inducing the production of dopamine-generating cells for Parkinson's disease, the conversion to progenitor cells that were able to region-specifically differentiate inside the brain into neurons and astrocytes, or to cardiomyocytes, enhancing cardiac repair after myocardial infarction.[230-236] Where the implantation of other cells inclusive of predifferentiated SC is to be avoided open scaffolds can give superior results and may provide differentiation in a more efficient and specific manner.

The combination of ESC and open scaffolds can be expected to be especially beneficial for engineering tissues where vascularisation is essential (muscle, liver, spleen tissues). However, the use of human embryonic stem cells in regenerative therapies raises intense bioethical debates as human embryos are their only source.[232] Biologically, at least to some extent, ESC can be replaced by human adult stem cells, many of which can be cultured *in vitro*. Cell transplants from umbilical-cord blood or bone marrow have already been found to cure certain cancers, to aid in haematopoietic reconstitution after local tissue removal or destruction or to support the construction of heart valves.[233,235]

Adult stem cells do have a number of drawbacks.[12,237] Not all tissues can be harvested (neural) and some are prone to ageing characterised by the depletion of cell numbers over time. Although there are a number of encouraging reports on developmental plasticity of adult stem cells, suggesting that a single autogenic type could serve as a source of any differentiated type,[238] the use of stem cells continues being a matter of compromise. In this vein, the main efforts are focused on issues that require immediate consideration and are prerequisite in constructing bioartificial tissues making the concept technically plausible. Principally, whether the used cells are stem or lineage, they all are to build into a complex 3D structure.[239] This is the key factor in normal tissue development.[240]

4.4.2   Accepting Inescapable

Starting with seminal works from Kleinman and Bissell laboratories,[241,242] it is now clear that cultured within 3D supporting matrices cells are exposed to physicochemical and mechanical environments maximally approximated to physiological conditions of native cellular surroundings.[243] The evidence for this is being accumulated from different studies ranging from tissue morphogenesis and migration, metastasis and matrix remodelling to tissue engineering.[55,242,244-247] In many cases, it has been shown that the outcomes in 3D models may be diametrically different from those obtained on 2D surfaces.

For instance, cell migration observed in 2D would be restricted to adhesion and translocation, whereas in 3D the remodelling of the matrix together with the changing behaviour of migrating cell populations or mechanics of growing tissues can be assessed.[20,239] As a result, signals created by the cells and fed to the remodelled matrices are systematically passed by the matrix back to the cells to direct morphological processes that eventually define tissue phenotypes.[243] Cell–matrix interactions are thus assessed as truly reciprocal, the way they occur in native tissues and that in 2D cannot be revealed unambiguously.[247]

One of the main forces identified to drive the 3D organisation of a tissue is the variational changes in physical properties observed in the matrix as a result of local and bulk contractions incited by cell spreading.[245] The extent to which these are expressed is defined by spatial parameters introduced by the matrix. Most designed 3D matrices are governed by the aim of achieving bulk properties that can provide sufficient strength to resist mechanical stress wielded by cells. The relative success in it though is marked by statistical and heterogeneous dispersion of cells within a given matrix. Encouragingly, some matrices are accommodatable to the stress, exhibiting good remodelling capacities, but many are not, and respond solely with increases in density.[248,249] In native tissues this is compensated by the secretion of ECM components and does not cause dramatic changes in the matrix volume. For example, density increases of collagen or fibrin fibres that ensue at the expense of increased protein concentrations are proximal to surrounding cells.[249,250] This provides necessary tensile strengthening of the matrix that develops mechanical loading or tension controlled locally (by restraining or aligning fibres) and without affecting its bulk morphological properties. By contrast, in synthetic matrices (unless the synthesis of the native ECM has been induced) such changes invariably lead to decreases in the matrix volume, hampering cell assembly into the desired tissue architecture. In turn, this gives rise to morphologically undefined materials with poor differentiation profiles.

The stumbling point here can therefore be emphasised as the lack of control over cell positioning and matrix composition. This is addressed at the nano-to-microscale where the matrix–tissue assembly and structure–mechanics relationship are set.[6,20] Designing synthetic matrices is one strategy to advance this. However, given never-certain outcomes of synthetic designs, many teams shift their main research focus into devising technologies that would promote the cell-induced production of the ECM. The approach is becoming increasingly popular and is showing the anticipated tendency to extend beyond the subject in a traditional sense to other independently defined biotechnologies.[251–253] Yet, the underlying principles of all such possible methods converge into the same set of assumptions that for the purpose of this chapter are gathered in the context of two mainstream approaches: nano-to-microscale patterning or 3D micropatterning and *in vivo* bioreactors.

4.4.2.1 Patterning Positional

Micropatterning is usually given as a collective definition for miniaturisation techniques, irrespective of whether these reflect on top-down or bottom-up,

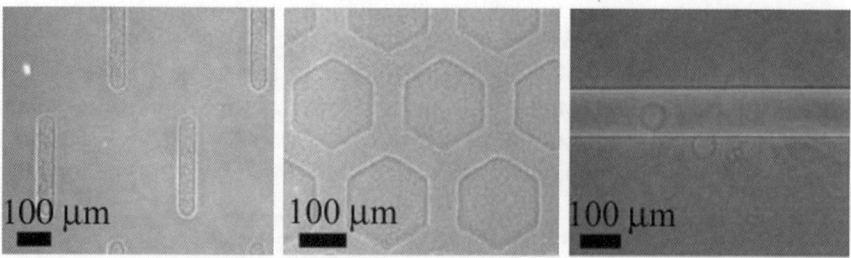

Figure 4.17   Micropatterned extracellular matrices; Arrays of posts (left) and
embedded matrigel (centre) in type I collagen, and a cylindrical channel
in fibrin (right), (reprinted from Nelson, C. M. and Tien, J. Microstructured extracellular matrices in tissue engineering and development.
Curr. Opinion Biotechnol., **17**, 518–523, Copyright 2006, with permission
from Elsevier).

2D- or 3D strategies.[254–256] However, as applicable to 3D tissue (re)construc-
tions in dynamically aggressive environments it can be bracketed out as a
bottom-up approach defined at the nano-to-microscale,[257] Figure 4.17.

This entails the use of temporary matrices that are, as described above, meant
to fully integrate into the host. Temporary matrices set up tissue architecture in
3D, induce the secretion and assembly of the native ECM and subsequently
deteriorate, leaving the maintenance of the architecture solely to the cell-
exerted ECM.

Surely, this is an idealised view. Nevertheless, any invention aiming at the
ideal temporary matrix would be focused on solving technical challenges
identical to those posed by the structural problem of an artificial matrix and on
implementing found solutions in the positional context within the living tissue.

For instance, in the application of micropatterning to 3D tissue culture
Bissell and colleagues demonstrated that 3D micropatterned matrices can
control the initial geometry of mammary epithelial tubules.[258]

A gel matrix was generated by moulding unpolymerised collagen around a
patterned elastomeric stamp. Epithelial cells embedded within it formed hollow
tubules conforming to the size and shape of the matrix cavities. When
branching was induced (by epidermal or hepatocyte growth factors) multicellular branches formed extending into the surrounding collagen. The branches were initiated only from the ends but not the sides of the tubules – a
pattern characteristic of the dichotomous branching of mammary end buds
in vivo. The tubules are thus polarised aiding in the directional cell invasion of
the surrounding tissues.

Interestingly, neutralising the activity of factors necessary for branching
morphogenesis such as matrix metalloproteinases or mesenchymal morphogens
could prevent the invasion, but failed to block the activation of branching. This
strongly suggests that the initiation of branching is spatially localised and that
the cells are instructed to branch according to their position within the tubule.
Being preset by the matrix this may be the factor that determines the invasive

phenotype. Indeed, by examining morphogenesis of tubules of varying geometry it was shown that the positional context of the matrix controls branch sites.[258] For instance, increasing the length of the tubules did not affect the polar character of branching even though increased branching density was observed; curved tubules were found to branch form the convex side of the curve, whereas bifurcated tubules branched from distal positions. The study demonstrates the role of the initial geometry of the tubule (seeded by the matrix) in dictating the positioning of epithelial branching, which in physiological terms is expressed in the transformation of epithelial tubule rudiments to mature ductal tree.[259]

The process of branching morphogenesis is universal across many cells types. In a series of studies concerning the role of dendritic branching in fibroblast motility, Grinnell and coworkers compared the behaviour of fibroblasts in 2D and 3D supports.[260,261]

It was initially observed that the flattened appearance of fibroblasts on planar surfaces is markedly different from that of connective tissue fibroblasts *in situ*. In contrast to the lamellar shape of the 2D fibroblast, the *in vivo* cells are usually characterised by the formation of dendritic networks with extended and slender branches. The researchers then hypothesised that such a difference in appearance should be due to the combination of topographic responsiveness of cells that are able to recognise nanometre features on the substratum surface,[262] and local surface stiffness that in one case – 2D surfaces – occurs and permits increased cell spreading and focal adhesion, but is not typical of the other – 3D matrices – where cells can be under isometric tension only during fibrotic conditions such as wound healing.[263]

With this in mind, a floating matrix – compliant or stress-accommodating 3D collagen gel with high cell-embedding capacity (typically $> 10^6/\text{ml}$) – was introduced. This is meant to help fibroblasts develop in a more *in situ*-like morphology compared with cells on surfaces. With no matrix contraction detected over the same incubation time as in planar surfaces, fibroblasts embedded in the matrix produced highly dynamic dendritic networks whose extensions appeared to project and retract.[264] In contrast, no dendritic projections were observed in collagen-coated surfaces. Consistent with this, the analysis of matured fibroblasts in restrained as opposed to floating matrices at low ($10^5/\text{ml}$) and high cell density ($10^6/\text{ml}$) revealed that matrix contraction was apparent only at high cell density, in which case dendritic networks appeared to be initial fibroblast morphology that over time converted to bipolar stellate structures.

The observation is intriguing as it tends to be typical for micropatterned and structurally restrained scaffolds but not for compliant and stress-responsive matrices.

Using the analogy of neuronal cells whose dendritic morphology is stabilised by actin and microtubules,[265] the dependence of the structure of the fibroblast dendritic networks on the cytoskeleton components was also assessed. It was found that cell extensions contained a core of microtubules with actin localised at the tips. The phenomenon of the formation and continued stabilisation of dendritic networks mediated by both actin microfilaments and microtubules was not observed in 2D and could only be mimicked using 3D matrices. In particular, the concentration of actin at the tips of the microtubule core can serve as another example of positional context in directing or setting-up branching morphogenesis.

4.4.2.2   Relating Interfacial

Most recently, Sia and coworkers extended positional setting into multiple micropatterning enabling interfacing multiple but not mixing 3D matrices in tissue modelling.[266]

The researchers argued that because native tissues are composed of multiple ECM types putting bulk and patterned natural ECM phases together would help in countering cellular contractile forces.

This, most of all technically challenging task, requires a construction capable of not only incorporating multiple microscale patterns as intended but also of interfacing individual phases within a sustained matrix geometry. It was therefore assumed that collagen fibres in a preformed ECM may act as nucleation sites for the assembly of collagen fibres in an adjacent gelling or bulk ECM. Adjacent matrices are thus to be used for the integration of adjacent phases; that is, matrices at the phase boundary would act as templates for the assembly of new matrices within the bulk ECM.

A proposed construct incorporating such specifications was based on the collagen fibres, patterned phases, flowing into microfluidic alginate hydrogels, gelling or bulk phases. It this design, the gelation of the patterned phase (ECM collagen I) occurred at the expense of collagen fibres (preseeded by collagen-enriched bulk phase) crossing from the bulk phase boundary, Figure 4.18(a). In contrast, when a collagen-free alginate was used as a bulk phase, patterned fibres formed exclusively in the flowing ECM solution and were not supplied by the bulk boundary. This supports the feeding model of new collagen fibres that tended to be assembled at the boundary by pre-existing collagen accumulated at the interface. Furthermore, it was found that in a collagen-containing bulk phases most of the patterned fibres appeared to concentrate at the interface, whereas in a pure alginate bulk phase patterned collagen was preferentially built via the channel.

The model was used to probe if collagen fibre assembly could link natural ECM phases together to resist cell contraction. In this study, fibroblasts and endothelial cells were seeded into alginate (with and without collagen, bulk phases) and channel-patterned collagen I matrices (patterned phases), respectively, Figure 4.18(b). Consistent with the design, fibroblasts appeared as uniformly distributed in the bulk phases, with the distribution of endothelial cells being spatially restricted to the micropatterned phase. In collagen-free bulk phases the patterned collagen was found to be detached from the bulk interface and significantly contracted after just a few hours of culturing. No further migration of endothelial cells was observed after the contraction. In marked contrast, continued migration and subsequent primary proliferation (upon the

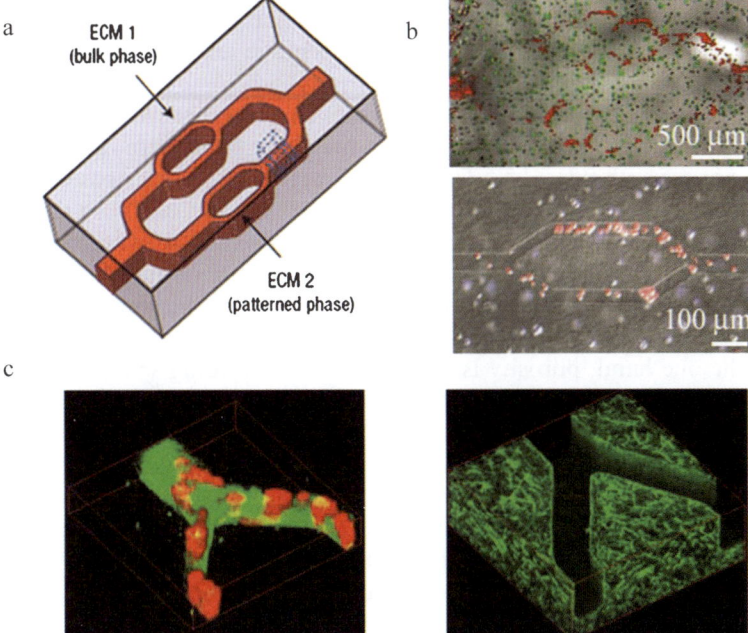

Figure 4.18 Microstructured multiple 3D matrices; (a) Schematic illustrating micro-
fluidically patterned and bulk hydrogel ECMs, (b) Micropatterned collagen imaged with seeded endothelial cells (red) localised in the collagen
channel and fibroblasts (green and blue) distributed in the bulk phase, (c)
3D reconstructions of a collagen channel (green) engineered with sharp
boundaries seeded with endothelial cells (red) in a pure alginate phase
(left) and a collagen-doped alginate bulk phase before patterning
showing a uniform distribution of collagen fibres (right), (reprinted by
permission from Macmillan Publishers Ltd: Gillette, B. M. *et al. In situ*
collagen assembly for integrating microfabricated three-dimensional cell-
seeded matrices. *Nature Mater.*, **7**, 636–640, Copyright (2008)).

addition of angiogenic growth factors) was observed in the collagen-loaded
bulk phases in which the phase interfaces remained stably integrated over the
course of several weeks, Figure 4.18(c).

The design exemplifies a number of observations made by other groups
based on different 3D micropatterned systems,[267–270] but its principle
distinction is high-resolution control over cell and ECM colocalisation and
positioning in 3D.

The method also represents an excellent interfacial model of diffusively
permeable phases that can be used for studying cell migration and proliferation
in response to designed external stimuli. But most importantly, designs of this
type hold promise for developing implementable 3D mimetics of the native
ECM via creating multiple ECM phases, the regularity and reconstruction of
which requires the flexibility of more than one assembly mode. The latter is

probably why most micropatterning techniques employ native ECM assemblies – collagen or fibrin matrices. The adaptability of artificial self-assembly systems based on helical, β-structured peptides or peptidic amphiphiles to 3D matrix micropatterning has yet to be demonstrated.

Alternative materials may derive from natural or synthetic polymers, in which case an appropriate 3D architecture to pattern against would be 3D porous scaffolds.[271,272] The main attraction of polymer-based materials is the relative ease with which the degree of their porosity and pore sizes can be determined and regulated.[273] For polymer-based scaffolds whose fabrication largely relies on solidification, crosslinking and thus on morphological pre-shaping, porosity becomes the prime property that puts this type into the category of open and temporary matrices, Figure 4.19.

On the one hand, porosity is responsible for providing spatial freedom for cells to infiltrate the matrix, migrate and proliferate. This freedom is more substantial than that provided by closed systems whose function in this regard may be limited to creating permeability sufficient for the inflow of nutrients and the elution of metabolic and biodegradation waste. On the other hand, porosity in conjunction with biodegradability is necessary for initiating the production of the native ECM. Equally importantly, once implemented this should be possible to be tailored to respond to changing proliferation rates of the seeded cells – an intrinsic characteristic of dynamically equilibrated self-assembled matrices. Altogether, this defines the main hurdle in engineering polymer-based scaffolds as providing reciprocal mechanotransduction between scaffolds and cells.[274,275]

Porous polymer scaffolds can be shown as ideal systems for engineering different types of tissues (*e.g.* cartilage, bone, muscle). However, to a large extent this can be attributed to their excellent grafting properties rather than to efficient nano-to-microscale patterning as imposed on living tissues they greatly lack.

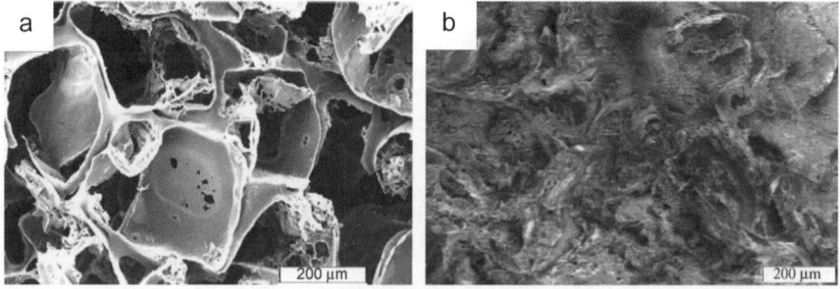

Figure 4.19   Electron micrographs of porous polymer 3D scaffolds; (a) before and (b) after cell-seeding, (reprinted from Zoldan, J. and Levenberg, S. Engineering Three-Dimensional Tissue Structures Using Stem Cells. *Methods Enzymol.*, **420**, 381–391, Copyright 2006, with permission from Elsevier).

Ironically, this turned around positively, prompting a comparable, if not more, efficient approach to bioartificial tissues – *in vivo* bioreactors.

4.4.2.3   Grafting Integral

Arguably, the idea of *in vivo* bioreactors and its experimental introduction was stimulated by the drawbacks of grafting strategies, in particular in relation to bone regeneration.[276] Traditional bone grafting is reliant on auto- and allografts that are utilised for bone losses, spinal fusions, long-bone fractures and nonunions or for bone reconstruction after tumour resection.[277–279] Although bone grafts have substantially advanced bone engineering their use still poses a number of complications mostly derived from the imperfections of the graft materials themselves that consequently lead to tissue devascularisation, risks of infections or morbidity at donor sites.[280–282]

Exemplified by the application in bone engineering, the grafting strategy has evolved into the concept of an ideal graft designer that can be formulated as the sum of three interrelated elements: cell-conductive matrix, cell-conductive proteins and committed cells.[276,283,284] All three being pooled together make up a simplified version of an *in vivo* bioreactor.

Originally, Holt *et al.*[276] showed that such a reactor can be made as a sealed hydroxyapatite scaffold or coralline cylinder (porous or conductive matrix) assembled with a vascular pedicle and impenetrable silicone shield. The bioreactor is introduced into a living tissue and coated with growth factors (conductive proteins), with the recruitment of osteogenic cells (committed cells) capitalising on the neovascularisation of the scaffold acting as a conduit. In this respect, the bioreactor is not a closed bioartificial tissue, as it may seem, but a sealed open tissue as its main purpose – emulation of the bone formation – occurs in an isolated environment and the surrounding tissue serves as a supply of cells invading the reactor. Indeed, it was shown that the cells not only successfully invade the scaffold but also spread and proliferate within with the subsequent production of the native ECM – the main characteristic of a living open tissue. The design gives a very good example of how vascularisation can be initiated and maintained within a short period of time but sufficient for the generation of living tissue. It also advocates for *in vivo* approach to engineering bone as opposed to *in vitro* techniques constantly failing to spatially relate forming blood vessels and bone-committed cells in initiating and maintaining architectural integrity of bone. With many unsuccessful stories of engineering functional bone *in vitro* stumbled on the pivotal reliance of bone tissue on vascularisation, it has become clear that the development of different cell types into one tissue architecture has to present a concomitant process.[285] The realisation of the problem prompted the emergence of approaches that emphasised using porous polymeric materials (ceramics, demineralised grafts) in combination with bone stem cells and morphogenetic proteins; that is, *in vivo* bioreactor designers. Considered as a straight route to the interplay between scaffolds, cells and growth factors and hence to creating environments promoting functional tissue regeneration, this format of *in vivo* bioreactors has been widely accepted and used.[286–288]

Yet, an intriguing concept was reported by Stevens *et al.*[289] who hypothesised that the transplantation formula, scaffold-cell growth factors, in bone engineering can be totally avoided. It was shown that large volumes of bone can be produced without the need for cell transplantation and the administration of growth factor. The essence of the strategy is not an introduction or injection but a creation of an *in vivo* bioreactor within the body. The bioreactor in this case is a space deliberately formed between the surface of a long bone and its membrane rich in stem cells such that the required cell population and all necessary growth signals are prelocalised or derived locally. The engineering process is thus underpinned by a provoked healing response within the created space. The idea of constructing a bioreactor by simply generating a space of a chosen geometry within the body is a very attractive alternative to exogenously introduced bioreactors. However, the technical realisation of the concept proved to require a comparable external intervention. Firstly, a hydraulic elevation procedure earlier developed by the group[290] was used to generate the space. Secondly, once made the space was filled with a cell-supporting matrix, alginate gel, followed by its crosslinking. Thus, the proposed bioreactor is at least a two-component system that cannot function without an injectable 3D matrix.

The role of injectable scaffolds and their impact on tissue development has formed a separate subject in regenerative medicine that proves to be extremely instrumental in the formulation of best-performing 3D matrices. This field is vast and demands much closer focus than permitted within the scope of a single section. More detailed accounts of this spectrum of tissue-engineering studies are given in several excellent reviews that provide a contemporary assessment of the pros and cons of injectable and in general polymer-based scaffolds as candidates for *in vivo* 3D matrices.[14,20,291,292]

Possibly due to their intrinsic conceptual limitations or relatively short history *in vivo* bioreactor designs have yet to progress beyond laboratory development. Distinctively supportive of the concept in this regard are investigations targeting other issues in cell and tissue biology;[253] for example, in providing 3D biomimetic models that can find use in establishing the effects of micro-environmental conditions and nanoscale patterning on tumour malignancy *in vivo* as compared to those established *in vitro*.[288]

4.5   Outlook

Assessing the progress of regenerative medicine through the prism of *in vivo* bioreactors provides valid reasoning for levelling up design directions towards an artificial matrix. By their complexity and promised efficiency *in vivo* bioreactors constitute the strategy to develop; by their sufficiently moderate technological advancement they reinstate the midway point between the chemical mimicry of the ECM and the clinical implementation of regenerative therapies. The key is, however, in that devising more efficient bioreactors this divulges the main challenges facing artificially guided regeneration as an integrated problem. As the very task of constructing synthetic substituents of the ECM the problem is multifold. Ultra- and macrostructural integrity of cell-supportive scaffolds, reliable cell sources and biocompatible mass transport, matrix remodelling and maintained vascularisation are those stumbling blocks that lay between the potential and the reality of induced regeneration, between "today's Frankenstein and tomorrow's benefactor".[229]

References

1. R. J. Goss, *Principles of Regeneration*, Academic Press, New York, 1969.
2. R. J. Goss, *The natural history (and mystery) of regeneration*, ed. C. E. Dinsmore, Cambridge University Press, Cambridge, 1991.
3. E. M. Tanaka, Regeneration: If They Can Do It Why Can't We? *Cell*, 2003, **113**, 559–562.
4. O. J. Reichman, Evolution of Regeneration Capabilities, *The American Naturalist*, 1984, **123**, 752–763.
5. R. J. Goss, Why Mammals Don't Regenerate—Or Do They? *News Physiol. Sci.*, 1987, **2**, 112–115.
6. R. Langer and C. A. Vacanti, Tissue engineering, *Science*, 1993, **260**, 920–926.
7. R. Langer and D. A. Tirrell, Designing materials for biology and medicine, *Nature*, 2004, **428**, 487–492.
8. A. S. Alvarado and P. A. Tsonis, Bridging the regeneration gap: genetic insights from diverse animal models, *Nat. Rev. Genet.*, 2006, **7**, 873–884.
9. P. A. Tsonis, Regenerative biology: the emerging field of tissue repair and restoration, *Differentiation*, 2002, **70**, 397–409.
10. P. A. Tsonis, Regeneration in Vertebrates, *Dev. Biol.*, 2000, **221**, 273–284.
11. G. C. Gurtner, S. Werner, Y. Barrandon and M. T. Longaker, Wound repair and regeneration, *Nature*, 2008, **453**, 314–321.
12. D. L. Stocum, Tissue restoration through regenerative biology and medicine, *Adv. Anat. Embryol. Cell Biol.*, 2004, **176**, 1–101.
13. S. Levenberg, R. Langer and P. S. Gerald, Advances in Tissue Engineering, *Curr. Top. Dev. Biol.*, 2004, **61**, 113–134.
14. B.-S. Kim and D. J. Mooney, Development of biocompatible synthetic extracellular matrices for tissue engineering, *Trends Biotechnol.*, 1998, **16**, 224–230.
15. T. Starborg, Y. Lu, R. S. Meadows, K. E. Kadler and D. F. Holmes, *Electron Microscopy in Cell–Matrix Research Methods*, 2008, **45**, 53–64.
16. P. D. Yurchenco, D. E. Birk and R. P. Mecham (eds.), *Extracellular Matrix Assembly and Structure*, Academic, San Diego, 1994.

17. E. A. Silva, D. J. Mooney and P. S. Gerald, Synthetic Extracellular Matrices for Tissue Engineering and Regeneration, *Curr. Top. Dev. Biol.*, 2004, **64**, 181–205.

18. H. K. Kleinman, D. Philp and M. P. Hoffman, Role of the extracellular matrix in morphogenesis, *Curr. Opin. Biotechnol.*, 2003, **14**, 526–532.

19. R. B. Vernon and E. H. Sage, Between molecules and morphology. Extracellular matrix and creation of vascular form, *Am. J. Pathol.*, 1995, **147**, 873–883.

20. M. P. Lutolf and J. A. Hubbell, Synthetic biomaterials as instructive extracellular microenvironments for morphogenesis in tissue engineering, *Nat. Biotechnol.*, 2005, **23**, 47–55.

21. C. M. Nelson and J. Tien, Microstructured extracellular matrices in tissue engineering and development, *Curr. Opin. Biotechnol.*, 2006, **17**, 518–523.

22. M. van der Rest and R. Garrone, Collagens as multidomain proteins, *Biochimie*, 1990, **72**, 473–484.

23. P. M. Cowan, S. McGavin and A. C. T. North, The Polypeptide Chain Configuration of Collagen, *Nature*, 1955, **176**, 1062–1064.

24. G. N. Ramachandran and G. Kartha, Structure of Collagen, *Nature*, 1954, **174**, 269–270.

25. G. N. Ramachandran and G. Kartha, Structure of Collagen, *Nature*, 1955, **176**, 593–595.

26. A. Rich and F. H. C. Crick, The Structure of Collagen, *Nature*, 1955, **176**, 915–916.

27. A. Steplewski, V. Hintze and A. Fertala, Molecular basis of organisation of collagen fibrils, *J. Struct. Biol.*, 2007, **157**, 297–307.

28. M. van der Rest and R. Garrone, Collagen family of proteins, *FASEB J.*, 1991, **5**, 2814–2823.

29. K. E. Kadler, D. F. Holmes, J. A. Trotter and J. A. Chapman, Collagen fibril formation, *Biochem. J.*, 1996, **316**, 1–11.

30. D. F. Holmes, H. K. Graham, J. A. Trotter and K. E. Kadler, STEM/TEM studies of collagen fibril assembly, *Micrometre*, 2001, **32**, 273–285.

31. D. F. Holmes, C. J. Gilpin, C. Baldock, U. Ziese, A. J. Koster and K. E. Kadler, Corneal collagen fibril structure in three dimensions: Structural insights into fibril assembly, mechanical properties, and tissue organization, *Proc. Natl. Acad. Sci. USA*, 2001, **98**, 7307–7312.

32. L. M. Shaw and B. R. Olsen, FACIT collagens: diverse molecular bridges in extracellular matrices, *Trends Biochem. Sci.*, 1991, **16**, 191–194.

33. C. Wang, H. Luosujärvi, J. Heikkinen, M. Risteli, L. Uitto and R. Myllylä, The third activity for lysyl hydroxylase 3: galactosylation of hydroxylysyl residues in collagens *in vitro*, *Matrix Biol.*, 2002, **21**, 559–566.

34. J. Baum and B. Brodsky, Folding of peptide models of collagen and misfolding in disease, *Curr. Opin. Struct. Biol.*, 1999, **9**, 122–128.

35. H. Herrmann and U. Aebi, Intermediate Filaments: Molecular Structure, Assembly Mechanism, and Integration Into Functionally Distinct Intracellular Scaffolds, *Annu. Rev. Biochem.*, 2004, **73**, 749–789.

36. A. M. Hindeleh and R. Hosemann, Paracrystals representing the physical state of matter, *J. Phys. C: Solid State Phys.*, 1988, **21**, 4155–4170.

37. A. Nagayama and S. Dales, Rapid purification and the immunological specificity of mammalian microtubular paracrystals possessing an ATPase activity, *Proc. Natl. Acad. Sci. USA*, 1970, **66**, 464–471.

38. M. K. Gardner, A. J. Hunt, H. V. Goodson and D. J. Odde, Microtubule assembly dynamics: new insights at the nanoscale, *Curr. Opin. Cell Biol.*, 2008, **20**, 64–70.

39. B. T. Helfand, L. Chang and R. D. Goldman, Intermediate filaments are dynamic and motile elements of cellular architecture, *J. Cell Sci.*, 2004, **117**, 133–141.

40. H. K. Graham, D. F. Holmes, R. B. Watson and K. E. Kadler, Identification of collagen fibril fusion during vertebrate tendon morphogenesis. The process relies on unipolar fibrils and is regulated by collagen–proteoglycan interaction, *J. Mol. Biol.*, 2000, **295**, 891–902.

41. D. F. Holmes, M. P. Lowe and J. A. Chapman, Vertebrate (chick) collagen fibrils formed *in vivo* can exhibit a reversal in molecular polarity, *J. Mol. Biol.*, 1994, **235**, 80–83.

42. K. E. Kadler, Y. Hojima and D. J. Prockop, Collagen fibrils *in vitro* grow from pointed tips in the C- to N-terminal direction, *Biochem. J.*, 1990, **268**, 339–343.

43. U. Hansen and P. Bruckner, Macromolecular Specificity of Collagen Fibrillogenesis: Fibrils of Collagens I and XI Contain a Heterotypic Alloyed Core and a Collagen I Sheath, *J. Biol. Chem.*, 2003, **278**, 37352–37359.

44. T. C. Baradet, J. C. Haselgrove and J. W. Weisel, Three-dimensional reconstruction of fibrin clot networks from stereoscopic intermediate voltage electron microscope images and analysis of branching, *Biophys. J.*, 1995, **68**, 1551–1560.

45. J. W. Weisel, The mechanical properties of fibrin for basic scientists and clinicians, *Biophys. Chem.*, 2004, **112**, 267–276.

46. J. W. Weisel, in *Adv. Prot. Chem.*, eds. D. A. D. Parry and J. M. Squire, Academic Press, 2005, 247–299.

47. J. W. Weisel, C. Nagaswami and L. Makowski, Twisting of fibrin fibers limits their radial growth, *Proc. Natl. Acad. Sci. USA*, 1987, **84**, 8991–8995.

48. L. Medved, T. Ugarova, Y. Veklich, N. Lukinova and J. Weisel, Electron microscope investigation of the early stages of fibrin assembly: Twisted protofibrils and fibers, *J. Mol. Biol.*, 1990, **216**, 503–509.

49. E. A. Ryan, L. F. Mockros, J. W. Weisel and L. Lorand, Structural Origins of Fibrin Clot Rheology, *Biophys. J.*, 1999, **77**, 2813–2826.

50. C. M. Kielty, M. J. Sherratt and C. A. Shuttleworth, Elastic fibres, *J. Cell Sci.*, 2002, **115**, 2817–2828.

51. C. Kielty, T. Wess, L. Haston, J. Ashworth, M. Sherratt and C. Shuttleworth, Fibrillin-rich microfibrils: elastic biopolymers of the extracellular matrix, *J. Mus. Res. Cell Motil.*, 2002, **23**, 581–596.

52. F. Ramirez, L. Y. Sakai, H. C. Dietz and D. B. Rifkin, Fibrillin micro-fibrils: multipurpose extracellular networks in organismal physiology, *Physiol. Genomics*, 2004, **19**, 151–154.

53. W. F. Daamen, J. H. Veerkamp, J. C. M. van Hest and T. H. van Kuppevelt, Elastin as a biomaterial for tissue engineering, *Biomaterials*, 2007, **28**, 4378–4398.

54. R. Q. Qian and R. W. Glanville, Alignment of Fibrillin Molecules in Elastic Microfibrils Is Defined by Transglutaminase-Derived Cross-Links, *Biochemistry*, 1997, **36**, 15841–15847.

55. K. Wolf, I. Mazo, H. Leung, K. Engelke, U. H. von Andrian, E. I. Deryugina, A. Y. Strongin, E.-B. Brocker and P. Friedl, Compensation mechanism in tumor cell migration: mesenchymal-amoeboid transition after blocking of pericellular proteolysis, *J. Cell Biol.*, 2003, **160**, 267–277.

56. Y. Hegerfeldt, M. Tusch, E.-B. Brocker and P. Friedl, Collective Cell Movement in Primary Melanoma Explants: Plasticity of Cell-Cell Inter-action, {beta}1-Integrin Function, and Migration Strategies, *Cancer Res.*, 2002, **62**, 2125–2130.

57. D. Schulz Torres, M. T. Freyman, I. V. Yannas and M. Spector, Tendon cell contraction of collagen-GAG matrices *In vitro*: effect of cross-linking, *Biomaterials*, 2000, **21**, 1607–1619.

58. R. J. Schlueter and A. Veis, The Macromolecular Organisation of Dentine Matrix Collagen. II. Periodate Degradation and Carbohydrate Cross-Linking, *Biochemistry*, 1964, **3**, 1657–1665.

59. J. M. Orban, L. B. Wilson, J. A. Kofroth, M. S. El-Kurdi, T. M. Maul and D. A. Vorp, Crosslinking of collagen gels by transglutaminase, *J. Biomed. Mater. Res. A*, 2004, **68**(a), 756–762.

60. H. P. Wiesmann, U. Meyer, U. Plate, H. J. Höhling and W. J. Kwang, Aspects of Collagen Mineralisation in Hard Tissue Formation, *Int. Rev. Cytology*, 2004, **242**, 121–156.

61. L. B. Rocha, R. L. Adam, N. J. Leite, K. Metze and M. A. Rossi, Bio-mineralisation of polyanionic collagen-elastin matrices during cavarial bone repair, *J. Biomed. Mater. Res. A*, 2006, **79**(a), 237–245.

62. S. D. Figueiró, A. A. M. Macêdo, M. R. S. Melo, A. L. P. Freitas, R. A. Moreira, R. S. de Oliveira, J. C. Góes and A. S. B. Sombra, On the dielectric behaviour of collagen-algal sulfated polysaccharide blends: Effect of glutaraldehyde crosslinking, *Biophys. Chem.*, 2006, **120**, 154–159.

63. S. Singh and J. Behari, Physical characteristics of bone composite mate-rials, *J. Biol. Phys.*, 1984, **12**, 121–128.

64. C. C. Silva, D. Thomazini, A. G. Pinheiro, N. Aranha, S. D. Figueiró, J. C. Góes and A. S. B. Sombra, Collagen-hydroxyapatite films: piezo-electric properties, *Mater. Sci. Eng. B*, 2001, **86**, 210–218.

65. S. Sun, T. Kou and H. Zhu, A study on bioelectret collagen, *J. Appl. Polym. Sci.*, 1997, **64**, 267–271.

66. M. Takeuchi, M. Sekino, N. Iriguchi and S. Ueno, Dependence of the Spin-Spin Relaxation Time of Water in Collagen Gels on Collagen Fiber Directions, *Magn. Reson. Med. Sci.*, 2004, **3**, 153–157.

67. N. Dubey, P. C. Letourneau and R. T. Tranquillo, Neuronal contact guidance in magnetically aligned fibrin gels: effect of variation in gel mechano-structural properties, *Biomaterials*, 2001, **22**, 1065–1075.

68. S. Sarkar, M. Dadhania, P. Rourke, T. A. Desai and J. Y. Wong, Vascular tissue engineering: microtextured scaffold templates to control organisation of vascular smooth muscle cells and extracellular matrix, *Acta Biomater.*, 2005, **1**, 93–100.

69. D. Olsen, C. Yang, M. Bodo, R. Chang, S. Leigh, J. Baez, D. Carmichael, M. Perälä, E.-R. Hämäläinen, M. Jarvinen and J. Polarek, Recombinant collagen and gelatin for drug delivery, *Adv. Drug Deliv. Rev.*, 2003, **55**, 1547–1567.

70. F. W. Kotch and R. T. Raines, Self-assembly of synthetic collagen triple helices, *Proc. Natl. Acad. Sci. USA*, 2006, **103**, 3028–3033.

71. M. Seligman, R. F. Eilberg and L. Fishman, Mineralisation of elastin extracted from human aortic tissues, *Calcif. Tissue Res.*, 1975, **17**, 229–234.

72. W. Hornebeck, J. M. Tixier and L. Robert, Inducible adhesion of mesenchymal cells to elastic fibers: elastonectin, *Proc. Natl. Acad. Sci. USA*, 1986, **83**, 5517–5520.

73. L. J. Currie, J. R. Sharpe and R. Martin, The use of fibrin glue in skin grafts and tissue-engineered skin replacements: a review, *Plast. Reconstr. Surg.*, 2001, **108**, 1713–1726.

74. A. Prado, P. Andrades, S. Danilla, S. Benitez and P. Wisnia, Use of Aerosolized Bovine-Prepared Fibrin Glue for Skin Fixation After Primary Open Rhinoplasty: A Prospective Randomized and Controlled Trial, *Aesthetic Plast. Surg.*, 2006, **30**, 568–573.

75. C. Lasa, J. Hollinger, W. Drohan and M. MacPhee, Delivery of demineralized bone powder by fibrin sealant, *Plast. Reconstr. Surg.*, 1995, **96**, 1409–1418.

76. D. J. Mackenzie, R. Sipe, D. Buck, W. Burgess and J. Hollinger, Recombinant human acidic fibroblast growth factor and fibrin carrier regenerates bone, *Plast. Reconstr. Surg.*, 2001, **107**, 989–996.

77. B. K. Mann, Biologic gels in tissue engineering, *Clin. Plast. Surg.*, 2003, **30**, 601–609.

78. T. Coviello, P. Matricardi and F. Alhaique, Drug delivery strategies using polysaccharidic gels, *Exp. Opinion Drug Deliv.*, 2006, **3**, 395–404.

79. G. Skjåk-Braek, Alginates: biosyntheses and some structure-function relationships relevant to biomedical and biotechnological applications, *Biochem. Soc. Trans.*, 1992, **20**, 27–33.

80. O. Smidsrød and G. Skjåk-Braek, Alginate as immobilisation matrix for cells, *Trends Biotechnol.*, 1990, **8**, 71–78.

81. W. Hashimoto, M. Okamoto, T. Hisano, K. Momma and K. Murata, Sphingomonas sp. A1 lyase active on both poly [beta]–mannuronate and heteropolymeric regions in alginate, *J. Ferment. Bioeng.*, 1998, **86**, 236–238.

82. Y. Mishima, K. Momma, W. Hashimoto, B. Mikami and K. Murata, Crystal Structure of AlgQ2, a Macromolecule (Alginate)-binding Protein

of Sphingomonas sp. A1, Complexed with an Alginate Tetrasaccharide at 1.6-A Resolution, *J. Biol. Chem.*, 2003, **278**, 6552–6559.

83. B. Thu, O. Smidsrød, G. Skjak-Brk, R. H. Wijffels, R. M. Buitelaar, C. Bucke and J. Tramper, Alginate gels – Some structure-function correlations relevant to their use as immobilisation matrix for cells, *Prog. Biotechnol.*, 1996, **11**, 19–30.

84. P. Gacesa, Alginates, *Carbohydr. Polym.*, 1988, **8**, 161–182.

85. K. I. Draget, G. Skjåk-Bræk and O. Smidsrød, Alginate based new materials, *Int. J. Biol. Macromol.*, 1997, **21**, 47–55.

86. N. Bhattarai, Z. Li, D. Edmondson and M. Zhang, Alginate-Based Nanofibrous Scaffolds: Structural, Mechanical, and Biological Properties, *Adv. Mater.*, 2006, **18**, 1463–1467.

87. J. Rowley, G. Madlambayan, J. Faulkner and D. J. Mooney, Alginate hydrogels as synthetic extracellular matrix materials, *Biomaterials*, 1999, **20**, 45–53.

88. J. L. Drury and D. J. Mooney, Hydrogels for tissue engineering: scaffold design variables and applications, *Biomaterials*, 2003, **24**, 4337–4351.

89. F. Brandl, F. Sommer and A. Goepferich, Rational design of hydrogels for tissue engineering: Impact of physical factors on cell behaviour, *Biomaterials*, 2007, **28**, 134–146.

90. S. R. Khetani and S. N. Bhatia, Engineering tissues for *In vitro* applications, *Curr. Opin. Biotechnol.*, 2006, **17**, 524–531.

91. A. Gutowska, B. Jeong and M. Jasionowski, Injectable gels for tissue engineering, *Anat. Rec*, 2001, **263**, 342–349.

92. V. Liu Tsang and S. N. Bhatia, Three-dimensional tissue fabrication, *Adv. Drug Delivery Rev.*, 2004, **56**, 1635–1647.

93. K. Y. Lee and D. J. Mooney, Hydrogels for Tissue Engineering, *Chem. Rev.*, 2001, **101**, 1869–1880.

94. K. H. Bouhadir, K. Y. Lee, E. Alsberg, K. L. Damm, K. W. Anderson and D. J. Mooney, Degradation of Partially Oxidized Alginate and Its Potential Application for Tissue Engineering, *Biotechnol. Prog.*, 2001, **17**, 945–950.

95. T. Boontheekul, H.-J. Kong and D. J. Mooney, Controlling alginate gel degradation utilising partial oxidation and bimodal molecular weight distribution, *Biomaterials*, 2005, **26**, 2455–2465.

96. H. J. Kong, D. Kaigler, K. Kim and D. J. Mooney, Controlling Rigidity and Degradation of Alginate Hydrogels via Molecular Weight Distribution, *Biomacromolecules*, 2004, **5**, 1720–1727.

97. J. L. Drury, T. Boontheekul and D. J. Mooney, Cellular cross-linking of peptide modified hydrogels, *J. Biomech. Eng.*, 2005, **127**, 220–228.

98. H. J. Kong and D. J. Mooney, The effects of poly(ethyleneimine) (PEI) molecular weight on reinforcement of alginate hydrogels, *Cell Transplant*, 2003, **12**, 779–785.

99. C. K. Kuo and P. X. Ma, Ionically crosslinked alginate hydrogels as scaffolds for tissue engineering: Part 1. Structure, gelation rate and mechanical properties, *Biomaterials*, 2001, **22**, 511–521.

100. B. Balakrishnan and A. Jayakrishnan, Self-cross-linking biopolymers as injectable in situ forming biodegradable scaffolds, *Biomaterials*, 2005, **26**, 3941–3951.

101. T. Majima, T. Funakosi, N. Iwasaki, S.-T. Yamane, K. Harada, S. Nonaka, A. Minami and S.-I. Nishimura, Alginate and chitosan polyion complex hybrid fibers for scaffolds in ligament and tendon tissue engineering, *J. Orthop. Sci.*, 2005, **10**, 302–307.

102. M. Goldberg, R. Langer and X. Jia, Nanostructured materials for applications in drug delivery and tissue engineering, *J. Biomater. Sci. Polym. Ed.*, 2007, **18**, 241–268.

103. H. Park, C. Cannizzaro, G. Vunjak-Novakovic, R. Langer, C. A. Vacanti and O. C. Farokhzad, Nanofabrication and Microfabrication of Functional Materials for Tissue Engineering, *Tissue Eng.*, 2007, **13**, 1867–1877.

104. D. Howard, L. D. Buttery, K. M. Shakesheff and S. J. Roberts, Tissue engineering: strategies, stem cells and scaffolds, *J. Anat.*, 2008, **213**, 66–72.

105. F. Cellesi, N. Tirelli and J. A. Hubbell, Materials for cell encapsulation via a new tandem approach combining reverse thermal gelation and covalent crosslinking, *Macromol. Chem. Phys.*, 2002, **203**, 1466–1472.

106. W. E. Hennink and C. F. van Nostrum, Novel crosslinking methods to design hydrogels, *Adv. Drug Delivery Rev.*, 2002, **54**, 13–36.

107. J. A. Hubbell, Bioactive biomaterials, *Curr. Opin. Biotechnol.*, 1999, **10**, 123–129.

108. K. T. Nguyen and J. L. West, Photopolymerisable hydrogels for tissue-engineering applications, *Biomaterials*, 2002, **23**, 4307–4314.

109. T. J. Sanborn, P. B. Messersmith and A. E. Barron, *In situ* crosslinking of a biomimetic peptide-PEG hydrogel via thermally triggered activation of factor XIII, *Biomaterials*, 2002, **23**, 2703–2710.

110. M. P. Lutolf, F. E. Weber, H. G. Schmoekel, J. C. Schense, T. Kohler, R. Muller and J. A. Hubbell, Repair of bone defects using synthetic mimetics of collagenous extracellular matrices, *Nat. Biotechnol.*, 2003, **21**, 513–518.

111. R. J. Christie, D. J. Findley, M. Dunfee, R. D. Hansen, S. C. Olsen and D. W. Grainger, Photopolymerized hydrogel carriers for live vaccine ballistic delivery, *Vaccine*, 2006, **24**, 1462–1469.

112. J. Elisseeff, K. Anseth, D. Sims, W. McIntosh, M. Randolph and R. Langer, Transdermal photopolymerisation for minimally invasive implantation, *Proc. Natl. Acad. Sci. USA*, 1999, **96**, 3104–3107.

113. K. Jurgensen, D. Aeschlimann, V. Cavin, M. Genge and E. B. Hunziker, A New Biological Glue for Cartilage-Cartilage Interfaces: Tissue Transglutaminase, *J. Bone Joint Surg. Am.*, 1997, **79**, 185–193.

114. D. J. Heath, P. Christian and M. Griffin, Involvement of tissue transglutaminase in the stabilisation of biomaterial/tissue interfaces important in medical devices, *Biomaterials*, 2002, **23**, 1519–1526.

115. E. A. Verderio, T. S. Johnson and M. Griffin, Transglutaminases in wound healing and inflammation, *Prog. Exp. Tumor Res.*, 2005, **38**, 89–114.

116. M. P. Lutolf, G. P. Raeber, A. H. Zisch, N. Tirelli and J. A. Hubbell, Cell-responsive synthetic hydrogels, *Adv. Mater.*, 2003, **15**, 888–892.

117. J. Kopecek and J. Yang, Hydrogels as Smart Biomaterials, *Polym. Int.*, 2007, **56**, 1078–1098.

118. J. Kopecek, Hydrogel biomaterials: A smart future?, *Biomaterials*, 2007, **28**, 5185–5192.

119. J. Kopecek, Smart and genetically engineered biomaterials and drug-delivery systems, *Eur. J. Pharm. Sci.*, 2003, **20**, 1–16.

120. W. A. Petka, J. L. Harden, K. P. McGrath, D. Wirtz and D. A. Tirrell, Reversible Hydrogels from Self-Assembling Artificial Proteins, *Science*, 1998, **281**, 389–392.

121. W. Shen, K. Zhang, J. A. Kornfield and D. A. Tirrell, Tuning the erosion rate of artificial protein hydrogels through control of network topology, *Nat. Mater.*, 2006, **5**, 153–158.

122. C. Wang, R. J. Stewart and J. Kopecek, Hybrid hydrogels assembled from synthetic polymers and coiled-coil protein domains, *Nature*, 1999, **397**, 417–420.

123. A. P. Nowak, V. Breedveld, L. Pakstis, B. Ozbas, D. J. Pine, D. Pochan and T. J. Deming, Rapidly recovering hydrogel scaffolds from self-assembling diblock-copolypeptide amphiphiles, *Nature*, 2002, **417**, 424–428.

124. E. Katchalski and M. Sela, Synthesis and chemical properties of poly alpha-amino acids, *Adv. Protein Chem.*, 1958, **13**, 243–492.

125. T. J. Deming, Methodologies for preparation of synthetic block-copoly-peptides: materials with future promise in drug delivery, *Adv. Drug Delivery Rev.*, 2002, **54**, 1145–1155.

126. S. Zhang, Fabrication of novel biomaterials through molecular self-assembly, *Nat. Biotechnol.*, 2003, **21**, 1171–1178.

127. M. J. Pandya, G. M. Spooner, M. Sunde, J. R. Thorpe, A. Rodger and D. N. Woolfson, Sticky-End Assembly of a Designed Peptide Fiber Provides Insight into Protein Fibrillogenesis, *Biochemistry*, 2000, **39**, 8728–8734.

128. T. Koide, D. L. Homma, S. Asada and K. Kitagawa, Self-complementary peptides for the formation of collagen-like triple helical supramolecules, *Bioorg. Med. Chem. Lett.*, 2005, **15**, 5230–5233.

129. R. Z. Kramer, L. Vitagliano, J. Bella, R. Berisio, L. Mazzarella, B. Brodsky, A. Zagari and H. M. Berman, X-ray crystallographic determination of a collagen-like peptide with the repeating sequence (Pro-Pro-Gly), *J. Mol. Biol.*, 1998, **280**, 623–638.

130. G. B. Fields, J. L. Lauer, Y. Dori, P. Forns, Y.-C. Yu and M. Tirrell, Proteinlike molecular architecture: Biomaterial applications for inducing cellular receptor binding and signal transduction, *Biopolymers*, 1998, **47**, 143–151.

131. R. Martin, L. Waldmann and D. L. Kaplan, Supramolecular assembly of collagen triblock peptides, *Biopolymers*, 2003, **70**, 435–444.

132. H. P. Bachinger, K. J. Doege, J. P. Petschek, L. I. Fessler and J. H. Fessler, Structural implications from an electronmicroscopic comparison of procollagen V with procollagen I, pC-collagen I, procollagen IV, and a Drosophila procollagen, *J. Biol. Chem.*, 1982, **257**, 14590–14592.

133. J. C. Myers, D. Li, P. S. Amenta, C. C. Clark, C. Nagaswami and J. W. Weisel, Type XIX Collagen Purified from Human Umbilical Cord Is Characterized by Multiple Sharp Kinks Delineating Collagenous Subdomains and by Intermolecular Aggregates via Globular, Disulfide-linked, and Heparin-binding Amino Termini, *J. Biol. Chem.*, 2003, **278**, 32047–32057.

134. M. D. Shoulders, J. A. Hodges and R. T. Raines, Reciprocity of Steric and Stereoelectronic Effects in the Collagen Triple Helix, *J. Am. Chem. Soc.*, 2006, **128**, 8112–8113.

135. F. W. Kotch, I. A. Guzei and R. T. Raines, Stabilisation of the Collagen Triple Helix by O-Methylation of Hydroxyproline Residues, *J. Am. Chem. Soc.*, 2008, **130**, 2952–2953.

136. J. C. Horng, A. J. Hawk, Q. Zhao, E. S. Benedict, S. D. Burke and R. T. Raines, Macrocyclic Scaffold for the Collagen Triple Helix, *Org. Lett.*, 2006, **8**, 4735–4738.

137. N. C. Yoder and K. Kumar, Fluorinated amino acids in protein design and engineering, *Chem. Soc. Rev.*, 2002, **31**, 335–341.

138. S. Rele, Y. Song, R. P. Apkarian, Z. Qu, V. P. Conticello and E. L. Chaikof, D-Periodic Collagen-Mimetic Microfibers, *J. Am. Chem. Soc.*, 2007, **129**, 14780–14787.

139. K. E. Kadler, Y. Hojima and D. J. Prockop, Assembly of type I collagen fibrils de novo. Between 37 and 41 degrees C the process is limited by micro-unfolding of monomers, *J. Biol. Chem.*, 1988, **263**, 10517–10523.

140. S. Leikin, D. C. Rau and V. A. Parsegian, Temperature-favoured assembly of collagen is driven by hydrophilic not hydrophobic interactions, *Nat. Struct. Biol.*, 1995, **2**, 205–210.

141. R. A. Gelman, D. C. Poppke and K. A. Piez, Collagen fibril formation *In vitro*. The role of the nonhelical terminal regions, *J. Biol. Chem.*, 1979, **254**, 11741–11745.

142. S. Zhang, T. Holmes, C. Lockshin and A. Rich, Spontaneous assembly of a self-complementary oligopeptide to form a stable macroscopic membrane, *Proc. Natl. Acad. Sci. USA*, 1993, **90**, 3334–3338.

143. S. Zhang and A. Rich, Direct conversion of an oligopeptide from a beta-sheet to an alpha-helix: A model for amyloid formation, *Proc. Natl. Acad. Sci. USA*, 1997, **94**, 23–28.

144. S. Zhang, T. C. Holmes, C. M. DiPersio, R. O. Hynes, X. Su and A. Rich, Self-complementary oligopeptide matrices support mammalian cell attachment, *Biomaterials*, 1995, **16**, 1385–1393.

145. A. Aggeli, M. Bell, N. Boden, J. N. Keen, P. F. Knowles, T. C. B. McLeish, M. Pitkeathly and S. E. Radford, Responsive gels formed by the

spontaneous self-assembly of peptides into polymeric beta-sheet tapes, *Nature*, 1997, **386**, 259–262.

146. N. L. Ogihara, G. Ghirlanda, J. W. Bryson, M. Gingery, W. F. DeGrado and D. Eisenberg, Design of three-dimensional domain-swapped dimers and fibrous oligomers, *Proc. Natl. Acad. Sci. USA*, 2001, **98**, 1404–1409.

147. R. P. Nagarkar, R. A. Hule, D. J. Pochan and J. P. Schneider, De Novo Design of Strand-Swapped beta-Hairpin Hydrogels, *J. Am. Chem. Soc.*, 2008, **130**, 4466–4474.

148. S. Kojima, Y. Kuriki, T. Yoshida, K. Yazaki and K. Miura, Fibril formation by an amphipathic alpha-helix-forming polypeptide produced by gene engineering, *Proc. Jpn. Acad. B Phys. Biol. Sci.*, 1997, **73**, 7–11.

149. H. Yokoi, T. Kinoshita and S. Zhang, Dynamic reassembly of peptide RADA16 nanofiber scaffold, *Proc. Natl. Acad. Sci. USA*, 2005, **102**, 8414–8419.

150. M. Altman, P. Lee, A. Rich and S. Zhang, Conformational behaviour of ionic self-complementary peptides, *Protein Sci.*, 2000, **9**, 1095–1105.

151. S. Zhang, C. Lockshin, R. Cook and A. Rich, Unusually stable beta-sheet formation in an ionic self-complementary oligopeptide, *Biopolymers*, 1994, **34**, 663–672.

152. A. Sponner, W. Vater, S. Monajembashi, E. Unger, F. Grosse and K. Weisshart, Composition and Hierarchical Organisation of a Spider Silk, *PLoS ONE*, 2007, **2**, e998.

153. J. M. Gosline, P. A. Guerette, C. S. Ortlepp and K. N. Savage, The mechanical design of spider silks: from fibroin sequence to mechanical function, *J. Exp. Biol.*, 1999, **202**, 3295–3303.

154. Y. Yokosaki, E. L. Palmer, A. L. Prieto, K. L. Crossin, M. A. Bourdon, R. Pytela and D. Sheppard, The integrin alpha 9 beta 1 mediates cell attachment to a non-RGD site in the third fibronectin type III repeat of tenascin, *J. Biol. Chem.*, 1994, **269**, 26691–26696.

155. A. L. Prieto, G. M. Edelman and K. L. Crossin, Multiple integrins mediate cell attachment to cytotactin/tenascin, *Proc. Natl. Acad. Sci. USA*, 1993, **90**, 10154–10158.

156. T. C. Holmes, S. de Lacalle, X. Su, G. Liu, A. Rich and S. Zhang, Extensive neurite outgrowth and active synapse formation on self-assembling peptide scaffolds, *Proc. Natl. Acad. Sci. USA*, 2000, **97**, 6728–6733.

157. R. G. Ellis-Behnke, Y.-X. Liang, S.-W. You, D. K. C. Tay, S. Zhang, K.-F. So and G. E. Schneider, Nano neuro knitting: Peptide nanofiber scaffold for brain repair and axon regeneration with functional return of vision, *Proc. Natl. Acad. Sci. USA*, 2006, **103**, 5054–5059.

158. P. B. Harbury, T. Zhang, P. S. Kim and T. Alber, A switch between two-, three-, and four-stranded coiled coils in GCN4 leucine zipper mutants, *Science*, 1993, **262**, 1401–1407.

159. E. K. O'Shea, R. Rutkowski and P. S. Kim, Evidence that the leucine zipper is a coiled coil, *Science*, 1989, **243**, 538–542.

160. J. H. Brown, Breaking symmetry in protein dimers: Designs and functions, *Protein Sci.*, 2006, **15**, 1–13.
161. F. H. C. Crick, The packing of -helices: simple coiled-coils, *Acta Crystallogr.*, 1953, **6**, 689–697.
162. A. N. Lupas and M. Gruber, in *Adv. Protein Chem.*, eds. D. A. D. Parry and J. M. Squire, Academic Press, 2005, 37–38.
163. P. Burkhard, J. Stetefeld and S. V. Strelkov, Coiled coils: a highly versatile protein-folding motif, *Trends Cell Biol.*, 2001, **11**, 82–88.
164. D. N. Woolfson, in *Adv. Protein Chem.*, eds. D. A. D. Parry and J. M. Squire, Academic Press, 2005, 79–112.
165. E. K. O'Shea, K. J. Lumb and P. S. Kim, Peptide "Velcro": Design of a heterodimeric coiled coil, *Curr. Biol.*, 1993, **3**, 658–667.
166. J. H. Brown, K.-H. Kim, G. Jun, N. J. Greenfield, R. Dominguez, N. Volkmann, S. E. Hitchcock-DeGregori and C. Cohen, Deciphering the design of the tropomyosin molecule, *Proc. Natl. Acad. Sci. USA*, 2001, **98**, 8496–8501.
167. A. Karabinos, J. Schünemann, M. Meyer, U. Aebi and K. Weber, The Single Nuclear Lamin of Caenorhabditis elegans Forms *In vitro* Stable Intermediate Filaments and Paracrystals with a Reduced Axial Periodicity, *J. Mol. Biol.*, 2003, **325**, 241–247.
168. D. Papapostolou, A. M. Smith, E. D. T. Atkins, S. J. Oliver, M. G. Ryadnov, L. C. Serpell and D. N. Woolfson, Engineering nanoscale order into a designed protein fiber, *Proc. Natl. Acad. Sci. USA*, 2007, **104**, 10853–10858.
169. M. G. Ryadnov and D. N. Woolfson, Self-Assembled Templates for Polypeptide Synthesis, *J. Am. Chem. Soc.*, 2007, **129**, 14074–14081.
170. D. Papapostolou, E. H. C. Bromley, C. Bano and D. N. Woolfson, Electrostatic Control of Thickness and Stiffness in a Designed Protein Fiber, *J. Am. Chem. Soc.*, 2008, **130**, 5124–5130.
171. K. Weber, N. Geisler, U. Plessmann, A. Bremerich, K. F. Lechtreck and M. Melkonian, SF-assemblin, the structural protein of the 2-nm filaments from striated microtubule associated fibers of algal flagellar roots, forms a segmented coiled coil, *J. Cell Biol.*, 1993, **121**, 837–845.
172. K.-F. Lechtreck, Analysis of striated fiber formation by recombinant SF-assemblin *In vitro*, *J. Mol. Biol.*, 1998, **279**, 423–438.
173. E. Fuchs, Intermediate filaments and disease: mutations that cripple cell strength, *J. Cell Biol.*, 1994, **125**, 511–516.
174. K. Haubold, H. Herrmann, S. J. Langer, R. M. Evans, L. A. Leinwand and M. W. Klymkowsky, Acute effects of desmin mutations on cytoskeletal and cellular integrity in cardiac myocytes, *Cell Motil. Cytoskelet.*, 2003, **54**, 105–121.
175. S. A. Potekhin, T. N. Melnik, V. Popov, N. F. Lanina, A. A. Vazina, P. Rigler, A. S. Verdini, G. Corradin and A. V. Kajava, De novo design of fibrils made of short alpha-helical coiled coil peptides, *Chem. Biol.*, 2001, **8**, 1025–1032.

176. J. Y. Su, R. S. Hodges and C. M. Kay, Effect of chain length on the formation and stability of synthetic alpha-helical coiled coils, *Biochemistry*, 1994, **33**, 15501–15510.

177. R. Lutgring and J. Chmielewski, General strategy for covalently stabilising helical bundles: A novel five-helix bundle protein, *J. Am. Chem. Soc.*, 1994, **116**, 6451–6452.

178. V. N. Malashkevich, R. A. Kammerer, V. P. Efimov, T. Schulthess and J. Engel, The Crystal Structure of a Five-Stranded Coiled Coil in COMP: A Prototype Ion Channel?, *Science*, 1996, **274**, 761–765.

179. D. E. Wagner, C. L. Phillips, W. M. Ali, G. E. Nybakken, E. D. Crawford, A. D. Schwab, W. F. Smith and R. Fairman, Toward the development of peptide nanofilaments and nanoropes as smart materials, *Proc. Natl. Acad. Sci. USA*, 2005, **102**, 12656–12661.

180. A. Lomander, W. Hwang and S. Zhang, Hierarchical self-assembly of a coiled-coil peptide into fractal structure, *Nano Lett.*, 2005, **5**, 1255–1260.

181. S. B. Prusiner, Prions. *Proc. Natl. Acad. Sci. USA*, 1998, **95**, 13363–13383.

182. B. C. May, C. Govaerts, S. B. Prusiner and F. E. Cohen, Prions: so many fibers, so little infectivity, *Trends Biochem. Sci.*, 2004, **29**, 162–165.

183. R. Sneer, M. J. Weygand, K. Kjaer, D. A. Tirrell and H. Rapaport, Parallel beta-Sheet Assemblies at Interfaces, *ChemPhysChem.*, 2004, **5**, 747–750.

184. A. V. Kajava, S. A. Potekhin, G. Corradin and R. D. Leapman, Organisation of designed nanofibrils assembled from alpha-helical peptides as determined by electron microscopy, *J. Pept. Sci.*, 2004, **10**, 291–297.

185. T. N. Melnik, V. Villard, V. Vasiliev, G. Corradin, A. V. Kajava and S. A. Potekhin, Shift of fibril-forming ability of the designed {alpha}-helical coiled-coil peptides into the physiological pH region, *Protein Eng.*, 2003, **16**, 1125–1130.

186. Y. Zimenkov, V. P. Conticello, L. Guo and P. Thiyagarajan, Rational design of a nanoscale helical scaffold derived from self-assembly of a dimeric coiled coil motif, *Tetrahedron*, 2004, **60**, 7237–7246.

187. Y. Zimenkov, S. N. Dublin, R. Ni, R. S. Tu, V. Breedveld, R. P. Apkarian and V. P. Conticello, Rational Design of a Reversible pH-Responsive Switch for Peptide Self-Assembly, *J. Am. Chem. Soc.*, 2006, **128**, 6770–6771.

188. S. N. Dublin and V. P. Conticello, Design of a Selective Metal Ion Switch for Self-Assembly of Peptide-Based Fibrils, *J. Am. Chem. Soc.*, 2008, **130**, 49–51.

189. M. G. Ryadnov and D. N. Woolfson, Engineering the morphology of a self-assembling protein fibre, *Nat. Mater.*, 2003, **2**, 329–332.

190. M. G. Ryadnov and D. N. Woolfson, Introducing Branches into a Self-Assembling Peptide Fiber, *Angew. Chem. Int. Ed.*, 2003, **42**, 3021–3023.

191. M. G. Ryadnov and D. N. Woolfson, Fiber Recruiting Peptides: Non-covalent Decoration of an Engineered Protein Scaffold, *J. Am. Chem. Soc.*, 2004, **126**, 7454–7455.

192. M. G. Ryadnov and D. N. Woolfson, MaP Peptides: Programming the Self-Assembly of Peptide-Based Mesoscopic Matrices, *J. Am. Chem. Soc.*, 2005, **127**, 12407–12415.

193. J. D. Hartgerink, E. Beniash and S. I. Stupp, Self-Assembly and Mineralisation of Peptide-Amphiphile Nanofibers, *Science*, 2001, **294**, 1684–1688.

194. J. D. Hartgerink, E. Beniash and S. I. Stupp, Peptide-amphiphile nanofibers: A versatile scaffold for the preparation of self-assembling materials, *Proc. Natl. Acad. Sci. USA*, 2002, **99**, 5133–5138.

195. G. A. Silva, C. Czeisler, K. L. Niece, E. Beniash, D. A. Harrington, J. A. Kessler and S. I. Stupp, Selective Differentiation of Neural Progenitor Cells by High-Epitope Density Nanofibers, *Science*, 2004, **303**, 1352–1355.

196. H. Mihara, S. Matsumura and T. Takahashi, Construction and Control of Self-Assembly of Amyloid and Fibrous Peptides, *Bull. Chem. Soc. Jpn.*, 2005, **78**, 572–590.

197. T. Scheibel, R. Parthasarathy, G. Sawicki, X.-M. Lin, H. Jaeger and S. L. Lindquist, Conducting nanowires built by controlled self-assembly of amyloid fibers and selective metal deposition, *Proc. Natl. Acad. Sci. USA*, 2003, **100**, 4527–4532.

198. V. Villard, O. Kalyuzhniy, O. Riccio, S. A. Potekhin, T. N. Melnik, A. V. Kajava, C. Rüegg and G. Corradin, Synthetic RGD-containing alpha-helical coiled coil peptides promote integrin-dependent cell adhesion, *J. Pept. Sci.*, 2006, **12**, 206–212.

199. J. L. Myles, B. T. Burgess and R. B. Dickinson, Modification of the adhesive properties of collagen by covalent grafting with RGD peptides, *J. Biomater. Sci. Polym. Ed.*, 2000, **11**, 69–86.

200. M. O. Guler, L. Hsu, S. Soukasene, D. A. Harrington, J. F. Hulvat and S. I. Stupp, Presentation of RGDS Epitopes on Self-Assembled Nanofibers of Branched Peptide Amphiphiles, *Biomacromolecules*, 2006, **7**, 1855–1863.

201. J. C. Tiller, G. Bonner, L.-C. Pan and A. N. Klibanov, Improving biomaterial properties of collagen films by chemical modification, *Biotechnol. Bioeng.*, 2001, **73**, 246–252.

202. M. J. B. Wissink, R. Beernink, J. S. Pieper, A. A. Poot, G. H. M. Engbers, T. Beugeling, W. G. van Aken and J. Feijen, Immobilisation of heparin to EDC/NHS-crosslinked collagen. Characterisation and *in vitro* evaluation, *Biomaterials*, 2001, **22**, 151–163.

203. A. Y. Wang, X. Mo, C. S. Chen and S. M. Yu, Facile Modification of Collagen Directed by Collagen Mimetic Peptides, *J. Am. Chem. Soc.*, 2005, **127**, 4130–4131.

204. A. Y. Wang, C. A. Foss, S. Leong, X. Mo, M. G. Pomper, S. M. Yu and M. Seungju, Spatio-Temporal Modification of Collagen Scaffolds Mediated by Triple Helical Propensity, *Biomacromolecules*, 2008, **9**, 1755–1763.

205. X. Mo, Y. An, C.-S. Yun and S. M. Yu, Nanoparticle-Assisted Visualisation of Binding Interactions between Collagen Mimetic Peptide and Collagen Fibers, *Angew. Chem., Int. Ed.*, 2006, **45**, 2267–2270.

206. J. H. Collier, B. H. Hu, J. W. Ruberti, J. Zhang, P. Shum, D. H. Thompson and P. B. Messersmith, Thermally and photochemically triggered self-assembly of peptide hydrogels, *J. Am. Chem. Soc.*, 2001, **123**, 9463–9464.

207. M. Zhou, D. Bentley and I. Ghosh, Helical Supramolecules and Fibers Utilising Leucine Zipper-Displaying Dendrimers, *J. Am. Chem. Soc.*, 2004, **126**, 734–735.

208. Y. Liu and D. Eisenberg, 3D domain swapping: As domains continue to swap, *Protein Sci.*, 2002, **11**, 1285–1299.

209. J. W. Bryson, S. F. Betz, H. S. Lu, D. J. Suich, H. X. Zhou, K. T. O'Neil and W. F. DeGrado, Protein Design: A Hierarchic Approach, *Science*, 1995, **270**, 935–941.

210. M. S. Lamm, K. Rajagopal, J. P. Schneider and D. J. Pochan, Laminated Morphology of Nontwisting beta-Sheet Fibrils Constructed via Peptide Self-Assembly, *J. Am. Chem. Soc.*, 2005, **127**, 16692–16700.

211. J. P. Schneider, D. J. Pochan, B. Ozbas, K. Rajagopal, L. Pakstis and J. Kretsinger, Responsive Hydrogels from the Intramolecular Folding and Self-Assembly of a Designed Peptide, *J. Am. Chem. Soc.*, 2002, **124**, 15030–15037.

212. I. Cherny and E. Gazit, Amyloids: Not Only Pathological Agents but Also Ordered Nanomaterials, *Angew. Chem., Int. Ed.*, 2008, **47**, 4062–4069.

213. A. T. Petkova, R. D. Leapman, Z. Guo, W.-M. Yau, M. P. Mattson and R. Tycko, Self-Propagating, Molecular-Level Polymorphism in Alzheimer's {beta}-Amyloid Fibrils, *Science*, 2005, **307**, 262–265.

214. M. Reches and E. Gazit, Casting Metal Nanowires Within Discrete Self-Assembled Peptide Nanotubes, *Science*, 2003, **300**, 625–627.

215. C. H. Görbitz, Nanotube Formation by Hydrophobic Dipeptides, *Chemistry*, 2001, **7**, 5153–5159.

216. C. H. Görbitz, Crystal and molecular structures of the isomeric dipeptides alpha-L-aspartyl-L-alanine and beta-L-aspartyl-L-alanine, *Acta Chem. Scand. B*, 1987, **41**, 679–685.

217. C. H. Görbitz, An NH3+ . . . phenyl interaction in L-phenylalanyl-L-valine, *Acta Crystallogr. C*, 2000, **56**, 1496–1498.

218. Y. Zhang, H. Gu, Z. Yang and B. Xu, Supramolecular Hydrogels Respond to Ligand-Receptor Interaction, *J. Am. Chem. Soc.*, 2003, **125**, 13680–13681.

219. V. Jayawarna, M. Ali, T. A. Jowitt, A. F. Miller, A. Saiani, J. E. Gough and R. V. Ulijn, Nanostructured Hydrogels for Three-Dimensional Cell Culture Through Self-Assembly of Fluorenylmethoxycarbonyl-Dipeptides, *Adv. Mater.*, 2006, **18**, 611–614.

220. A. M. Smith, R. J. Williams, C. Tang, P. Coppo, R. F. Collins, M. L. Turner, A. Saiani and R. V. Ulijn, Fmoc-Diphenylalanine Self Assembles to a Hydrogel via a Novel Architecture Based on pi-pi Interlocked beta-Sheets, *Adv. Mater.*, 2008, **20**, 37–41.

221. R. P. Lanza, R. Langer and J. P. Vacanti (eds.), *Principles of Tissue Engineering*, Academic Press, New York, 2000.

222. R. M. Nerem, Tissue Engineering: The Hope, the Hype, and the Future, *Tissue Eng.*, 2006, **12**, 1143–1150.

223. J. J. Marler, J. Upton, R. Langer and J. P. Vacanti, Transplantation of cells in matrices for tissue regeneration, *Adv. Drug Delivery Rev.*, 1998, **33**, 165–182.

224. A. Lichtenberg, S. Cebotari, I. Tudorache, A. Hilfiker and A. Haverich, Biological scaffolds for heart valve tissue engineering, *Methods Mol. Med.*, 2007, **140**, 309–317.

225. K. Mendelson and F. Schoen, Heart Valve Tissue Engineering: Concepts, Approaches, Progress, and Challenges, *Ann. Biomed. Eng.*, 2006, **34**, 1799–1819.

226. J. R. Fuchs, B. A. Nasseri and J. P. Vacanti, Tissue engineering: a 21st century solution to surgical reconstruction, *Ann. Thorac. Surg.*, 2001, **72**, 577–591.

227. J. I. Dawson and R. O. C. Oreffo, Bridging the regeneration gap: Stem cells, biomaterials and clinical translation in bone tissue engineering, *Arch. Biochem. Biophys.*, 2008, **473**, 124–131.

228. R. M. Nerem, Tissue engineering: confronting the transplantation crisis, *Adv. Exp. Med. Biol.*, 2003, **534**, 1–9.

229. D. L. Stocum, *Regenerative biology and medicine*, Elsevier, Oxford, 2006.

230. A. Vats, N. S. Tolley, A. E. Bishop and J. M. Polak, Embryonic stem cells and tissue engineering: delivering stem cells to the clinic, *J. Roy. Soc. Med.*, 2005, **98**, 346–350.

231. H. J. Rippon and A. E. Bishop, Embryonic stem cells, *Cell Prolif.*, 2004, **37**, 23–34.

232. H. M. Blau, T. R. Brazelton and J. M. Weimann, The Evolving Concept of a Stem Cell: Entity or Function? *Cell*, 2001, **105**, 829–841.

233. P. Bianco, M. Riminucci, S. Gronthos and P. G. Robey, Bone Marrow Stromal Stem Cells: Nature, Biology, and Potential Applications, *Stem Cells*, 2001, **19**, 180–192.

234. P. Bianco and P. G. Robey, Stem cells in tissue engineering, *Nature*, 2001, **414**, 118–121.

235. S. Lane, H. J. Rippon and A. E. Bishop, Stem cells in lung repair and regeneration, *Regen. Med.*, 2007, **2**, 407–415.

236. A. E. Bishop, L. D. Buttery and J. M. Polak, Embryonic stem cells, *J. Pathol.*, 2002, **197**, 424–429.

237. I. L. Weissman, Translating Stem and Progenitor Cell Biology to the Clinic: Barriers and Opportunities, *Science*, 2000, **287**, 1442–1446.

238. I. L. Weissman, Stem Cells: Units of Development, Units of Regeneration, and Units in Evolution, *Cell*, 2000, **100**, 157–168.

239. C. M. Nelson and M. J. Bissell, Of Extracellular Matrix, Scaffolds, and Signaling: Tissue Architecture Regulates Development, Homeostasis, and Cancer, *Annu. Rev. Cell Dev. Biol.*, 2006, **22**, 287–309.

240. J. Zoldan and S. Levenberg, Engineering Three-Dimensional Tissue Structures Using Stem Cells, *Methods Enzymol.*, 2006, **420**, 381–391.

241. H. K. Kleinman, M. L. McGarvey, J. R. Hassell, V. L. Star, F. B. Cannon, G. W. Laurie and G. R. Martin, Basement membrane complexes with biological activity, *Biochemistry*, 1986, **25**, 312–318.

242. V. M. Weaver, O. W. Petersen, F. Wang, C. A. Larabell, P. Briand, C. Damsky and M. J. Bissell, Reversion of the malignant phenotype of human breast cells in three-dimensional culture and in vivo by integrin blocking antibodies, *J. Cell Biol.*, 1997, **137**, 231–245.

243. E. Cukierman, R. Pankov and K. M. Yamada, Cell interactions with three-dimensional matrices, *Curr. Opin. Cell Biol.*, 2002, **14**, 633–640.

244. K. M. Yamada and E. Cukierman, Modelling Tissue Morphogenesis and Cancer in 3D, *Cell*, 2007, **130**, 601–610.

245. F. Grinnell, C.-H. Ho, E. Tamariz, D. J. Lee and G. Skuta, Dendritic Fibroblasts in Three-dimensional Collagen Matrices, *Mol. Biol. Cell*, 2003, **14**, 384–395.

246. G. Y. Lee, P. A. Kenny, E. H. Lee and M. J. Bissell, Three-dimensional culture models of normal and malignant breast epithelial cells, *Nat. Methods*, 2007, **4**, 359–365.

247. E. Cukierman, R. Pankov, D. R. Stevens and K. M. Yamada, Taking Cell-Matrix Adhesions to the Third Dimension, *Science*, 2001, **294**, 1708–1712.

248. M. Miron-Mendoza, J. Seemann and F. Grinnell, Collagen Fibril Flow and Tissue Translocation Coupled to Fibroblast Migration in 3D Collagen Matrices, *Mol. Biol. Cell*, 2008, **19**, 2051–2058.

249. S. Rhee and F. Grinnell, Fibroblast mechanics in 3D collagen matrices, *Adv. Drug Deliv. Rev.*, 2007, **59**, 1299–1305.

250. H. Jiang, S. Rhee, C.-H. Ho and F. Grinnell, Distinguishing fibroblast promigratory and procontractile growth factor environments in 3-D collagen matrices, *FASEB J.*, 2008, **22**, 2151–2160.

251. W. M. Saltzman and W. L. Olbricht, Building drug delivery into tissue engineering design, *Nat. Rev. Drug Discov.*, 2002, **1**, 177–186.

252. M. Heyde, K. A. Partridge, R. O. Oreffo, S. M. Howdle, K. M. Shakesheff and M. C. Garnett, Gene therapy used for tissue-engineering applications, *J. Pharm. Pharmacol.*, 2007, **59**, 329–350.

253. G. Chan and D. J. Mooney, New materials for tissue engineering: towards greater control over the biological response, *Trends Biotechnol.*, 2008, **26**, 382–392.

254. T. H. Park and M. L. Shuler, Integration of Cell Culture and Micro-fabrication Technology, *Biotechnol. Prog.*, 2003, **19**, 243–253.

255. M. Geissler and Y. X. Patterning: Principles, Some New Developments, *Adv. Mater.*, 2004, **16**, 1249–1269.

256. S. Raghavan and C. S. Chen, Micropatterned Environments in Cell Biology, *Adv. Mater.*, 2004, **16**, 1303–1313.

257. A. Folch and M. Toner, Microengineering of cellular interactions, *Annu. Rev. Biomed. Eng.*, 2000, **2**, 227–256.

258. C. M. Nelson, M. M. VanDuijn, J. L. Inman, D. A. Fletcher and M. J. Bissell, Tissue Geometry Determines Sites of Mammary

Branching Morphogenesis in Organotypic Cultures, *Science*, 2006, **314**, 298–300.

259. B. S. Wiseman, M. D. Sternlicht, L. R. Lund, C. M. Alexander, J. Mott, M. J. Bissell, P. Soloway, S. Itohara and Z. Werb, Site-specific inductive and inhibitory activities of MMP-2 and MMP-3 orchestrate mammary gland branching morphogenesis, *J. Cell Biol.*, 2003, **162**, 1123–1133.

260. S. Rhee, H. Jiang, C.-H. Ho and F. Grinnell, Microtubule function in fibroblast spreading is modulated according to the tension state of cell-matrix interactions, *Proc. Natl. Acad. Sci. USA*, 2007, **104**, 5425–5430.

261. F. Grinnell, B. L. Rocha, C. Iucu, S. Rhee and H. Jiang, Nested collagen matrices: A new model to study migration of human fibroblast populations in three dimensions, *Exp. Cell Res.*, 2006, **312**, 86–94.

262. A. Curtis and C. Wilkinson, New depths in cell behaviour: reactions of cells to nanotopography, *Biochem. Soc. Symp.*, 1999, **65**, 15–26.

263. J. J. Tomasek, G. Gabbiani, B. Hinz, C. Chaponnier and R. A. Brown, Myofibroblasts and mechano-regulation of connective tissue remodeling, *Nat. Rev. Mol. Cell Biol.*, 2002, **3**, 349–363.

264. E. Tamariz and F. Grinnell, Modulation of Fibroblast Morphology and Adhesion during Collagen Matrix Remodelling, *Mol. Biol. Cell*, 2002, **13**, 3915–3929.

265. E. K. Scott and L. Luo, How do dendrites take their shape?, *Nat. Neurosci.*, 2001, **4**, 359–365.

266. B. M. Gillette, J. A. Jensen, B. Tang, G. J. Yang, A. Bazargan-Lari, M. Zhong and S. K. Sia, In situ collagen assembly for integrating micro-fabricated three-dimensional cell-seeded matrices, *Nat. Mater.*, 2008, **7**, 636–640.

267. T. R. Sodunke, K. K. Turner, S. A. Caldwell, K. W. McBride, M. J. Reginato and H. M. Noh, Micropatterns of Matrigel for three-dimensional epithelial cultures, *Biomaterials*, 2007, **28**, 4006–4016.

268. E. Figallo, M. Flaibani, B. Zavan, G. Abatangelo and N. Elvassore, Micropatterned Biopolymer 3D Scaffold for Static and Dynamic Culture of Human Fibroblasts, *Biotechnol. Prog.*, 2007, **23**, 210–216.

269. J. Y. Shen, M. B. Chan-Park, B. He, A. P. Zhu, X. Zhu, R. W. Beuerman, E. B. Yang, W. Chen and V. Chan, Three-Dimensional Microchannels in Biodegradable Polymeric Films for Control Orientation and Phenotype of Vascular Smooth Muscle Cells, *Tissue Eng.*, 2006, **12**, 2229–2240.

270. J. Le Beyec, R. Xu, S.-Y. Lee, C. M. Nelson, A. Rizki, J. Alcaraz and M. J. Bissell, Cell shape regulates global histone acetylation in human mammary epithelial cells, *Exp. Cell Res.*, 2007, **313**, 3066–3075.

271. J. Vacanti and R. Langer, Tissue engineering: the design and fabrication of living replacement devices for surgical reconstruction and transplantation, *Lancet*, 1999, **354**, 32–34.

272. L. G. Griffith, Emerging Design Principles in Biomaterials and Scaffolds for Tissue Engineering, *Ann. NY Acad Sci.*, 2002, **961**, 83–95.

273. O. R. Davies, A. L. Lewis, M. J. Whitaker, H. Tai, K. M. Shakesheff and S. M. Howdle, Applications of supercritical CO_2 in the fabrication of

polymer systems for drug delivery and tissue engineering, *Adv. Drug Delivery Rev.*, 2008, **60**, 373–387.

274. D. E. Ingber, Tensegrity II. How structural networks influence cellular information processing networks, *J. Cell Sci.*, 2003, **116**, 1397–1408.

275. S. Levenberg, N. F. Huang, E. Lavik, A. B. Rogers, J. Itskovitz-Eldor and R. Langer, Differentiation of human embryonic stem cells on three-dimensional polymer scaffolds, *Proc. Natl. Acad. Sci. USA*, 2003, **100**, 12741–12746.

276. G. E. Holt, J. L. Halpern, T. T. Dovan, D. Hamming and H. S. Schwartz, Evolution of an in vivo bioreactor, *J. Orthop. Res.*, 2005, **23**, 916–923.

277. T. W. Bauer and G. F. Muschler, Bone graft materials. An overview of the basic science, *Clin. Orthop. Relat. Res.*, 2000, **371**, 10–27.

278. Y. Khan, M. J. Yaszemski, A. G. Mikos and C. T. Laurencin, Tissue Engineering of Bone: Material and Matrix Considerations, *J. Bone Joint Surg. Am.*, 2008, **90**, 36–42.

279. S. S. Tseng, M. A. Lee and A. H. Reddi, Nonunions and the Potential of Stem Cells in Fracture-Healing, *J. Bone Joint Surg. Am.*, 2008, **90**, 92–98.

280. J. Nishida and T. Shimamura, Methods of reconstruction for bone defect after tumor excision: a review of alternatives, *Med. Sci. Monit.*, 2008, **14**, RA107–113.

281. M. V. Risbud, I. M. Shapiro, A. Guttapalli, A. Di Martino, K. G. Danielson, J. M. Beiner, A. Hillibrand, T. J. Albert, D. G. Anderson and A. R. Vaccaro, Osteogenic potential of adult human stem cells of the lumbar vertebral body and the iliac crest, *Spine*, 2006, **31**, 83–89.

282. R. Zeiser, A. Beilhack and R. S. Negrin, Acute graft-versus-host disease-challenge for a broader application of allogeneic hematopoietic cell transplantation, *Curr. Stem Cell Res. Ther.*, 2006, **1**, 203–212.

283. W. Paul and C. P. Sharma, Ceramic Drug Delivery: A Perspective, *J. Biomater. Appl.*, 2003, **17**, 253–264.

284. H. Petite, V. Viateau, W. Bensaid, A. Meunier, C. de Pollak, M. Bourguignon, K. Oudina, L. Sedel and G. Guillemin, Tissue-engineered bone regeneration, *Nat. Biotech.*, 2000, **18**, 959–963.

285. H. J. Kong and D. J. Mooney, Microenvironmental regulation of bio-macromolecular therapies, *Nat. Rev. Drug Discov.*, 2007, **6**, 455–463.

286. R. W. Sands and D. J. Mooney, Polymers to direct cell fate by controlling the microenvironment, *Curr. Opin. Biotechnol.*, 2007, **18**, 448–453.

287. C. A. Simmons, E. Alsberg, S. Hsiong, W. J. Kim and D. J. Mooney, Dual growth factor delivery and controlled scaffold degradation enhance in vivo bone formation by transplanted bone marrow stromal cells, *Bone*, 2004, **35**, 562–569.

288. C. Fischbach, R. Chen, T. Matsumoto, T. Schmelzle, J. S. Brugge, P. J. Polverini and D. J. Mooney, Engineering tumors with 3D scaffolds, *Nat. Methods*, 2007, **4**, 855–860.

289. M. M. Stevens, R. P. Marini, D. Schaefer, J. Aronson, R. Langer, V. P. Shastri and V. Prasad, In vivo engineering of organs: The bone bioreactor, *Proc. Natl. Acad. Sci. USA*, 2005, **102**, 11450–11455.

290. R. P. Marini, M. M. Stevens, R. Langer and V. P. Shastri, Hydraulic Elevation of the Periosteum: A Novel Technique for Periosteal Harvest, *J. Invest. Surg.*, 2004, **17**, 229–233.
291. X. Liu and P. X. Ma, Polymeric Scaffolds for Bone Tissue Engineering, *Ann. Biomed. Eng.*, 2004, **32**, 477–486.
292. Q. P. Hou, P. A. De Bank and K. M. Shakesheff, Injectable scaffolds for tissue regeneration, *J. Mater. Chem.*, 2004, **14**, 1915–1923.

CHAPTER 5
Concluding Remarks

5.1 Learning Fluent

The process of self-assembly is central to the organisation of matter, with its notable role played in biological systems. Despite Nature's flawless performance and the wealth of methods in constructing large monomolecular structures such as proteins and DNA, spontaneous noncovalent multicopy associations of separate subunits is evidently preferred to covalent bonding in generating functional objects ranging from organelles to cells and tissues.

From a designer's perspective the choice does not require additional reasoning as current synthetic approaches are well behind natural examples. From Nature's perspective the wealth dwells on molecular synthesis and self-assembly being reciprocal and complementary; for instance, ribosomes synthesise proteins from amino-acid monomers but themselves assemble from proteins.

In light of this, employing self-assembly becomes stipulated by the dynamic nature of biological components that in most, if not all, cases have to be architecturally fictile, *i.e.* capable of undergoing stimuli-responsive assembly disassembly cycles. This is an intrinsic property of molecular encoding – the mechanism of establishing processing templates or codes and reading from these in building supramolecular structures.

That said, self-assembly can be viewed as a language allowing programming ordered hierarchical structures. Similarly and inherently, the language itself manifests several hierarchical levels. It can be simple or advanced. It can contain very few or general operators (*e.g.* hydrophobic effect or van der Waals interactions), but it can also encompass more definitive interrelations coding specific assembly pathways, already known (*e.g.* nucleic-acid base-pairings or protein-folding motifs) and to be discovered.

The complexity and refinement of the resulting structures is thus strongly correlated with the level at which language is applied that makes the success of

RSC Nanoscience and Nanotechnology No. 7
Bionanodesign
By M Ryadnov
© Maxim Ryadnov 2009
Published by the Royal Society of Chemistry, www.rsc.org

bionanodesign, both particular and general, contingent on one's "linguistic" proficiency.

5.2 Parsing Semantic

Our ability to predict and prescribe desired assemblies is avowedly low, so is the current choice of instruments and approaches, with the main efforts being concentrated on compiling toolkits of building blocks that may be instructive on the nanoscale. The building of such a dictionary is addressed by developing principles of *de novo* design, linking this with the natural expressions of self-assembling motifs is viewed by many as the main strategy.

To a large extent it is an empirical approach underlined by a historically approved tendency of revealing molecular processes with their subsequent probing in proof-of-principle designs. This can be related to memorising what is and becomes available, *i.e.* to enriching the existing vocabulary.

This is both the beginning of and the lifetime continuous-learning process.

However, mastering a language lies, first of all, in its structuring – formulating a systematic and orderly arrangement, a matrix that would support and provide readable periodicities connecting molecules (words) with nanostructures (sentences) and *vice versa*. This may spread out to further structural variations or even different languages. Curiously enough, in this respect today's use and study of self-assembly is very group specific, and presents rather a series of remotely related "dialects" than one common language.

We know, for example, that desired DNA sequences can be written to program complex structures and unknown DNA or RNA sequences can be divided and read to a single nucleotide. But we hardly know how and why some folds of nucleic acids template reproducible nucleoproteid complexes and others do not, and why the mutation of a single nucleotide has to compromise the integrity of an entire nanostructure. In this regard, current theories are componental but not interchangeable. This may explain why elucidating the key principles of self-assembly has proved to be strictly coherent with their application to constructing nanostructured materials to have gradually evolved into a number of subfields that address incisive technological aspects. Pursuing novel technologies remains to be the main focus.

Inexplicitly stated, this arguably leads to a pragmatic inference of systematising the established self-assembly phenomena into one universal protocol.

5.3 Drawing Pragmatic

This path is challenging but the choice is inevitable. Learning from experience is probably the only process of universalising the never-ceasing stream of information acquired from different subfields with different patterns and degrees of relatedness. Therefore, following basics seeding all self-assembly processes assumes to be ubiquitous in the successful completion of an intended design.

Vaguely, these can be formulated as three diverging but correlative characteristics.

One deals with molecular constituents themselves and the interactions these rely upon in ascending from amorphous structural expressions to cooperative periodicities. These are set by primary covalent structures that are mainly responsible for the types of perceptual assemblies and by the environmental conditions that either favour or disfavour a given assembly plan of transcribing unordered structures to ordered complexes.

Another derives from the first and perceives the responsive properties of the assembled structures that converge into three main points; reversibility, supramolecular complementarity and motion. All of these are defined and assured by inter- and intramolecular dynamics that serve as semantic categories in the classification of assemblies in accordance with their environment and the level of hierarchical accomplishments. For example, if reversibility is an equal prerequisite for all assembly steps, the complementarity of supramolecular components is of intermediate importance and predominantly reflects localised adjustments, whereas motion differs by type for initial and resultant forms and depends on the equilibria reached.

The main distinction of the third is the adherence of probable and apparent error variations to multiplicity of noncovalent interactions in assembly. Specifically, these include defects at various assembly stages, their multiplication with increasing complexity and means of their removal or circumvention. In natural systems this problem is not of immediate importance, with perfection settled by various proofreading mechanisms. Understandably, artificial systems lacking compensatory means suffer from myriads of imperfections. Unlike the previous two this approves the characteristic discriminative, but equally influential in the prescriptive self-assembly. Therefore, with it viewed in the context of variational competitions, emulating biomolecular self-assembly claims to provide efficient factors for limiting imperfections. However, these constitute the morphology of an advanced language, a version of which has yet to be translated into a system of meaningful parameters amendable for design.

Current interest in designing self-assembling nanosystems is tremendous, with an emerging tendency focusing away from sole applications. This feeds the next phase in bionanodesign, which might not be final but the one that will install a connecting bridge between molecules and autonomously operating systems constituting words and texts in the deciphered language of molecular self-assembly.

CHAPTER 6
Revealing Contributory

To assist an interested reader in gaining a broader view of current trends in bionanodesign and to encourage the formation of a more critical view on the given presentation a short extraction of laboratories whose research have shaped the field is presented.

The collection is arranged in alphabetical order and evidently is not complete as it serves the purpose of familiarisation and reflects the author's subjective perception of the problem rather than giving a full coverage of ongoing research.

However, in preparing the list a key assumption was made to accentuate the integral purpose of the book confirmed by the research commitment of included groups working on nanoscale design employing natural biomolecular motifs or templates.

By unwritten convention, each laboratory is represented by a group leader that again in terms of personal contributions may be incomplete and subjective but is irrelevant within the bigger picture of pursuing the mutually shared goal of bionanodesign – deciphering the language of biomolecular self-assembly.

Adleman Leonard (University of Southern California)
Seeded the idea of and provided the proof-of-principle for DNA computing
Aebi Ueli (University of Basel)
Solved key structural parameters driving the assembly of α-helical filaments
Aoyama Yasuhiro (Kyoto University)
Proposed a "glyco" approach for complexation of nucleic acids into nanosized particles
Belcher Angela (Massachusetts Institute of Technology)
Put forward phage coding for specifying the selection of inorganic nanomaterials
Behr Jean-Paul (Louis Pasteur University Strasbourg)

RSC Nanoscience and Nanotechnology No. 7
Bionanodesign
By M Ryadnov
© Maxim Ryadnov 2009
Published by the Royal Society of Chemistry, www.rsc.org

Formulated the concept of "molecular galenics" – multicomponent one-constituent-one-function assemblies – in designing synthetic viruses

Bissell Mina (Lawrence Berkeley National Laboratory)
Pioneered 3D cell culture
Burkhard Peter (University of Connecticut)
Reported the first polyhedra design based on de novo peptide sequences
Chmielewski Jean (Purdue University)
Explored the concept of self-replicating peptides for stimuli-modulated nanomaterials
Conticello Vincent (Emory University)
Extended the strategy of staggered fibrillar assembly to different protein folds
Cozarrelli Nicholas (University of California, Berkeley)
Set fundamental principles for DNA topology
DeGrado William (University of Pennsylvania)
Established key conventions of protein design applicable to nanoscale engineering
Deming Timothy (University of California, Los Angeles)
Engineered self-healing hydrogels based on dynamically restructuring peptide matrices
Dobson Chris (Cambridge University)
Laid the foundations of the assembly of amyloid-like nanofibrils
Douglas Trevor (Montana State University)
Led (with Mark Young) studies on nanoscale imparting of protein cages and viruses
Dragnea Bogdan (Indiana University)
Exploited core-directed polymorphism in the assembly of viral shells
Fairman Robert (Haverford College)
Proposed hydrophobic axial stagger for preferential network assembly of protein fibres
Fertala Andrzej (Thomas Jefferson University)
Identified key characteristics in designing structurally functional collagen mimetics
Gazit Ehud (Tel Aviv University)
Elucidated a minimum amyloid sequence as a core motif in the synthesis of nanomaterials –
Ghadiri Reza (Scripps Research Institute)
Introduced the strategy of alternating chirality in the assembly of peptide nanotubes
Ghosh Indraneel (University of Arizona)
Described a supramolecular translation of connate peptide helices into helical nanofibres
Görbitz Carl (University of Oslo)
Predicted and solved the structure of nanotubes assembled from hydrophobic dipeptides
Harrison Stephen (Harvard University)
Determined structural parameters of protein nanocage assemblies

Hartgerink Jeffrey (Rice University)
Pursued a common rationale of nanofibre designs derived from different protein folds
Heddle Jonathan (Tokyo Institute of Technology)
Probed ring-shaped proteins as templates for size- and shape-defined nanostructures
Howdle Steven (Nottingham University)
Elaborated microporous polymer scaffolds as 3D open matrices
Huang Leaf (University of North Carolina)
Proposed "ternary complexes" as minimal formulations for dynamic nonviral vectors
Hubbell Jeffrey (Swiss Federal Institute of Technology)
Demonstrated the necessity of "living" biomaterials instructive on the nanoscale
Johnson John (Scripps Research Institute)
Established structure–assembly relationship in engineering isosahedral viruses
Kadler Karl (Manchester University)
Described the fundamental mechanisms of collagen fibrillar self-assembly
Kajava Andrej (University of Montpellier)
Formulated the rationale of axial mismatching in designing helix-based fibrillar matrices
Kopecek Jindrich (University of Utah)
Postulated the hybrid nature of self-sustainable and stimuli-responsive hydrogels
Kostarelos Kostas (Imperial College London)
Addressed the toxicity and tropism of viral vectors through artificial enveloping
LaBean Thomas (Duke University)
Developed DNA lattices as assembly templates for hybrid nanomaterials
Lakey Jeremy (Newcastle University)
Applied native membrane-active protein scaffolds to patterning nanoscale interfaces
Langer Robert (Massachusetts Institute of Technology)
Postulated key criteria in designing biological substitutes
Lindquist Susan (Whitehead Institute for Biomedical Research)
Proposed selective deposition on nanofibre surfaces for metal nanowires
Manchester Marianne (Scripps Research Institute)
Investigated viruses as nanodefined MRI contrast-enhancing agents for tumour targeting
Mann Stephen (Bristol University)
Initiated the application of protein cages and viruses in nanomaterial synthesis
Messersmith Phillip (Northwestern University)
Devised locally triggered gelation of protein fibrillar matrices
Mihara Hisakazu (Tokyo Institute of Technology)
Explored the impact of kinetic and thermodynamic factors on morphological preferences of peptide nanofibre designs
Miller Andrew (Imperial College London)

Introduced the strategy of modular upgrading to liposome-mediated DNA delivery
Mooney David (Harvard University)
Led physicochemical studies on sustained biological response of cell-supporting scaffolds
Naik Rajesh (US Air Force Research Laboratory)
Exploited nanomaterial synthesis directed by solid-binding peptide sequences
Niemeyer Christof (Dortmund University)
Applied principles of positional conjugation to stoichiometric nanostructured devices
Nussinov Ruth (Tel Aviv University)
Led efforts in computational nanostructure design reliant on protein folding
Pierce Niles (California Institute of Technology)
Developed arbitrary algorithms for programming DNA self-assembly pathways
Raines Ronald (University of Wisconsin-Madison)
Proposed topologically staggered collagen assembly in engineering 3D matrices
Rothemund Paul (California Institute of Technology)
Authored perhaps the most elegant concept in nanodesign – DNA origami
Safinya Cyrus (University of California, Santa Barbara)
Solved and rationalised the structure of lipoplexes
Sanford John (Cornell University)
The inventor of the gene gun
Schneider Joel (University of Delaware)
Linked reversible sol-gel transition of protein hydrogels with conformational switching
Shakesheff Kevin (Nottingham University)
Elucidated design principles for injectable scaffolds
Shih William (Harvard University)
Described probably the most exquisite nanoscale 3D design – DNA octahedron
Seeman Nadrian (New York University)
The founder and inspirer of DNA nanodesign
Seymour Leonard (Oxford University)
Proposed "stealth viruses", chemically coated viruses, as alternative vectors for gene transfer
Sleytr Uwe (University of Agricultural Sciences, Vienna)
Identified crystalline bacterial cell surface layers (S-layer) as nanopatterning platforms
Stevens Molly (Imperial College London)
Explored in vivo bioreactors as a transplantation-free approach to tissue repair
Stocum David (Indiana University)
Postulated key aspects and research foci of regenerative biology and medicine
Stupp Samuel (Norhtwestern University)
Described the most comprehensively studied fibrous analogue of the ECM
Tame Jeremy (Yokohama State University)
Specified tailoring of nanomaterials using native and mutant protein shapes
Tirrell David (California Institute of Technology)

Defined convergence points of protein design and polymer chemistry as a design paradigm to smart hydrogels

Turberfield Andrew (Oxford University)

Generated geometrically defined 3D DNA nanostructures as precursors for DNA motors

Ulijn Rein (Strathclyde University)

Exploited hydrophobic tuning as a minimalist approach to programmable peptide self-assembly

Vologodskii Alexander (New York University)

Adapted the theory of knots in DNA supercoiling and assembly

Wagner Carston (University of Minnesota)

Reported a chemically controlled approach to discrete protein nanorings

Wagner Ernst (Ludwig Maximilian University of Munich)

Described bioresponsive shielding as an optimisation model to synthetic viruses

Weisel John (University of Pennsylvania)

Established fundamentals of matrix formation exemplified by fibrin assembly

Whitesides George (Harvard University)

Postulated molecular self-assembly as a nanoscale synthetic strategy in generating structures

Winfree Eric (California Institute of Technology)

Invented algorithmic DNA self-assembly

Woolfson Derek (Bristol University)

Translated the sticky-ended cohesion of DNA hybridisation into peptide fibrillogenesis

Xu Bing (Hong Kong University)

Explored correlation between gelation and primary structure of peptide hydrogelators

Yeates Todd (University of California, Los Angeles)

Applied protein folding topologies to the construction of spatially fixed cages ("nanohedra"), filaments and primitive organelles

Young Mark (Montana State University)

Led (with Trevor Douglas) studies on nanoscale imparting of protein cages and viruses

Yu Michael (Johns Hopkins University)

Reported a strand-invasion approach for the facile modification of collagen fibres

Yurke Bernard (Bell Laboratories)

Introduced and experimentally followed the notion of DNA-fuelled DNA nanomachines

Zhang Shounguang (Massachusetts Institute of Technology)

Pioneered peptide self-assembly as an explicitly separate and self-sufficient discipline

Zlotnick Adam (Oklahoma University)

Proposed approximation equilibrium models for polyhedral protein assemblies

Subject Index